Annals of Mathematics Studies

Number 123

Automorphic Representations of Unitary Groups in Three Variables

by

Jonathan D. Rogawski

PRINCETON UNIVERSITY PRESS

PRINCETON, NEW JERSEY

1990

Princeton University Press books are printed on acid-free paper, and meet the guidelines for permanence and durability of the Committee on Production Guidelines for Book Longevity of the Council on Library Resources

Printed in the United States of America
by Princeton University Press, 41 William Street
Princeton, New Jersey

Library of Congress Cataloging-in-Publication Data

Rogawski, Jonathan David.
Automorphic representations of unitary groups in three variables / by Jonathan D. Rogawski.
p. cm. – (Annals of mathematics studies ; no. 123)
Includes bibliographical references (p.) and index.
ISBN 0-691-08586-2 (cloth: acid-free paper)
ISBN 0-691-08587-0 (paper: acid-free paper)
1. Unitary groups. 2. Trace formulas. 3. Representations of groups. 4. Automorphic forms. I. Title. II. Series.
QA171.R617 1990
512'.2–dc20 90-38535

To Julie, with love

Table of Contents

It is a basic problem in the theory of automorphic representations to compare the trace formulas of reductive groups G' and G over a global field F in various situations. For example, G' may be an inner form of G, or the set of fixed points of an automorphism of G. The base change problem for a cyclic extension E/F, in which $G = \mathrm{Res}_{E/F}(G')$, is an example of the latter situation. Such a comparison was carried out in the representation-theoretic context for the first time in the case of GL(2) ([JL]), resulting in the Jacquet-Langlands correspondence between automorphic representations of GL(2) and unit groups of quaternion algebras. The base change problem for GL(n) and cyclic extensions was treated in the work of Saito, Shintani, and Langlands for $n = 2$ and was generalized to $n > 2$ by Arthur-Clozel ([L_1], [AC]). To deal with more general reductive groups, it is necessary to "stabilize" the trace formula, in the sense of Langlands ([L_2], [L_6]). A comparison of trace formulas involves a correspondence between the conjugacy classes of the two groups, and, in general, such a correspondence must be phrased in terms of stable conjugacy. The stable trace formula is also an essential ingredient in the comparison of the trace formula with the Lefschetz trace formula in the theory of Shimura varieties ((L_7], [Mo]).

The first part of this book (§1-§10) is devoted to developing a stable trace formula for the group $G = U(3)$, the quasi-split unitary group in three variables defined with respect to a quadratic extension E/F. At the same time, the comparison of trace formulas for G and $\mathrm{Res}_{E/F}(G)$ is made. The stabilization procedure is described in §5 and is carried out in §10 (Theorem 10.3.1). The combination of the stable trace formula and the base change comparison yields a great deal of information about automorphic representations of $U(3)$. These consequences are worked out in the second part of the book (§13-15).

Stabilization involves several concepts and ingredients due to Langlands which are known under the general heading of "endoscopy". These include stable conjugacy, endoscopic groups, and transfer of orbital integrals. The definitions are reviewed in §3 and §4. At the level of representations, endoscopy shows up in the theory of L-packets, i.e., the partition of the set of irreducible admissible representations of a connected reductive group over a local F into finite sets Π called L-packets which satisfy certain natural properties. Such a partition has been shown to exist in the archimedean case ([L_8]), but not yet, in general, in the p-adic case. For GL(n), locally and globally, all L-packets are singletons. Once local L-packets are defined at all places, global L-packets are defined as restricted tensor products $\Pi = \otimes\Pi_v$, where Π_v is a local L-packet which, for almost all v, contains an unramified representation. A restricted tensor product $\otimes\pi_v$ belongs to Π if $\pi_v \in \Pi_v$ for all v and π_v is unramified for almost all v.

The first example of a stable trace formula was obtained for the group SL(2) by Labesse and Langlands. In this case, two representations are defined to lie in the same L-packet (locally and globally) if they are conjugate under the adjoint action of $\mathrm{PGL}_2(F)$. Although this definition also works for SL(n), using conjugation by $\mathrm{PGL}_n(F)$, in general L-packets cannot be defined in terms of the action of the adjoint group or any other larger group. Let Π be a discrete L-packet on SL(2), i.e., Π contains elements which occur discretely in $L^2(SL_2(F)\backslash SL_2(\mathbf{A}))$. Any two members of Π have the same local components almost everywhere. However, they may occur in the discrete spectrum with different multiplicities. Let us call Π stable if all members of Π have the same multiplicity. The stable trace formula for SL(2) leads to a classification of the set of Π which are not stable in terms of automorphic characters θ of groups $T(F)\backslash T(\mathbf{A})$, where T is the torus defined by the norm one group E^1 of a quadratic extension E/F. By the Langlands classification for tori, θ is associated to a homomorphism $\theta^* : W_F \to {}^LT$ where W_F is the Weil group of F. There is a natural homomorphism $\varphi : {}^LT \to {}^L\mathrm{SL}(2)$ and the composition $\varphi \circ \theta^* : W_F \to {}^L\mathrm{SL}(2)$ defines an L-packet $\Pi(\theta)$ which is known to be cuspidal ([L$_2$]). By the results of [LL], Π is stable unless it is of the form $\Pi(\theta)$ and, furthermore, there is a formula for the multiplicities of the representations belonging to the packets $\Pi(\theta)$. The tori T are endoscopic groups for SL(2), and the results just quoted illustrate the pattern that is expected to hold in general, that the L-packets which are not stable will occur as functorial transfers of L-packets on endoscopic groups.

The group $G = U(3)$ has only one proper elliptic endoscopic group, namely, the group $H = U(2) \times U(1)$. By the general theory (§4), there is an embedding of L-groups $\xi_H\colon {}^LH \to {}^LG$. We also have the embedding $\psi_G\colon {}^LG \to {}^L\widetilde{G}$, where $\widetilde{G} = \mathrm{Res}_{E/F}(G)$, corresponding to the base change lifting from $U(3)$ to $\mathrm{GL}(3)_{/E}$. In §13, as a consequence of the global theory, we define local L-packets for G and we prove the existence of the transfers of automorphic L-packets associated to ξ_H and ψ_G by the principle of functoriality. This leads to a classification of automorphic L-packets of G (§3.13). Let $\Pi_s(G)$ be the set of discrete L-packets on G whose base change transfer to $\mathrm{GL}(3)_{/E}$ is discrete. The elements of $\Pi_s(G)$ are stable in the above sense (Theorem 13.3.3). Furthermore, the base change transfer defines a bijection between $\Pi_s(G)$ and the set of discrete representations of $\mathrm{GL}(3)_{/E}$ which are invariant under the automorphism $g \to {}^t\bar{g}^{-1}$, where the bar denotes conjugation with respect to E/F, and whose central character, viewed as a character of I_E, has trivial restriction to I_F. Let $\Pi_e(G)$ be the set of cuspidal L-packets which are transfers of cuspidal L-packets on H. As in the case of SL(2), there is a multiplicity formula for the elements of L-packets belonging to $\Pi_e(G)$ (**Theorem 13.3.7**). The union $\Pi_s(G) \cup \Pi_e(G)$ is disjoint and accounts for most of the discrete spectrum of G. To describe the remaining discrete representations, we define an enlarged L-packet $\Pi(\rho)$, which we call an "A-packet", for each one-dimensional automorphic representation ρ of H. Locally, if v is a place of F, the L-packet $\xi_H(\rho_v)$ consists of a single representation $\pi^n(\rho_v)$ (§12). The A-packet $\Pi(\rho_v)$ contains $\pi^n(\rho_v)$ and it contains an additional representation $\pi^s(\rho_v)$ precisely when v remains prime in E. Every discrete representation of G which does not belong to an L-packet in $\Pi_s(G)$ or $\Pi_e(G)$ belongs to a unique A-packet $\Pi(\rho)$. The existence of these A-packets is predicted by the conjectures of J. Arthur ([A$_4$], [R$_5$]), as is the structure of the multiplicity formula for them (Theorem 13.3.7). The representations $\pi^n(\rho_v)$ are non-tempered and the philosophy of [A$_4$] suggests that the only cuspidal representations of G with a non-tempered component are those belonging to the A-packets $\Pi(\rho)$.

If E_v/F_v is $\mathbf{C}/\mathbf{R}$, then G_v is the real group $U_{2,1}(\mathbf{R})$ and the representations of the form $\pi^n(\rho_v)$ include all those such that $H^1(\mathfrak{g}, K, F \otimes \pi) \neq 0$ for some finite dimensional representation F of G_v. The representation $\pi^s(\rho_v)$ is either square-integrable or is a component of a certain reducible principal series representation. In particular, we obtain examples where the latter representations occur as local components of cuspidal representations.

By Theorem 13.3.1, the multiplicity of an automorphic representation of G which occurs in the discrete spectrum is always equal to one. In addition, two L-packets which are equal locally almost everywhere coincide.

In §14, the comparison between $U(3)$ and its inner forms is carried out. This leads to a generalization (Theorem 15.3.1) of the vanishing theorem of Rapoport-Zink ([RZ]), according to which $H^1(\Gamma, F) = 0$ for all congruence subgroups Γ of $U(2,1)$ which arise from global inner forms of $U(3)$ associated to division algebras over an imaginary quadratic extension of $\mathbf{Q}$ with an involution of the second kind.

The oscillator representation and L-function methods have been used to investigate automorphic representations of $U(3)$. The existence of cuspidal representations of G with local component of the form $\pi^n(\rho_v)$ was first discovered by Howe and Piatetski-Shapiro ([HP]). Results on the transfer from H as well as base change to E for $U(3)$ were obtained by Gelbart and Piatetski-Shapiro ([GP]). In most cases, the trace formula approach leads to a more precise and complete formulation of the results. This approach has also been considered in papers of Flicker.

The results of the present work have been applied to determine the zeta-functions of the Shimura varieties associated to unitary groups in three variables for the case E quadratic imaginary ([Mo]). As a consequence, ℓ-adic representations are associated to cohomological automorphic representations on $U(3)$. This theory has been applied in [BR_2] to obtain some cases of the Tate conjectures for these Shimura varieties. Furthermore, the endoscopic transfer from H leads to a construction of ℓ-adic representations for Hilbert modular forms ([BR_3]).

The present work relies on the general theory of endoscopy, as begun by Langlands and developed and expanded in the papers of Kottwitz and Shelstad. Results due to Langlands-Shelstad on the transfer of orbital integrals for $U(3)$ ([LS_2]) play an important role. In addition, we rely on the lectures presented in two seminars held at the Institute for Advanced Study during the year 1983-1984. The general twisted form of the Arthur trace formula developed in [M] is used for the base change comparison of $U(3)$ with GL(3). The approach to the stable trace formula taken here is based on the lectures presented in [A]. My thanks are due to R. Langlands for help and encouragement on numerous occasions.

I wish to thank The Institute for Advanced Study, the National Science Foundation and the Sloan Foundation for support at various times during the preparation of this book.

Automorphic Representations

of Unitary Groups

in Three Variables

CHAPTER 1

Preliminary definitions and notation

1.1. The symbol F will be used to denote a local field or global field of characteristic zero. The ring of integers of F will be denoted by $\mathcal{O}_F$. If F is a number field, let $\mathbf{A}_F$, I_F, and C_F be the adeles, ideles, and idele classes of F, respectively, and if F is local, let $C_F = F^*$. Let W_F be the absolute Weil group of F and if L/F is a finite extension, let $W_{L/F}$ be the Weil group of L/F. If L/F is Galois, let $\Gamma(L/F)$ be the Galois group. The symbol Γ will denote $\Gamma(\overline{F}/F)$, where $\overline{F}$ is an algebraic closure of F. The norm and trace maps will be denoted by $N_{L/F}$ and $\mathrm{Tr}_{L/F}$, respectively and we write L^1 and L° for their kernels. We write I_L^1 for the norm one elements in I_L If F is global, v will denote a place of F. Let F_v and $\mathcal{O}_v$ be the completions of F and $\mathcal{O}_F$ at v, and let $L_v = L \otimes_F F_v$.

1.2. Let $\underline{G}$ be an algebraic group over F. We will write G for the group $\underline{G}(F)$ of F-points of $\underline{G}$. If F is global, bold $\mathbf{G}$ will denote the group of $\mathbf{A}_F$-points of $\underline{G}$, and if v is a place of F, we set $G_v = \underline{G}(F_v)$. Let $\underline{Z}_G$ be center of $\underline{G}$. Let $X^*(G) = \mathrm{Hom}(\underline{G}, \mathbf{G}_m)$ be the lattice of characters of $\underline{G}$ and let $X^*(G)_F$ be the sublattice of F-rational characters. Let $X_*(G)$ be the lattice dual to $X^*(G)$. If F is global, let $\mathbf{G}^1 = \{g \in \mathbf{G} : |\chi(g)| = 1$ for all $\chi \in X^*(G)_F\}$.

1.3. The symbol $\underline{G}$ will henceforth denote a connected reductive group over F. Fix a minimal F-parabolic subgroup P_0 of G and let M_0 be a Levi factor of P_0. A subgroup M containing M_0 is called a Levy subgroup if it is a Levi factor of a parabolic subgroup. Let $\mathcal{L}(G)$ be the set of Levi subgroups of G. A parabolic subgroup of G will be called standard if it contains P_0. Let $\mathcal{P}$ be the set of standard parabolic subgroups of G. If $P \in \mathcal{P}$, the unique Levi factor of P containing M_0 will be denoted by M_P and N_P will denote the unipotent radical of P. Set $N_0 = N_{P_0}$.

When F is global, we fix a maximal compact subgroup $\mathbf{K} = \Pi_v K_v$ of $\mathbf{G}$ as follows. We assume that K_v is special for all v and that for almost all finite v, $K_v = G(\mathcal{O}_v)$, where $G(\mathcal{O}_v) = i(G_v) \cap \mathrm{GL}_n(\mathcal{O}_v)$ for some fixed embedding i of G in $\mathrm{GL}_{n/F}$. The Iwasawa decomposition $\mathbf{G} = \mathbf{PK}$ holds.

1.4. Let ε be a (possibly trivial) automorphism of $\underline{G}$ of finite order. Elements $x, y \in G$ are called ε-conjugate if $g^{-1}x\varepsilon(g) = y$ for some $g \in G$. For $\gamma \in \underline{G}$, the subgroup $\{g \in \underline{G} : g^{-1}\gamma\varepsilon(g) = \gamma\}$ is called the ε-centralizer of γ in $\underline{G}$ and will be denoted by $\underline{G}_{\gamma\varepsilon}$ or $\underline{G}(\gamma\varepsilon)$.

Let $D_\varepsilon(g)$ be the coefficient of the lowest degree term in the polynomial $p(g) = \det(T + 1 - ad(g \times \varepsilon))$ which is not identically zero. Here $\mathrm{ad}(g \times \varepsilon)$ is the automorphism $X \longrightarrow g\varepsilon(X)g^{-1}$. An element $g \in G$ is called ε-regular if $D_\varepsilon(g) \neq 0$. Let $G^{\varepsilon r}$ denote the set of ε-regular elements in G. If γ is ε-regular, then $\underline{G}(\gamma\varepsilon)^\circ$ is a torus ([C$_2$], §2). An element $\gamma \in G$ is called ε-semisimple if $\gamma \times \varepsilon$ is semisimple as an element of the non-connected algebraic group $G^* = G \rtimes \langle\varepsilon\rangle$. There is a Jordan decomposition $\gamma \times \varepsilon = s'u$ where s' is a semisimple element of G^* of the form $s \times \varepsilon$ for some $s \in G$ and u is a unipotent element in G which commutes with s'. We call s the ε-semisimple part of γ.

A superscript ε will denote "ε-fixed points". Thus, $\mathcal{P}^\varepsilon$ is the set of $P \in \mathcal{P}$ such that $\varepsilon(P) = P$. In general, when ε is trivial, the ε will be deleted from the notation. For example, we write $\underline{G}_\gamma$ for $\underline{G}_{\gamma\varepsilon}$ when ε is trivial.

1.5. Let F be a local field and let ω be a character of a subgroup Z of the center of G. Let $C(G, \omega)$ denote the space of smooth functions f on G such that $\mathrm{supp}(f)$ is compact modulo Z, $f(zg) = \omega^{-1}(z)f(g)$ for $z \in Z$, and if F is archimedean, such that f is K-finite (where K is a fixed maximal compact subgroup of G). If F is p-adic and K is a hyperspecial maximal compact subgroup of G, let $\mathcal{H}(G) = \mathcal{H}(G, \omega)$ be the Hecke algebra of bi-K-invariant functions in $C(G, \omega)$. This Hecke algebra is non-zero only if ω is trivial on $Z \cap K$. If F is global, we denote $\mathcal{H}(G_v)$ by $\mathcal{H}_v$.

1.6. Representations. All representations of reductive groups over local field are assumed to be admissible. By abuse of notation, we will not distinguish between an irreducible representation and its isomorphism class. If π is any representation, let $JH(\pi)$ denote the set of irreducible constituents of π.

The set of irreducible admissible representations of a reductive group G will be denoted by $E(G)$. According to the local Langlands correspondance (which is known, in general, if F is archimedean, and in special cases if F

is p-adic), $E(G)$ is partitioned into finite subsets called L-packets. The set of L-packets will be denoted by $\Pi(G)$.

Let $P \in \mathcal{P}$ and let σ be a representation of a Levi factor M_P of P, regarded as a representation of P on which N_P acts trivially. We denote by $i_{G,P}(\sigma)$, the representation of G unitarily induced from σ, acting by right translation on smooth functions φ such that $\varphi(pg) = \delta_P^{1/2}\sigma(p)\varphi(g)$, where δ_P is the modulus function $\det(p|\,\mathrm{Lie}(N_P)|$. In some situations, $i_{G,P}(\sigma)$ will be denoted by $i_P(\sigma)$, $i_G(\sigma)$ or I_σ.

Let P be a Borel subgroup. If σ is a unitary character or if $i_P(\sigma)$ is irreducible, then the set $JH(i_P(\sigma))$ is an L-packet. We will call such an L-packet a principal series (abbreviated p.s.) L-packet. In §11 and §12, a p.s. L-packet consisting of more that one element (i.e., when σ is unitary and $i_P(\sigma)$ is reducible) will be called an l.d.s. L-packet.

The term square-integrable (resp., supercuspidal, over a p-adic field), will be used to mean square-integrable (resp., supercuspidal) modulo the center and irreducible. An L-packet Π will be called square-integrable (resp. supercuspidal, tempered, unitary) if each element of Π is square-integrable (resp. supercuspidal, tempered, unitary). The set of square-integrable L-packets will be denoted by $\Pi^2(G)$.

If $\pi \in E(G)$ satisfies $\varepsilon(\pi) = \pi$, there exists an operator $\pi(\varepsilon)$ such that $\pi(\varepsilon)\pi(g)\pi(\varepsilon)^{-1} = \pi(\varepsilon(g))$. Suppose that the restriction of π to Z is ω (where Z is some fixed subgroup of the center of G). The twisted character $\chi_{\pi\varepsilon}$ of π is the distribution on $C(G,\omega)$ defined by $\chi_{\pi\varepsilon}(\phi) = \mathrm{Tr}(\pi(\phi)\pi(\varepsilon))$, where

$$\pi(\phi) = \int_{Z\backslash G} \phi(g)\pi(g)dg \ .$$

The distribution $\chi_{\pi\varepsilon}$ depends on the choice of $\pi(\varepsilon)$. There exists a locally constant function on $G^{\varepsilon r}$, which we also denote by $\chi_{\pi\varepsilon}$, such that

$$\chi_{\pi\varepsilon}(f) = \int_{Z\backslash G} f(g)\chi_{\pi\varepsilon}(g)dg.$$

This is due to Harish-Chandra when ε is trivial or G is a real group and was extended by Clozel ([C$_2$]) to the p-adic case for non-trivial ε.

For $\gamma \in G$ and $f \in C(G,\omega)$, we will consider the (twisted) orbital integral:

$$\Phi_{G,\varepsilon}(\gamma, f) = \Phi_\varepsilon(\gamma, f) = \int_{G(\gamma\varepsilon)'\backslash G} f(g^{-1}\gamma\varepsilon(g))dg \ .$$

where $G(\gamma\varepsilon)' = \{g \in G : g^{-1}\gamma\varepsilon(g)\gamma^{-1} \in Z'\}$. Here Z' is a subgroup of the center which is trivial if ε is trivial. In the cases under consideration where ε is non-trivial, we specify Z' (§4.10 and §4.11).

1.7. Measures. Assume that F is global. An additive character $\psi = \Pi\psi_v$ of $F\backslash\mathbf{A}_F$ defines a Haar measure dx_v on F_v for all places v, namely, dx_v is the Haar measure on F_v which is self-dual with respect to the Fourier transform defined by ψ_v. We also obtain a measure $\otimes dx_v$ on $\mathbf{A}_F$. The local and global Haar measures corresponding to $\psi = \psi_0 \circ \mathrm{Tr}_{F/\mathbf{Q}}$, where ψ_0 is the unique additive character of $\mathbf{Q}\backslash\mathbf{A}_Q$ such that $\psi_0(x_\infty) = e^{2\pi i x_\infty}$ for all $x_\infty \in \mathbf{R}$, will be called the standard additive measures.

An algebraic differential form Ω of degree n on an n-dimensional variety over F_v defines a measure $|\Omega|$ in a standard way. In local coordinates, $\Omega = \varphi(x^1, \dots, x^n)dx^1 \wedge \dots \wedge dx^n$ and we define

$$|\Omega| = |\varphi(x^1, \dots, x^n)| dx_v^1 \cdots dx_v^n \ .$$

A measure obtained in this way will be called an algebraic measure. When the variety and the form are defined globally, this gives a measure locally for all places. In particular, if Ω is an invariant form of degree $\dim(G)$ on G, we obtain a Haar measures $|\Omega|_v$ on G_v for all places v.

Recall that an $\overline{F}$-isomorphism $\psi : \underline{G}' \to \underline{G}$ between F-groups $\underline{G}$ and $\underline{G}'$ is called an inner twisting if for all $\tau \in \Gamma$, there exists $a_\tau \in \underline{G}$ such that $\psi^\tau = ad(a_\tau) \circ \psi$ for all $\tau \in \Gamma$, where $\psi^\tau(g) = \tau(\psi(\tau^{-1}(g)))$. In this case, $\underline{G}'$ is said to be an inner form of $\underline{G}$. The pull-back $\Omega' = \psi^*(\Omega)$ is an invariant form of degree $\dim(G')$, defined over F since Ω is invariant under conjugation. Measures dg and dg' in G and G', respectively, are said to be compatible if $dg = c|\Omega|_v$ and $dg' = c|\Omega'|_v$ for some constant c. All measures on groups are assumed to be Haar measures.

Let λ be the representation of Γ on the lattice $X^*(G)$. The local Tamagawa measure with respect to Ω is defined by: $dg_v = L(1, \lambda_v)|\Omega|_v$ where λ_v is the restriction of λ to the decomposition group at v. The product measure Πdg_v will be called the (unnormalized) Tamagawa measure.

If ω is a character of a subgroup $\mathbf{Z}$ of the center of $\mathbf{G}$, we denote by $C(\mathbf{G}, \omega)$ the space of smooth compactly supported functions f on $\mathbf{G}$ which are linear combinations of functions of the form $f = \Pi f_v$, where $f_v \in C(G_v, \omega_v)$ for all v and f_v is the unit in $\mathcal{H}_v$ for almost all finite v.

We will use unnormalized Tamagawa measures globally and Tamagawa measures locally, unless otherwise stated. With this convention, we define, the global orbital integral:

$$\Phi_\varepsilon(\gamma, f) = \int\limits_{\mathbf{G}(\gamma\varepsilon)\backslash\mathbf{G}} f(g^{-1}\gamma\varepsilon(g))dg = \prod \Phi_\varepsilon(\gamma, f_v) .$$

for $f \in C(G,\omega)$ and ε-semisimple $\gamma \in G$.

1.8 *L*-groups. Let $(\underline{B}, \underline{T})$ be a Borel pair in $\underline{G}$, i.e., $\underline{T}$ is a maximal torus in $\underline{G}$ and $\underline{B}$ is a Borel subgroup containing $\underline{T}$. A based root datum for $\underline{G}$ is a quadruple $\psi(G) = (X^*(T), \Delta^*, X_*(T), \Delta_*)$ where Δ^* and Δ_* are the sets of simple roots and co-roots which are positive with respect to $\underline{B}$, respectively. Up to canonical isomorphism, $\psi(G)$ is independent of the choice of $(\underline{B}, \underline{T})$. If $(\underline{B}', \underline{T}')$ is another Borel pair, then there exists $g \in \underline{G}$ such that $\mathrm{ad}(g)$ carries $(\underline{B}', T')$ to $(\underline{B}, \underline{T})$. The induced isomorphism on root data is independent of g.

Let $\mathrm{Out}(\underline{G})$ denote the quotient of $\mathrm{Aut}(\underline{G})$ by the group of inner automorphisms. Then $\mathrm{Out}(\underline{G})$ acts on $\psi(G)$ and is isomorphic to the group of automorphisms of $\psi(G)$. ([Bo]). The dual root data $\psi(G)^\wedge$ is the quadruple $\psi(G)^\wedge = (X_*(T), \Delta_*, X^*(T), \Delta^*)$.

The Galois group Γ acts on $\psi(G)$. If $\tau \in \Gamma$, then τ defines an isomorphism between the root data of $(\underline{B}, \underline{T})$ and $(\tau(\underline{B}), \tau(\underline{T}))$, and hence induces an automorphism of $\psi(\underline{G})$. We obtain a homomorphism $\Gamma \to \mathrm{Out}(\underline{G})$.

The dual group $\widehat{G}$ of $\underline{G}$ is the complex, connected reductive group whose root data $\psi(\widehat{G})$ is isomorphic to $\psi(G)^\wedge$ ([Bo], [Kt3]). Let $(B, T, \{X_\alpha\})$ be a splitting for $\widehat{G}$, i.e., (B, T) is a Borel pair and $\{X_\alpha\}$ is a set of basis elements for the root spaces in $\mathrm{Lie}(\widehat{G})$ associated to the set of simple roots of T with respect to B. The splitting defines a section s of the map $\mathrm{Aut}(\widehat{G}) \to \mathrm{Out}(\widehat{G})$. For $\delta \in \mathrm{Out}(\widehat{G})$, $s(\delta)$ is the unique automorphism of $\widehat{G}$ in the preimage of δ which fixes (B, T) and permutes the elements of $\{X_\alpha\}$. Fix an isomorphism of $\psi(G)^\wedge$ with $\psi(\widehat{G})$. This induces an isomorphism of $\mathrm{Out}(G)$ with $\mathrm{Out}(\widehat{G})$. We obtain a homomorphism $\Gamma \to \mathrm{Out}(\widehat{G})$, and, via s, an action of Γ on $\widehat{G}$ which preserves the splitting. The L-group of G is the semi-direct product ${}^LG = \widehat{G} \rtimes W_F$, where W_F acts on $\widehat{G}$ through its projection to Γ. Observe that although the action of Γ on $\widehat{G}$ depends on the choice of $(B, T, \{X_\alpha\})$, the restriction of the action to $Z(\widehat{G})$ is independent of this choice.

A map between L-groups, or a map from W_F to an L-group will be called an L-map if it commutes with the natural projections to W_F.

1.9 Unitary groups. Let E/F be a quadratic extension and let σ be the conjugation of E with respect to F. We also write $\sigma(x) = \overline{x}$. The character of order two of I_F associated to E/F by class field theory will be denoted by $\omega_{E/F}$. If χ is a character of E^* or I_E, $\overline{\chi}$ will denote the character $\overline{\chi}(x) = \chi(\overline{x})$.

Let $\Phi \in \mathrm{GL}_n(E)$ be a Hermitian matrix, i.e., ${}^t\overline{\Phi} = \Phi$. Let $\underline{U}_\Phi$ be the unitary group defined by Φ. Then

$$U_\Phi = \{g \in \mathrm{GL}_n(E)\colon g\Phi\, {}^t\overline{g} = \Phi\}.$$

Let $\Phi_n = (\Phi_{ij})$ where $\Phi_{ij} = (-1)^{i-1}\delta_{i,n+1-j}$ and δ_{ab} is the Kronecker delta. Then Φ_n is Hermitian if n is odd and $\xi\Phi_n$ is Hermitian if n is even, where $\xi \in E^*$ is an element of trace 0. We denote by $\underline{U}(n)$ the unitary group with respect to Φ_n or $\xi\Phi_n$, in the two cases, respectively. It is quasi-split and the subgroup of upper triangular matrices is a Borel subgroup. A group $\underline{G}$ will be called a unitary group if it is an inner form of $\underline{U}(n)$.

Let D be a finite-dimensional semisimple algebra over E. We call (D, α) a pair of the second kind if α is an involution of the second kind, i.e., α is an anti-automorphism of D whose restriction to E is σ. If D is a simple algebra with center E, then there is an algebraic group $\underline{G}$ over F such that $G = \{g \in D^* : \alpha(g)g = 1\}$. The group $\underline{G}$ is a unitary group and all unitary groups are obtained in this way from pairs (D, α). Over E, $\underline{G}$ is isomorphic to the algebraic group defined by D^*. If $D = M_n(E)$, then $\alpha(g) = \Phi\, {}^t\overline{g}\Phi^{-1}$ for some Hermitian form Φ and $\underline{G} = \underline{U}_\Phi$.

Suppose that E/F is a quadratic extension of number fields and let v a place of F. We have $G_v = \{g \in (D \otimes_F F_v)^* : \alpha(g)g = 1\}$. If v splits into two places w, w' of E, then $D \otimes_F F_v = D_w \times D_{w'}$ and α induces an anti-automorphism of D_w with $D_{w'}$. In this case, projection onto the first or second factors induces an isomorphism of G_v onto D_w^* or $D_{w'}^*$, respectively (these groups are isomorphic via $g \to \alpha(g)^{-1}$). If v remains prime, then α induces an isomorphism of D_v with its opposite algebra. This shows that the class of D_v in the Brauer group has order two. In particular, D_v is the split algebra if $\dim_E(D)$ is odd. In this case, G_v is isomorphic to U_Φ for some Hermitian matrix $\Phi \in \mathrm{GL}_n(E_v)$. By a theorem of Landherr ([L]), if v is finite, then the equivalence class of a Hermitian matrix Φ is determined by the class of $\det(\Phi)$ in $F_v^*/N_{E/F}(E_v^*)$. Observe that $\underline{U}_\Phi$ is isomorphic to $\underline{U}_{\lambda\Phi}$ for all $\lambda \in F_v^*$ and hence isomorphism class of $\underline{U}_\Phi$ depends only on $F_v^*/(F_v^*)^n N_{E/F}(E_v^*)$. In particular, if n is odd, there is a unique isomorphism class of unitary groups with respect to E_v/F_v. In the global case, the equivalence class of Φ is determined by the class of $\det(\Phi)$ in $F^*/N_{E/F}(F^*)$ and the signatures of Φ at archimedean places v such that E_v/F_v is isomorphic to $\mathbf{C}/\mathbf{R}$.

1.10. Unitary groups is three variables. We fix the following notation. If $G = U(n)$ or $\mathrm{GL}(n)$, B will denote the Borel subgroup of upper-triangular matrices, N will denote its unipotent radical, and M will denote the diagonal subgroup of B. A diagonal matrix with diagonal entries $a_1, \ldots, a_n$ will be denoted by $d(a_1, \ldots, a_n)$. Let α_j be the simple root of M defined by $\alpha_j(a_1, \ldots, a_n)) = a_j/a_{j+1}$ for $j = 1, \ldots, n-1$.

For $x, z \in E$ such that $x\overline{x} = z + \overline{z}$, set

$$u(x,z) = \begin{pmatrix} 1 & x & z \\ 0 & 1 & \overline{x} \\ 0 & 0 & 1 \end{pmatrix}.$$

If $G = U(3)$, then $N = \{u(x,z) : x, z \in E,\ x\overline{x} = z + \overline{z}\}$ and $M = \{d(\alpha, \beta, \overline{\alpha}^{-1}) : \alpha \in E^*,\ \beta \in E^1\}$. We will often let δ_0 denote a fixed non-zero element of E°. For $x \in E$, let $u(x) = u(x, \frac{x\overline{x}}{2})$ and for $w \in E^\circ$, let $n(w) = u(0, w)$.

If $G = U(2)$, then $M = \{d(\alpha, \overline{\alpha}^{-1}) : \alpha \in E^*\}$ and $N = \{n(t) : t \in F\}$, where $n(t) = \left(\begin{smallmatrix} 1 & t \\ 0 & 1 \end{smallmatrix}\right)$. Note that the derived group of G is isomorphic to $\mathrm{SL}(2)_{/F}$. The root of M in N will be denoted by α.

Let $G = U(n)$, where $n = 2$ or 3, and assume that E/F is global. We fix the following choices of maximal compact subgroups K_v of G_v for all places v. If v is p-adic, let K_v be the subgroup of integer matrices in G_v, except if $G = U(3), E_v/F_v$ is a quadratic extension, and the residual characteristic of v is two. In this case, choose $\lambda \in E_v$ such that $\mathrm{Tr}_{E/F}(\lambda) = 1$ and $\|\lambda\| = \min\{\|x\| : x \in E\ \mathrm{Tr}_{E/F}(x) = 1\}$. Let K_v be the stabilizer in G_v of the lattice

$$\left\{ \begin{pmatrix} a \\ b \\ c \end{pmatrix} \in E^3 : a, b, \lambda c \in \mathcal{O}_v \right\}.$$

Then K_v is a special maximal compact subgroup ([Ti]). Set

$$U_n(\mathbf{R}) = \{g \in \mathrm{GL}_n(\mathbf{C}) :\ {}^t\overline{g}g = 1\}$$
$$O_n(\mathbf{R}) = \{g \in \mathrm{GL}_n(\mathbf{R}) :\ {}^t\overline{g}g = 1\}.$$

If $E_v/F_v = \mathbf{C}/\mathbf{R}$, let $K_v = G_v \cap U_n(\mathbf{R})$. If v is archimedean and splits in E, then G_v is isomorphic to $\mathrm{GL}_n(E_w)$, where w is a place of E dividing v. Let K_v be the subgroup of G_v whose image in $\mathrm{GL}_n(E_w)$ is $U_n(\mathbf{R})$ (resp., $O_n(\mathbf{R})$), if $F_v = \mathbf{C}$ (resp., $F_v = \mathbf{R}$). Note that K_v is independent of the choice of w. In general, if f is a function on G_v and K is a compact subgroup of G_v, f^K will denote the averaged function $f^K(x) = \int_K f(kxk^{-1})dk$.

CHAPTER 2

The trace formula

The purpose of this chapter is to review the twisted version of Arthur's trace formula as it is presented in [M] (cf. also [La]).

2.1 The twisted trace formula. Let $\underline{Z}$ be a subgroup of $\underline{Z}_G$ such that $\underline{Z}\backslash\underline{Z}_G$ is anisotropic and let ω be a unitary character of $Z\backslash\mathbf{Z}$. Let $L(G) = L(G,\omega)$ be the space of measurable functions φ on $G\backslash\mathbf{G}$ such that $\varphi(zg) = \omega(z)\varphi(g)$ for all $z \in \mathbf{Z}$ and

$$\int_{\mathbf{Z}G\backslash\mathbf{G}} |\varphi(g)|^2 dg < \infty.$$

Let ρ be the representation of $\mathbf{G}$ on $L(G)$ by right translation. The space $L(G)$ decomposes, under the action of G, into a direct sum $L_c(G) \oplus L_d(G)$, where $L_d(G)$ is the closed span of the irreducible closed invariant subspaces of $L(G)$ and $L_c(G)$ is the orthogonal complement of $L_d(G)$ in $L(G)$. Let ρ_d be the restriction of ρ to $L_d(G)$.

Let P be a parabolic subgroup of G. If φ is a function on G which is left-invariant by P, set

$$\varphi_P(g) = \int_{N_P\backslash\mathbf{N}_P} \varphi(ng)\, dn\ .$$

A function $\varphi \in L(G)$ is called a cusp form if for all P, $\varphi_P(g) = 0$ for almost all $g \in \mathbf{G}$. The subspace $L_0(G)$ of cusp forms is a closed $\mathbf{G}$-invariant subspace which, by a theorem of Gelfand and Piatetski-Shapiro, is contained in $L_d(G)$. An irreducible representation of $\mathbf{G}$ is called cuspidal (resp., discrete) if it occurs in $L_0(G)$ (resp., $L_d(G)$).

Let ε be an F-automorphism of $\underline{G}$ of finite order. We assume that ε normalizes M_0, P_0, and Z and that $\omega(\varepsilon^{-1}(z)) = \omega(z)$ for $z \in \mathbf{Z}$. Define an operator $\rho(\varepsilon)$ on $L(G)$ by $(\rho(\varepsilon)\varphi)(g) = \varphi(\varepsilon^{-1}(g))$. Then $\rho(\varepsilon)$ defines an extension of ρ to the semi-direct product $\mathbf{G} \rtimes \langle\varepsilon\rangle$. Furthermore, $\rho(\varepsilon)$

preserves $L_0(G)$ and $L_d(G)$. Denote the restriction of $\rho(\varepsilon)$ to $L_d(G)$ by $\rho_d(\varepsilon)$.

Let $f \in C(G,\omega)$. The operator $\rho(f)$ is defined and $\rho_d(f)\rho_d(\varepsilon)$ is known to be of trace class ([Mu]). The twisted trace formula will be applied exclusively to functions of this type.

Let (M,σ) be a pair consisting of a Levi subgroup M and a unitary cuspidal representation σ of M. We will tacitly assume that the subgroup $\mathbf{Z}$ acts via ω on all representations of $\mathbf{G}$ and its Levi subgroups. Two pairs (M,σ) and (M',σ') are said to be equivalent if (M,σ) is G-conjugate to $(M',\sigma'\otimes\alpha)$ for some character α of $\mathbf{M}^1\backslash\mathbf{M}$. A cuspidal datum is an equivalence class of pairs $\chi = \{(M,\sigma)\}$.

The trace formula gives two expressions for the integral

$$T_G(f) = \int\limits_{\mathbf{Z}G\backslash\mathbf{G}} k^T(x)dx$$

where $k^T(x)$ is a function defined below. We write the trace formula as an equality

$$\sum J_{\mathcal{O}}^T(f) = \sum J_\chi^T(f) \,. \tag{2.1.1}$$

where both sides are equal to $T_G(f)$. The left-hand side is called the $\mathcal{O}$-expansion and is a sum over a set of representatives $\{\mathcal{O}\}$ for the set of ε-semisimple conjugacy classes in G modulo Z. The right-hand side is called the χ-expansion and is a sum over the set of cuspidal data. Each term in (2.1.1) is a polynomial in T.

The version of the twisted trace formula presented in [M] will be used, with a minor change. In this work, $k^T(x)$ is integrated over $\mathbf{Z}G\backslash\mathbf{G}$, whereas in [M], it is integrated over $G\backslash\mathbf{G}^1$.

2.2 The $\mathcal{O}$-expansion. To define $k^T(x)$, we assume for simplicity, that $F = \mathbf{Q}$ (of course, this entails no loss of generality). For $P \in \mathcal{P}$, let A_P be the maximal $\mathbf{Q}$-split torus in the center of M_P and let $\mathcal{A}_P = \mathrm{Lie}(A_P(\mathbf{R}))$. We also denote $\mathcal{A}_P$ by $\mathcal{A}_{M_P}$. An element $g \in \mathbf{G}$ can be written in the form $g = amnk$ where $a \in A_P(\mathbf{R})^\circ$, $m \in \mathbf{M}_P^1$, $n \in \mathbf{N}$, and $k \in \mathbf{K}$. Let $H_P(g)$ denote the logarithm of a. Set $\mathcal{A}_0 = \mathcal{A}_{P_0}$ and let $H(g) = H_{P_0}(g)$. We identify $\mathcal{A}_0$ with its dual by means of a positive-definite inner product which is invariant under ε and the Weyl group. Let $\Delta_0 = \{\alpha\}$ be the set of simple roots of A_0 in N_0 and let $\widehat{\Delta}_0 = \{\omega_\alpha\}$ be the dual basis of the span of Δ_0 in $\mathcal{A}_0$. Let $\widehat{\Delta}_P$ be the set of ω_α such that α does not occur in M_P. For $\alpha \in \Delta_0$, define

$$\overline{\omega}_\alpha = \frac{1}{\ell}\sum_{j=0}^{\ell-1} \varepsilon^j(\omega_\alpha)$$

where ℓ is the order of ε.

Assume that $P \in \mathcal{P}^\varepsilon$. Let $a(P,\varepsilon) = \dim(\mathcal{A}_P^\varepsilon)$ and let $\hat{\tau}_{P,\varepsilon}$ be the characteristic function of the set

$$\{H \in \mathcal{A}_0 : \langle H, \overline{\omega}_\alpha \rangle > 0 \quad \text{for} \quad \omega_\alpha \in \widehat{\Delta}_P\} .$$

Define:

$$K_P(x,y) = \sum_{\gamma \in Z\backslash M_P} \int_{\mathbf{N}_P} f(x^{-1}\gamma n \varepsilon(y))dn .$$

Then $K_P(x,y)$ is the kernel of the operator $\rho_P(f)\rho_P(\varepsilon)$, where ρ_P is the representation of $\mathbf{G} \rtimes \langle\varepsilon\rangle$ on the space of functions φ on $\mathbf{G}$ such that $\varphi(zmng) = \omega(z)\varphi(g)$ for $z \in \mathbf{Z}$, $m \in M_P$, $n \in \mathbf{N}_P, g \in \mathbf{G}$, and

$$\int_{\mathbf{K}} \int_{\mathbf{Z}M_P\backslash \mathbf{M}_P} |\varphi(mk)|^2 dmdk < \infty .$$

The function $k^T(x)$ is defined by

$$k^T(x) = \sum_{P \in \mathcal{P}^\varepsilon} (-1)^{a(P,\varepsilon)} \sum_{\delta \in P\backslash G} K_P(\delta x, \delta x)\hat{\tau}_{P,\varepsilon}(H(\delta x) - T) .$$

The sums over δ and γ in the definitions of $k^T(x)$ and $K_P(x,y)$ are finite. For T sufficiently regular, $k^T(x)$ is integrable over $\mathbf{Z}G\backslash\mathbf{G}$.

Elements $\gamma_1, \gamma_2 \in G$ will be called ε-semisimply conjugate if the semisimple parts of the Jordan decompositions of γ_1 and γ_2 are ε-conjugate in G. If $\mathcal{O}$ is an ε-semisimple conjugacy class in G, we denote by $\underline{\mathcal{O}}$ the set of elements in G which are ε-semisimply conjugate to an element of $\mathcal{O}$.

The $\mathcal{O}$-expansion is derived directly from the definition of $k^T(x)$. Let $\{\mathcal{O}\}$ be a set of representatives for the set of ε-semisimple conjugacy classes in G modulo the action of Z. For each $\mathcal{O}$, let $\underline{\mathcal{O}}'$ be a set of elements in $\underline{\mathcal{O}}$ such that for all $\gamma \in \underline{\mathcal{O}}$, there is a unique $z \in Z$ such that $z\gamma \in \underline{\mathcal{O}}'$. Define:

$$K_{P,\mathcal{O}}(x,y) = \sum_{M_P \cap \underline{\mathcal{O}}'} \int_{\mathbf{N}_P} f(x^{-1}\gamma n \varepsilon(y))dn .$$

Clearly, $K_P = \sum_{\{\mathcal{O}\}} K_{P,\mathcal{O}}$. Set

$$k_{\mathcal{O}}^T(f) = \sum_{P \in \mathcal{P}^\varepsilon} (-1)^{a(P,\varepsilon)} \sum_{\delta \in P \backslash G} K_{P,\mathcal{O}}(\delta x, \delta x)\hat{\tau}_{P,\varepsilon}(H(\delta x) - T) \, .$$

Then $k^T(x) = \sum_{\{\mathcal{O}\}} k_{\mathcal{O}}^T(x)$. Furthermore, $k_{\mathcal{O}}^T$ is integrable over $\mathbf{Z}G\backslash\mathbf{G}$. Let $J_{\mathcal{O}}^T(f)$ denote its integral. Then (2.1.1) is equal to $\sum J_{\mathcal{O}}^T(f)$.

There is a more useful expression for $J_{\mathcal{O}}^T(f)$. Let $P \in \mathcal{P}^\varepsilon$ and for $\gamma \in M_P \cap \mathcal{O}'$, let γ_s be the ε-semisimple part of γ. Set

$$N(x, \gamma, f) = \int_{\mathbf{N}_P(\gamma_s \varepsilon)} f(x^{-1} n \gamma \varepsilon(x)) dn$$

$$j_{P,\mathcal{O}}(x) = \sum_{\gamma \in M_P \cap \underline{\mathcal{O}}'} \sum_{\eta \in N_P(\gamma_s \varepsilon) \backslash N_P} N(\eta x, \gamma, f)$$

$$J_{\mathcal{O}}^T(x) = \sum_{P \in \mathcal{P}^\varepsilon} \sum_{\delta \in P \backslash G} (-1)^{a(P,\varepsilon)} j_{P,\mathcal{O}}(\delta x) \hat{\tau}_{P,\varepsilon}(H(\delta x) - T).$$

Then $J_{\mathcal{O}}^T$ is integrable over $\mathbf{Z}G\backslash\mathbf{G}$ and its integral is equal to $J_{\mathcal{O}}^T(f)$ ([M]).

2.3 The fine χ-expansion. The space $L(G)$ decomposes into an orthogonal direct sum of subspaces $L_\chi(G)$, indexed by the set $\{\chi\}$ of cuspidal data. Let (M_1, ρ) be a representative for χ and let P_1 be the standard parabolic subgroup containing M_1 as a Levi subgroup. Let ρ^1 be the restriction of ρ to $\mathbf{M}_1^1$ and let ψ be a smooth function on $\mathbf{G}$ such that

(i) $\psi(zmng) = \omega(z)\psi(g)$ for $n \in \mathbf{N}_{P_1}$, $m \in M_1$ and $z \in \mathbf{Z}$.

(ii) For all $g \in \mathbf{G}$, the function $m \to \psi(mg)$ belongs to the ρ^1-isotypic component of the space $L_0^2(M_1 \backslash \mathbf{M}_1^1)$ of cusp forms on $\mathbf{M}_1^1$.

(iii) The set of $a \in A_{P_1}(\mathbf{R})^\circ$ such that $\psi(amnk) \neq 0$ for some $m \in \mathbf{M}_1^1$, $n \in \mathbf{N}_{P_1}, k \in K$, is compact modulo $A_G(\mathbf{R})^\circ$.

By a basic result ([L$_9$]), the series

$$\varphi(g) = \sum_{\delta \in P \backslash G} \psi(\delta g)$$

converges and φ belongs to $L(G)$. Let $L_\chi(G)$ be the closed span of functions of this form. More generally, let $P \in \mathcal{P}$, set $M = M_P, N = N_P$, and suppose that $M_1 \subset M$. Then a subspace $L_\chi(M)$ of $L(M, \omega')$ is defined, where ω' is the restriction to $\mathbf{Z}_M$ of the central character of ρ.

Let σ be a unitary discrete representation of $\mathbf{M}$ and let $\hat{\sigma}$ be the direct sum of the irreducible subspaces of $L_\chi(M)$ isomorphic to σ. Extend σ to a representation of $\mathbf{P}$ trivial on $\mathbf{N}$ and let I_σ^P be the representation of $\mathbf{G}$ induced from $\hat{\sigma}$. We realize I_σ^P on the space of functions ϕ on $\mathbf{G}$ such that

(i) $\phi(\gamma n g) = \phi(g)$ for $\gamma \in P, n \in \mathbf{N}$.
(ii) For all $g \in \mathbf{G}$, the function $m \to \phi(mg)$ belongs to $\hat{\sigma}$.
(iii) $\int_{\mathbf{Z}_M M \backslash \mathbf{M} \times \mathbf{K}} |\phi(mk)|^2 dm dk < \infty$.

For $\lambda \in \mathcal{A}_P$, let χ_λ be the character of $\mathbf{P}$ defined by $\chi_\lambda(mn) = \exp(\langle \lambda, H_P(m) \rangle)$ for $m \in \mathbf{M}$, $n \in \mathbf{N}$. Let $I_{\sigma,\lambda}^P$ denote the representation induced from $\sigma \otimes \chi_\lambda$. We realize $I_{\sigma,\lambda}^P$ on the same space as I_σ^P, defining the action by

$$(I_{\sigma,\lambda}^P(g)\phi)(x) = \chi_\lambda(H_P(xg))\chi_\lambda(H_P(x))^{-1}\phi(xg) .$$

Let Ω_0 be the Weyl group of M_0. If M' is another Levi subgroup, let $\Omega(M, M')$ be the set of maps from $\mathcal{A}_M$ to $\mathcal{A}_{M'}$ obtained by restricting elements $s \in \Omega_0$ such that $s(M) = M'$. Let P' be a parabolic subgroup with Levi factor M'. For $s \in \Omega(M, M')$, there is an intertwining operator

$$M_{P'|P}(s, \lambda) : I_{\sigma,\lambda}^P \to I_{s(\sigma),s(\lambda)}^{P'} \cdot$$

For λ sufficiently positive with respect to P, $M_{P'|P}(s, \lambda)\varphi(g)$ is defined by the integral:

$$\int_{\mathbf{N}_s(P',P)\backslash \mathbf{N}_{P'}} \varphi(w^{-1}ng)\exp((\lambda+\rho_P)(H_P(w^{-1}ng))-(s(\lambda)+\rho_{P'})(H_{P'}(g)))dn.$$

Here w is a representative for s, $\mathbf{N}_s(P', P) = \mathbf{N}_{P'} \cap w\mathbf{N}_P w^{-1}$ and ρ_P is the half-sum of the roots of A_P in N_P. By Langlands' theory of Eisenstein series, $M_{P'|P}(s, \lambda)$ has a meromorphic continuation in λ such that $M_{P'|P}(s, \lambda)$ is unitary if $\mathrm{Re}(\lambda) = 0$.

Assume that $\mathcal{A}_G^\varepsilon = \{0\}$. The distribution $J_\chi(f)$ is equal to a sum of terms indexed by quintuples $(M, L, \mathcal{A}, \sigma, s)$ consisting of Levi subgroups M, L such that $M \subset L$, a subspace $\mathcal{A} \subset \mathcal{A}_M$, an element $s \in \Omega(M, \varepsilon(M))$, and a discrete representation σ of M which occurs in $L_\chi(M)$ such that

(1) $\mathcal{A}$ is the set of fixed points of $s\varepsilon$ in $\mathcal{A}_M$.
(2) The pair $(L, \mathcal{A})$ is conjugate to a pair $(L', \mathcal{A}')$ where L' is an ε-stable Levi factor and $\mathcal{A}' = \mathcal{A}_{L'}^\varepsilon$.
(3) σ is trivial on $\exp(\mathcal{A}_M)$ and its restriction to $\mathbf{Z}$ is ω.

(4) σ is equivalent to $s\varepsilon(\sigma)$.

Let Δ be the determinant of $s\varepsilon - 1$ acting on $\mathcal{A}_M/\mathcal{A}$ and set

$$c = (|\Delta| \cdot |\mathcal{P}(M)| \cdot |\Omega_0|)^{-1} |\Omega(M)| .$$

Here $\Omega(M)$ denotes the Weyl group of M_0 in M.

Since $\mathcal{A}_G^\varepsilon = \{0\}$, $\mathcal{A} = \{0\}$ for any quintuple such that $L = G$. Let $M_{P'|P}(s) = M_{P'|P}(s,0)$. Let $\mathcal{P}(M)$ be the set of parabolic subgroups containing M as a Levi subgroup. The term associated to a quintuple of the form $(M, G, \{0\}, \sigma, s)$ is

$$c \sum_{P \in \mathcal{P}(M)} \operatorname{Tr}(I_\sigma^P(f) M_{P|\varepsilon(P)}(s) I_\sigma^P(\varepsilon)). \tag{2.3.1}$$

Here $I_\sigma^P(\varepsilon)$ is the map from I_σ^P to $I_{\varepsilon(\sigma)}^{\varepsilon(P)}$ sending $\varphi(g)$ to $\varphi(\varepsilon^{-1}(g))$. This term is an ε-invariant distribution and is equal to a sum of twisted traces. Terms of this type are called discrete.

If $L \neq G$, the term associated to $(M, L, \mathcal{A}, \sigma, s)$ no longer defines an invariant distribution and it is more complicated to describe. We first define a subset $\mathcal{P}_\varepsilon(L)$ of $\mathcal{P}(L)$. Assume first that L is the Levi factor of an ε-stable standard parabolic subgroup. If $\alpha \in \Delta_0$ does not occur in L, then the restriction α' of α to $\mathcal{A}_L^\varepsilon$ is non-zero and the set of α' divides $\mathcal{A}_L^\varepsilon$ into chambers. To each chamber W is associated a parabolic subgroup $Q = LN$, where a root α occurs in N if and only if α' is positive on W. Let $\Delta_{Q,\varepsilon}$ be the set of α' which are positive on W. If $(L', \mathcal{A}')$ is conjugate to $(L, \mathcal{A})$ by an element $s \in \Omega_0$, let $\mathcal{P}_\varepsilon(L') = \{wQw^{-1} : Q \in \mathcal{P}_\varepsilon(L)\}$, where w is a representative for s, and if $Q' = wQw^{-1}$, let $\Delta_{Q',\varepsilon} = \{s(\alpha) : \alpha \in \Delta_{Q,\varepsilon}\}$. Let $v_{Q'}$ be the volume of the parallelepiped spanned by $\Delta_{Q',\varepsilon}$ and for $\Lambda \in \mathcal{A}$, define:

$$\theta_{Q',\varepsilon}(\Lambda) = v_{Q'}^{-1} \prod_{\alpha \in \Delta_{Q',\varepsilon}} \langle \Lambda, \alpha \rangle .$$

There exists $T_0 \in \mathcal{A}_0$ such that $H(w) = T_0 - sT_0$ for all w representing an element $s \in \Omega_0$. For $Q \in \mathcal{P}_\varepsilon(L)$, let $Y_Q(\Lambda)$ be the projection of $w(T - T_0) + T_0$ on $\mathcal{A}$, where w is chosen so that $w^{-1}Qw$ is standard.

Now let $P \in \mathcal{P}(M)$. For $Q \in \mathcal{P}_\varepsilon(L)$, there exists $R \in \mathcal{P}(M)$ such that $R \cap L$ is a parabolic subgroup of L and $R \subset Q$. The operator $M_{R|P}(1,\lambda)^{-1} M_{R|P}(1, \lambda + \Lambda)$ is shown to be independent of the choice of

R. For $\Lambda \in \mathcal{A}$ such that $\theta_{Q,\varepsilon}(\Lambda) \neq 0$ for $Q \in \mathcal{P}_\varepsilon(L)$, set

$$\mathcal{M}_L^T(P,\lambda,\Lambda) = \sum_{Q \in \mathcal{P}_\varepsilon(L)} \theta_{Q,\varepsilon}(\Lambda)^{-1} \exp(Y_Q(T)) M_{R|P}(1,\lambda)^{-1} M_{R|P}(1,\lambda+\Lambda)\,.$$

A suitable limit as $\Lambda \to 0$ exists (cf. [A5] for details) and we set $\mathcal{M}_L^T(P,\lambda) = \mathcal{M}_L^T(P,\lambda,0)$. The term in $J_\chi^T(f)$ associated to $(M,L,\mathcal{A},\sigma,s)$ is

(2.3.2)
$$c \sum_{P \in \mathcal{P}(M)} (2\pi)^{-a(P,\varepsilon)} \int_{i\mathcal{A}} \mathrm{Tr}(\mathcal{M}_L^T(P,\lambda) I_{\sigma,\lambda}^P(f) M_{P|\varepsilon(P)}(s,\varepsilon(\lambda)) I_{\sigma,\lambda}^P(\varepsilon)) d\lambda.$$

Let $\theta_G(f)$ denote the sum of all discrete terms (2.3.1) for all χ and all quintuples of the form $(M,G,\{0\},\sigma,s)$. For $L \neq G$, let $\theta_{G,L}(f)$ denote the sum over all χ and all quintuples $(M,L,\mathcal{A},\sigma,s)$. The terms $\theta_{G,L}(f)$ will arise only in §10.3 and §14.5. Their specific form will play essentially no role. All that is needed is a certain simple property of (2.3.2), which we now describe.

Let w be a finite place of F such that G_w is unramified (i.e., G_w is quasi-split and split over an unramified extension). Let B be a Borel subgroup of G_w contained in P_{0w} and let T be a maximal torus of B contained in M_{0w}. Let $\mathcal{H}_w$ be the Hecke algebra of G_w with respect to a hyperspecial maximal compact subgroup K_w and for $f_w \in \mathcal{H}_w$, let $f_w^\wedge$ denote the Satake transform of f_w (cf. §4.5). Thus $f_w^\wedge(\chi) = \mathrm{Tr}(i_{G_w,B}(\chi)(f_w))$ for χ an unramified character of T. Assume that σ_w is unramified. Then σ_w corresponds to class of unramified characters of T under conjugacy by the Weyl group of T in M. Let ψ be a representative of this class. For $\lambda \in i\mathcal{A}$, let χ'_λ be the restriction of χ_λ to T. The map $\lambda \to \chi'_\lambda$ defines a homomorphism from $i\mathcal{A}$ the group of unitary unramified characters of T. Its kernel is a lattice $\mathcal{Z}$. For $P \in \mathcal{P}(M)$, we have $\mathrm{Tr}(I_{\sigma_w,\lambda}^P(f_w)) = f_w^\wedge(\psi\chi'_\lambda)$.

Regard f_v as fixed for $v \neq w$ and consider the integral

(2.3.3)
$$\int_{i\mathcal{A}} \mathrm{Tr}(\mathcal{M}_L^T(P,\lambda) I_{\sigma,\lambda}^P(f) M_{P|\varepsilon(P)}(s,\lambda) I_{\sigma,\lambda}^P(\varepsilon)) d\lambda$$

as a distribution on f_w varying in $\mathcal{H}_w$. The result we need is that there exists an integrable function $\alpha(\lambda)$ on $i\mathcal{A}/\mathbb{Z}$ such that (2.3.3) is equal to:

$$\int_{i\mathcal{A}/\mathbb{Z}} f_w^{\wedge}(\psi\chi'_\lambda)\alpha(\lambda)d\lambda. \tag{2.3.4}$$

Let f_w^0 be the unit element in $\mathcal{H}_w$ and let $f^0 = \left(\prod_{v\neq w} f_v\right) f_w^0$. Define

$$\beta(\lambda) = \mathrm{Tr}(\mathcal{M}_L^T(P,\lambda)I_{\sigma,\lambda}^P(f^0)M_{P|\varepsilon(P)}(s,\lambda)I_{\sigma,\lambda}^P(\varepsilon))\ .$$

For $f_w \in \mathcal{H}_w$, we have an equality of operators

$$I_{\sigma_w,\lambda}^P(f_w) = f_w^{\wedge}(\psi\chi'_\lambda)I_{\sigma_w,\lambda}^P(f_w^0)$$

from which it follows that (2.3.3) is equal to

$$\int_{i\mathcal{A}} f_w^{\wedge}(\psi\chi'_\lambda)\beta(\lambda)d\lambda.$$

The function $\beta(\lambda)$ is integrable on $i\mathcal{A}$ (since (2.3.3) converges absolutely), the series $\alpha(\lambda) = \sum_{z\in\mathbb{Z}} \beta(\lambda + z)$ converges to an integrable function on $i\mathcal{A}/\mathbb{Z}$ and (2.3.4) holds with this choice of $\alpha(\lambda)$.

CHAPTER 3

Stable conjugacy

In the first part of this chapter, we review some general notions connected with stable conjugacy. The conjugacy classes and stable conjugacy classes that contribute to the $\mathcal{O}$-expansion of the trace formula for the unitary groups in three variables and the associated twisted trace formula for GL_3 are then classified and some cohomological data are calculated.

3.1. Stable conjugacy. In §3.1 – 3.3, $\underline{G}$ will denote a connected reductive group over F such that the derived group $\underline{G}_{\text{der}}$ is simply-connected (for definition of stable conjugacy when $\underline{G}_{\text{der}}$ is not simply-connected, see [Kt$_2$] and [KS] for the ordinary and twisted cases, respectively). Let ε be an automorphism of $\underline{G}$ of finite order. Let $\gamma \in G$ and set $\underline{I} = \underline{G}_{\gamma\varepsilon}$. An element $\delta \in G$ is said to be stably-ε-conjugate to γ if there exists $g \in G(\overline{F})$ such that $\delta = g^{-1}\gamma\varepsilon(g)$. In this case, $\tau(g)g^{-1} \in I(\overline{F})$ for all $\tau \in \Gamma$, and the image of the cocycle $\{\tau(g)g^{-1}\}$ in $H^1(F, I)$ belongs to $\mathfrak{D}(I/F)$, where

$$\mathfrak{D}(I/F) = \mathrm{Ker}\{H^1(F, G_{\gamma\varepsilon}) \longrightarrow H^1(F, G)\}.$$

Conversely, if $g \in G(\overline{F})$ and $\tau(g)g^{-1} \in G_{\gamma\varepsilon}(\overline{F})$ for all $\tau \in \Gamma$, then $g^{-1}\gamma\varepsilon(g)$ belongs to G. It follows that $\mathfrak{D}(I/F)$ parametrizes the set of ε-conjugacy classes within the stable-ε-conjugacy class of γ. Let $I^d = I \cap G_{\text{der}}$. Then $\mathfrak{D}(I/F)$ is contained in the image of $H^1(F, I^d)$ in $H^1(F, I)$ since g can be chosen inside $G_{\text{der}}(\overline{F})$. Depending on the context, $\mathfrak{D}(I/F)$ will also be denoted by $\mathfrak{D}_G(I/F)$, or $\mathfrak{D}(\gamma/F)$. We denote the ε-stable conjugacy class of γ by $\mathcal{O}_{\varepsilon-\text{st}}(\gamma)$. If $\delta \in \mathfrak{D}(\gamma)F)$ is represented by $\{\tau(g)g^{-1}\}$, γ^δ will denote an element in the conjugacy class of $g^{-1}\gamma g$ (thus γ^δ is only well-defined up to G-conjugacy).

The map $\mathrm{ad}(g^{-1})$ induces an $\overline{F}$-isomorphism between $\underline{G}_{\gamma\varepsilon}$ and $\underline{G}_{\delta\varepsilon}$. In fact, if $x \in G_{\gamma\varepsilon}(\overline{F})$, then $\tau(g^{-1}xg) = a_\tau g^{-1}\tau(x)ga_\tau^{-1}$, where $a_\tau = \tau(g)^{-1}g \in G_{\delta\varepsilon}(\overline{F})$. Hence $ad(g^{-1})$ defines an inner twisting. In particular, $\underline{G}_{\gamma\varepsilon}$ and $\underline{G}_{\delta\varepsilon}$ are isomorphic if $\underline{G}_{\gamma\varepsilon}$ is abelian.

Let $\mathcal{O}_{\varepsilon-\mathrm{st}}(G)$ denote the set of stable ε-semisimple conjugacy classes in G. We will drop the subscript ε when ε is trivial.

3.2. Some results of Kottwitz. In this section, we assume that ε is trivial. Let $\gamma \in G$ be a semisimple element and let $I = G_\gamma$. By a theorem of Steinberg ([St], pg. 57), the assumption that G_{der} is simply connected implies that I is connected.

THEOREM 3.2.1 (*Kottwitz-Steinberg,* [Kt$_2$], *Theorem* 4.1): *If $\underline{G}$ is quasi-split and $\underline{G}_{\mathrm{der}}$ is simply connected, then every conjugacy class in $G(\overline{F})$ which is defined over F contains an element of G.*

Suppose that $\underline{G}'$ is an inner form of $\underline{G}$ and let $\psi : \underline{G} \longrightarrow \underline{G}'$ be an inner twisting. By definition, for $\tau \in \Gamma$ there exist elements $x_\tau \in G(\overline{F})$ such that $\tau(\psi(g)) = x_\tau \psi(\tau(g)) x_\tau^{-1}$ for all $g \in G$. The $G'(\overline{F})$-conjugacy class of $\psi(\gamma)$ is therefore defined over F. If $\underline{G}'$ is quasi-split, then the intersection of this class with G' consists of a stable conjugacy class $\mathcal{O}_{\mathrm{st}}(\gamma')$ for some $\gamma' \in G$ by Theorem 3.2.1. In this case, we obtain a map

$$\mathcal{O}_{\varepsilon-\mathrm{st}}(G) \longrightarrow \mathcal{O}_{\varepsilon-\mathrm{st}}(G')$$

sending $\mathcal{O}_{\mathrm{st}}(\gamma)$ to $\mathcal{O}_{\mathrm{st}}(\gamma')$. We will say that a class $\mathcal{O}_{\mathrm{st}}(\gamma')$ in $\mathcal{O}_{\varepsilon-\mathrm{st}}(G')$ occurs in G if it is in the image of the above map.

As observed in [Kt$_3$], the map $H \longrightarrow Z(\widehat{H})$ defines an exact functor from the category of connected reductive groups over F with respect to normal F-homomorphisms (F-homomorphisms whose images are normal subgroups) and the category of diagonalizable groups over $\mathbf{C}$ with a Γ-action. Let $\underline{V}$ denote the torus $\underline{G}/\underline{G}_{\mathrm{der}}$. Then the sequence

$$1 \longrightarrow \widehat{V} \longrightarrow Z(\widehat{G}) \longrightarrow Z(\widehat{G}_{\mathrm{der}}) \longrightarrow 1$$

is exact. Hence $\widehat{V} = Z(\widehat{G})$ since G_{der} is simply connected and $Z(\widehat{G_{\mathrm{der}}})$ is trivial. Similarly, we have a Γ-equivariant exact sequence

$$1 \longrightarrow Z(\widehat{G}) \longrightarrow Z(\widehat{I}) \longrightarrow Z(\widehat{I^d}) \longrightarrow 1$$

and the associated long exact cohomology sequence yields:

$$\pi_0(Z(\widehat{G})^\Gamma) \longrightarrow \pi_0(Z(\hat{I})^\Gamma) \longrightarrow \pi_0(Z(\widehat{I^d})^\Gamma) \stackrel{b}{\longrightarrow} H^1(F, Z(\widehat{G})) \tag{3.2.1}$$

where, for a complex algebraic group H, $\pi_0(H)$ is the finite abelian group of connected components of H. Let $\mathcal{R}(I/F)$ denote the kernel of b if F is local and let $\mathcal{R}(I/F)$ denote the set of $x \in \pi_0(Z(\widehat{I^d})^\Gamma)$ such that $b(x)$ is locally trivial if F is global. In the global case, $\mathcal{R}(I/F)$ maps to $\mathcal{R}(I/F_v)$ for all v.

Let $A(G)$ be the dual of the abelian group $\pi_0(Z(\widehat{G})^\Gamma)$. Dual to (3.2.1), we have

$$A(I^d) \longrightarrow A(I) \longrightarrow A(G)\ .$$

Let $\mathcal{E}(I/F)$ be the image $A(I^d)$ in $A(I)$. If F is local, then $\mathcal{R}(I/F)$ and $\mathcal{E}(I/F)$ are dual abelian groups.

Assume that F is local for the rest of this section. By [Kt$_4$], Lemma 4.3, there is a canonical morphism $H^1(F,I) \longrightarrow A(I)$ such that the right square of the following diagram commutes:

$$\begin{array}{ccccc} H^1(F,I^d) & \longrightarrow & H^1(F,I) & \longrightarrow & H^1(F,G) \\ \downarrow & & \downarrow & & \downarrow \\ A(I^d) & \longrightarrow & A(I) & \longrightarrow & A(G)\ . \end{array}$$

The left square also commutes because the map $H^1(F,I) \longrightarrow A(I)$ is functorial with respect to normal homomorphisms of connected reductive group ([Kt$_4$], Theorem 1.2). The bottom arrows are group homomorphisms while the top arrows are maps of pointed sets.

Suppose that $\gamma' \in \mathcal{O}_{\mathrm{st}}(\gamma)$ and let α be the element of $\mathcal{D}(\gamma/F)$ corresponding to the conjugacy class of γ'. Let $\mathrm{inv}(\gamma,\gamma')$ be the image of α in $A(I)$. Then $\mathrm{inv}(\gamma,\gamma')$ lies in $\mathcal{E}(I/F)$ and each element of $\mathcal{R}(I/F)$ can be viewed as a function κ $(\mathrm{inv}(\gamma,\gamma'))$ on $\mathcal{O}_{\mathrm{st}}(\gamma)$ which is constant on conjugacy classes. The κ-orbital integrals described in §4 are built using elements of $\mathcal{R}(I/F)$. In particular, the role of a semisimple conjugacy class $\mathcal{O}_{\mathrm{st}}(\gamma)$ within the endoscopic analysis of G is governed by $\mathcal{R}(I/F)$. If F is p-adic or if F is local and γ is regular, the map $H^1(F,I) \longrightarrow A(I)$ is an isomorphism ([Kt$_4$], Lemma 1.2), and $\mathrm{inv}(\gamma,\gamma')$ is trivial if and only if γ and γ' are ε-conjugate.

Let T be an F-torus. Let $\overline{\mathbf{A}}$ denote the adele ring of $\overline{F}$, i.e., the direct limit of the rings $\mathbf{A}_K$ where K ranges over the finite extensions of F in $\overline{F}$. It is a consequence of Tate-Nakayama duality ([Kt$_3$]) that:

$$\text{(3.2.2)} \qquad A(T) = \begin{cases} H^1(F,T) & \text{if } F \text{ is local} \\ H^1(F,T(\overline{\mathbf{A}})/T(\overline{F})) & \text{if } F \text{ is global} \end{cases}$$

To calculate $A(T)$, it is often convenient to use that $A(T)$ is isomorphic to the Tate cohomology group $\widehat{H}^{-1}(F,X_*(T))$. Recall that $\widehat{H}^{-1}(F,X_*(T))$ is defined as follows ([Se]). Let K/F be a Galois extension over which T splits and let $N_{K/F}$ be the norm map on $X_*(T)$, i.e., $N_{K/F}(\lambda) = \sum \tau(\lambda)$, where the sum is over $\mathrm{Gal}(K/F)$. Then $\widehat{H}^{-1}(F, X_*(T))$ is canonically isomorphic to $\mathrm{Ker}(N_{K/F})/\mathcal{I}X_*(T)$ where $\mathcal{I}$ is the ideal in $\mathbf{Z}[\mathrm{Gal}(K/F)]$ generated by elements of the form $(\tau - 1)$.

The next result is a special case of [Kt$_3$], Lemma 2.2.

LEMMA 3.2.2. *Let T be an F-torus. Then $\pi_0(\hat{T}^\Gamma)$ is canonically isomorphic to $H^1(F, X^*(T))$.*

3.3. An obstruction. Let F be a global field and assume that G_{der} satisfies the Hasse principle. We also assume, for simplicity, that G is quasi-split (see [Kt$_3$] for the general case). Let $\gamma \in G$ be a semisimple element and set

$$\mathcal{O}_{\mathrm{st}}(\gamma/\mathbf{A}) = \{\gamma' \in \mathbf{G} : \gamma'_v \text{ is stably conj. to } \gamma \text{ in } G_v \text{ for all } v\}.$$

If $\gamma' \in \mathcal{O}_{\mathrm{st}}(\gamma/\mathbf{A})$, then γ'_v is conjugate to γ by an element of K_v for almost all v. Indeed, by [Kt$_4$], Proposition 7.1, this is the case for all v such that (i) G_v is unramified and K_v is hyperspecial, (ii) γ and γ'_v belong to K_v, and (iii) $1 - \alpha(\gamma)$ is either 0 or a unit for every root α of G. It follows that $g\gamma' g^{-1} = \gamma$ for some $g \in G(\overline{\mathbf{A}})$, where $\overline{\mathbf{A}}$ is the union of $\mathbf{A}_L$ over all finite extensions L/F.

We can suppose that $g \in G_{\mathrm{der}}(\overline{\mathbf{A}})$. Then the cocycle $\{a_\tau = \tau(g)g^{-1}\}$ takes values in $I^d(\overline{\mathbf{A}})$. Observe that γ' is $\mathbf{G}_{\mathrm{der}}$-conjugate to an element of G if and only if there is a choice of g such that $\{a_\tau\}$ takes values in $I^d(\overline{F})$. In fact, if $a_\tau \in I^d(\overline{F})$ for all τ, then the image of $\{a_\tau\}$ in $H^1(F, G_{\mathrm{der}})$ is locally trivial. By the Hasse principle, there exists $x \in G_{\mathrm{der}}(\overline{F})$ such that $\tau(g)g^{-1} = \tau(x)x^{-1}$ and γ' is $\mathbf{G}_{\mathrm{der}}$-conjugate to $x^{-1}\gamma x$. According to [Kt$_4$], §2.6, there is an exact sequence (for any connective reductive group)

$$H^1(F, I^d) \longrightarrow H^1(F, I^d(\overline{\mathbf{A}})) \xrightarrow{f} A(I^d) \tag{3.3.1}$$

Set $\mathrm{obs}(\gamma') = f(\alpha)$, where α is the class in $H^1(F, I^d(\overline{\mathbf{A}}))$ defined by $\{a_\tau\}$. We see that $\mathrm{obs}(\gamma')$ is trivial if and only if γ' is $\mathbf{G}_{\mathrm{der}}$-conjugate to an element of G.

The following proposition, due to Kottwitz ([Kt$_4$]), gives the condition for γ' to be $\mathbf{G}$-conjugate to an element of G in terms of $\mathrm{obs}(\gamma')$.

PROPOSITION 3.3.1: *Assume that G_{der} is simply connected and satisfies the Hasse principle. Let $\gamma' \in \mathcal{O}_{\mathrm{st}}(\gamma/\mathbf{A})$. Then γ' is $\mathbf{G}$-conjugate to an element of G if and only if $\kappa(\mathrm{obs}(\gamma')) = 1$ for all $\kappa \in \mathcal{R}(I/F)$.*

COROLLARY 3.3.2: *Let $\gamma' \in \mathcal{O}_{st}(\gamma/\mathbf{A})$. Then*

$$|\mathfrak{K}(I/F)|^{-1} \sum_{\kappa \in \mathfrak{K}(I/F)} \kappa(obs(\gamma'))$$

is equal to 1 or 0 according as γ' is or is not **G***-conjugate to an element of G.*

3.4. Cartan subgroups of unitary groups. Let G be a unitary group defined by a pair (D, α) consisting of a division algebra D of rank n over E and an involution α of the second kind. Let T be a Cartan Subgroup of G. Then α stabilizes the centralizer L' of T in D and (L', α') is a pair of the second kind, i.e., L' is a finite-dimensional, semisimple commutative algebra over E and α' induces σ on E. Let $T_{(L',\alpha')}$ be the F-torus such that

$$T_{(L',\alpha')}(F) = \{x \in L'^* : x\alpha'(x) = 1\} .$$

Then T is isomorphic to $T_{(L',\alpha')}$.

For any extension K/L, let $(\mathbf{G}_m)_{K/L}$ denote the L-torus $\mathrm{Res}_{K/L}(\mathbf{G}_m)$. If (L', α') is an irreducible pair of the second kind, i.e., is not isomorphic to a direct sum of more than one non-trivial pair, then either L' is a field or $L' = L'' \oplus L''$, where L'' is a field extension of E and $\alpha'(x, y) = (\alpha''(y), \alpha''(x))$, where α'' is an involution of the second kind on L''. In the latter case, $T_{(L',\alpha')}$ is isomorphic to $(\mathbf{G}_m)_{L''/F}$. If L' is a field and L is the fixed field of α, then $T_{(L',\alpha')}$ is isomorphic to $\ker(N_{L'/L})$, where $N_{L'/L}$ is the norm map from $(\mathbf{G}_m)_{L'/F}$ to $(\mathbf{G}_m)_{L/F}$. Clearly, every pair is isomorphic to a direct sum of irreducible pairs and T is isomorphic to a product of tori of the form $T_{(L',\alpha')}$ where (L', α') is irreducible.

We will say that Cartan subgroups T_1 and T_2 of G are stably conjugate if there exist regular elements $\gamma_j \in T_j$ such that γ_1 is stably conjugate to γ_2. This is the case if and only if there exists $g \in G(\overline{F})$ such that $gT_1g^{-1} = T_2$ and the isomorphism $t \longrightarrow gtg^{-1}$ is defined over F.

PROPOSITION 3.4.1: (a) *Two Cartan subgroups of G are stably conjugate if and only if they are isomorphic as F-tori.*

(b) *Assume that F is a global and that n is odd. Let T be an F-torus of dimension n of the form $T_{(L',\alpha')}$. Then T embeds in G if and only if T_v embeds in G_v for all places v of F.*

Proof: If T_1 and T_2 are Cartan subgroups of G which are isomorphic as F-tori, then $T_1(E)$ and $T_2(E)$ isomorphic Cartan subgroups of D^* and hence $gT_1(E)g^{-1} = T_2(E)$ for some $g \in D^*$. The map $t \longrightarrow \mathrm{ad}(g)t$ from T_1 to T_2 is defined over E and it is defined over F if $\alpha(g)g$ centralizes T_1. Let L' be the centralizer of T_1 in D, and let α' be the automorphism of L' defined by $\alpha'(\ell) = g^{-1}\alpha(g\ell g^{-1})g$. Since T_1 and T_2 are isomorphic as F-tori, the pairs (L', α) and (L', α') are isomorphic and there exists an automorphism β of L' such that $\alpha'(\ell) = \beta(\alpha(\beta^{-1}(\ell)))$. As is well-known (cf. [We], page 301), β is induced by $\mathrm{ad}(n)$ for some n in the normalizer of L' in D. We obtain $g^{-1}\alpha(g\ell g^{-1})g = n\alpha(n^{-1}\ell n)n^{-1}$. It follows that $\alpha(g)gn\alpha(n)$ and hence $\alpha(gn)gn$, centralize L' and we can replace g by gn. This proves (a).

We now prove (b). If T_v embeds in G_v for all places v of F, then $L' \otimes_F F_v$ embeds in $D \otimes_F F_v$ for all v. Hence L' embeds in D, and T embeds in G if and only if an embedding $i : L' \longrightarrow D$ exists such that $i(\alpha'(\ell)) = \alpha(\ell)$ for $\ell \in L'$. Fix any embedding of L' into D and regard L' as a subalgebra of D. There exists $x \in D^*$ such that $\alpha'(\ell) = \alpha(x^{-1}\ell x)$ for $\ell \in L'$ since the embeddings $\ell \longrightarrow \alpha(\ell)$ and $\ell \longrightarrow \alpha'(\ell)$ are conjugate in D^*. The problem is to show that there exists $g \in D^*$ such that $g^{-1}\alpha'(\ell)g = \alpha(g^{-1}\ell g)$ for $\ell \in L'$. This is the case if and only if the equation $g\alpha(g) = tx$ has a solution for some $g \in D^*$ and $t \in L'$. Observe that if (g, t) is a solution, then $\alpha(tx) = tx$.

A solution exists locally everywhere by hypothesis. Suppose that $g_1 \in (D \otimes \mathbf{R})^*$ and $t_1 \in (L' \otimes \mathbf{R})^*$ give an archimedean solution. If $\ell \in L' \otimes \mathbf{R}$, then $(\ell g_1)\alpha(\ell g_1) = \ell\alpha'(\ell)t_1 x$ and hence $(\ell g_1, \ell\alpha'(\ell)t_1)$ is also a solution. The map $\ell \longrightarrow \ell\alpha'(\ell)$ sends $(L' \otimes \mathbf{R})^*$ onto an open neighborhood of 1 in the group of α'-invariants in $(L' \otimes \mathbf{R})^*$. Since L'^* is dense in $(L' \otimes \mathbf{R})^*$, we can choose ℓ such that $t_2 = \ell\alpha'(\ell)t_1$ belongs to L'. We then obtain $g_2\alpha(g_2) = t_2 x$, where $g_2 = \ell g_1$.

Let

$$X = \{(y, z) \in D^* \times E^* : \alpha(y) = y, N_{D/E}(y) = N_{E/F}(z)\}.$$

The map sending g to $(g\alpha(g), N_{D/E}(g))$ gives rise to an exact sequence

$$D^* \longrightarrow X \longrightarrow H^1(F, G_0)$$

where $G_0 = \{g \in G : N_{D/E}(g) = 1\}$ is the special unitary group associated to G (cf. [Kn]). Since G_0 is simply connected, $H^1(F_v, G_{0v}) = 0$ for all finite v by Kneser's theorem. The Hasse principle holds for G_0 ([Kn]), and hence $(y, z) \in X$ is in the image of D^* if it is in the image of $(D \otimes \mathbf{R})^*$.

Let $m = N_{D/E}(t_2x) = N_{E/F}(N_{D/E}(g_2)) \in F^*$ and set $y = m^{-1}t_2x$. Now observe that $N_{D/E}(y) = m^{1-n}$ and m^{1-n} is a norm from E^* since n is odd. Hence $(y, m^{(1-n)/2})$ belongs X. Let $d = N_{D/E}(g_2)$. Then $(y, m^{(1-n)/2})$ is the image of $\lambda d^{-1}g_2 \in (D \otimes \mathbf{R})^*$ for $\lambda \in (E \otimes \mathbf{R})^*$ if $N_{E/F}(\lambda) = 1$ and $\lambda^n d^{1-n} = N_{E/F}(d)^{(1-n)/2}$, or equivalently, if $\lambda^n = (d/\bar{d})^{(1-n)/2}$. Such a λ exists since n is odd, and it follows that $g\alpha(g) = m^{-1}t_2x$ for some $g \in D^*$.

3.5. Stable conjugacy in unitary groups. We continue with the notation of the previous section.

LEMMA 3.5.1: (a) *$A(G)$ has order 2.*

(b) *If F is global, then $\ker^1(F, Z(\widehat{G}))$ is trivial.*

(c) *If I is the centralizer of a semisimple element of G, then $\mathfrak{R}(I/F)$ and $\mathcal{E}(I/F)$ are dual abelian groups.*

Proof: The action of Γ on $Z(\widehat{G})$ factors through $\Gamma(E/F)$. We have $Z(\widehat{G}) = \mathbf{C}^*$ and σ acts by $x \to x^{-1}$. Hence $Z(\widehat{G})^\Gamma = \{\pm 1\}$ and (a) follows. If F is global, then $H^1(E, Z(\widehat{G})) = \mathrm{Hom}(\Gamma_E, \mathbf{C}^*)$ and the restriction to E of an element $x \in \ker^1(F, Z(\widehat{G}))$ to E is trivial by the Chebotarev density theory. Hence x defines an element of $H^1(\Gamma(E/F), Z(\widehat{G}))$. This implies (b) since $H^1(\Gamma(E/F), Z(\widehat{G}))$ coincides with its localization at any place which remains prime in E. Part (c) is immediate from (b) and the definitions.

PROPOSITION 3.5.2: *Let T be a Cartan subgroup of G. Suppose that T is isomorphic to $T_{(L',\alpha')} \oplus T_{(L'',\alpha'')}$ where (L', α') is isomorphic to a direct sum of r irreducible pairs (L_i, α_i) such that L_i is a field, and (L'', α'') is a direct sum of irreducible pairs (L''', α''') such that L''' is not a field. Let L be the subalgebra of L' fixed α'.*

(a) *$H^1(F, T)$ is canonically isomorphic to $L/N_{L'/L}(L'^*)$.*

(b) *$A(T)$ is naturally isomorphic to $(\mathbf{Z}/2)^r$.*

(c) *If T is anisotropic, then $\mathcal{E}(T/F)$ is isomorphic to the subgroup $\{(\varepsilon_j) \in (\mathbf{Z}/2)^r : \Sigma\varepsilon_j = 0\}$. The order of $\mathfrak{R}(T/F)$ is 2^{r-1}.*

(d) *$\mathfrak{R}(T/F)$ is isomorphic to the image of $H^1(F, X^*(T))$ in $H^1(F, X^*(T^d))$*

(e) *$\widehat{H}^{-1}(F, X_*(T))$ is isomorphic to $(\mathbf{Z}/2)^r$. The image of $\widehat{H}^{-1}(F, X_*(T^d))$ in $\widehat{H}^{-1}(F, X_*(T))$ has index two.*

Proof: To prove (a) and (b), we can assume that T corresponds to an irreducible pair of the second kind. If $T = T_{(L'',\alpha'')}$, where L'' is not a field, then T is isomorphic to $(\mathbf{G}_m)_{K/E}$ for some extension K/E. In this case,

$H^1(F,T) = H^1(E,(\mathbf{G}_m)_{K/E})$ by Shapiro's lemma, and $H^1(E,(\mathbf{G}_m)_{K/E}) = 0$ by Theorem 90. It is clear that $N_{L'/L}$ is surjective in this case, and (a) follows. By (3.2.2), the triviality of $A(T)$ follows by Theorem 90 in the local case and by the vanishing of H^1 for the idele class group in the global case.

If $T = T_{(L',\alpha')}$, where L' is a field, then, then (a) follows from theorem 90 and the cohomology sequence associated to the exact sequence defined by the norm:

$$1 \longrightarrow T \longrightarrow (\mathbf{G}_m)_{L'/F} \longrightarrow (\mathbf{G}_m)_{L/F} \longrightarrow 1 \; .$$

To prove (b), we use that $A(T)$ is isomorphic to $\widehat{H}^{-1}(F,X_*(T))$. Let $\{\rho^j\}$ be the set of F-embeddings of L' into $\overline{F}$ and let $\{\rho_j\}$ be the dual basis of $X_*((\mathbf{G}_m)_{L'/F})$. As a right Γ_F-module, $X_*(T)$ is isomorphic to the submodule of $\oplus\mathbf{Z}\rho_j$ spanned by the set $\{(\alpha'-1)\rho_j\}$. Since T is anisotropic, $\ker(N_{L'/F})$ coincides with $X_*(T)$ and $\widehat{H}^{-1}(F,X_*(T))$ is isomorphic to the quotient of $X_*(T)$ by the span of $\{(\alpha'-1)\rho_j(\rho-1) : \rho \in \Gamma_F\}$. For all j,k there exists $\rho \in \Gamma_F$ such that $\rho_j\rho = \rho_k$ or $\alpha'\rho_k$ and hence

$$(\alpha'-1)\rho_j(\rho-1) = (\alpha'-1)(\pm\rho_k - \rho_j).$$

The span of $\{(\alpha'-1)\rho_j(\rho-1)\}$ is therefore the set of linear combinations $\sum a_j(\alpha'-1)\rho_j$ such that $\sum a_j$ is even. It follows that $\widehat{H}^{-1}(F,X_*(T))$ has order 2 and (b) follows

We now prove the remaining assertions. If (L',α') is a pair of the second kind, then the norm map $N_{L'/E}$ from L' to E induces a map from $T_{(L',\alpha)}$ to E^1. The restriction of the determinant map $\det_G$ of G to T is given by the product of the norm maps on each of the irreducible factors of T and T^d is its kernel. Let $\{\tau^{i,k}\}$ be a set of representatives for Γ_E/Γ_{L_i} and let ξ^i be a fixed embedding of L_i into $\overline{F}$. Then every embedding of L_i into $\overline{F}$ over F is of the form $\tau^{i,k}\xi^i$ or $\tau^{i,k}\xi^i\alpha'$. An element $\lambda \in X_*(T)$ can be written uniquely as $\lambda = \sum a_{ik}\cdot(\alpha'-1)\xi_i\tau^{ik}$ where $a_{ik} \in \mathbf{Z}$, and λ belongs to $X_*(T^d)$ if and only if $\sum_{i,k} a_{ik} = 0$. Under the isomorphism of $\widehat{H}^{-1}(F,X_*(T))$ with $(\mathbf{Z}/2)^r$, the image of λ is the vector whose i^{th} component is $\Sigma_k a_{ik}$. This proves (c) and (e). Part (d) follows from Lemma 3.2.2 and Lemma 3.5.1(b).

Let $\gamma \in G$ be a semisimple element. The following diagram commutes:

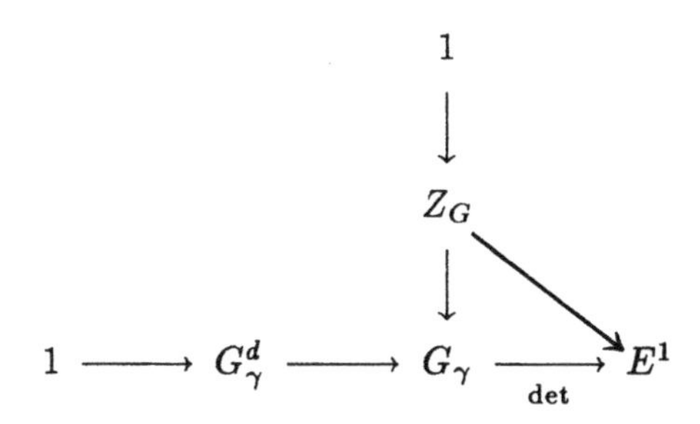

where $G_\gamma^d = \{g \in G_\gamma : \det(g) = 1\}$ and the diagonal arrow is the map $x \longrightarrow x^n$. The group $H^1(F, E^1)$, which we will identify with F^*/NE^*, is annihilated by 2. Hence, if n is odd, the diagonal arrow above induces an isomorphism $H^1(F, Z_G) \longrightarrow H^1(F, E^1)$. In this case, $H^1(F, Z_G)$ embeds in $H^1(F, G_\gamma)$ and $H^1(F, G_\gamma) \longrightarrow H^1(F, E^1)$ is surjective. This gives a splitting:

$$H^1(F, G_\gamma) \xrightarrow{\sim} Im[H^1(F, G_\gamma^d)] \times H^1(F, Z_G). \tag{3.5.1}$$

The subset $\mathfrak{D}(G_\gamma/F)$ is contained in $Im[H^1(F, G_\gamma^d)]$.

LEMMA 3.5.3: *Assume that n is odd.*

(a) *The map $G \to G_{\text{ad}}$ is surjective. In particular, G_{ad}-conjugacy coincides with conjugacy in G.*

(b) *The natural map $\psi : H^1(F, G_{\text{der}}) \longrightarrow H^1(F, G_{\text{ad}})$ is an isomorphism.*

Proof: The cokernel of $G \to G_{\text{ad}}$ is isomorphic to the kernel of the map $H^1(F, Z_G) \to H^1(F, G)$, and this is injective by the above remarks. This proves (a).

The cokernel of $\det_G : G \longrightarrow E^1$ is trivial and hence, $H^1(F, G_{\text{der}}) \longrightarrow H^1(F, G)$ is injective by the cohomology exact sequence. It follows from (3.5.1) for the case $\gamma = 1$ that ψ is injective. If T is any anisotropic F-torus, then $H^0(F, X^*(T)) = X^*(T)^\Gamma = \{0\}$. If F is local, this implies that $H^2(F, T)$ is also trivial by Tate-Nakayama duality ([Kt$_3$], §3). Since Z_G is anisotropic, this shows that ψ is surjective in the local case. Suppose that F is global. The Hasse principle holds for G_{der} by Landherr's theorem ([Kn]) and also for G_{ad} ([Ha], Satz 4.3.2). The upper horizontal and right vertical arrows in the following diagram are isomorphisms and the lower horizontal arrow is injective.

$$\begin{array}{ccc} H^1(F, G_{\text{der}}) & \longrightarrow & \Pi H^1(F_v, G_{\text{der}}) \\ \psi \downarrow & & \downarrow \\ H^1(F, G_{\text{ad}}) & \longrightarrow & \Pi H^1(F_v, G_{\text{ad}}) \end{array}$$

Hence ψ is also an isomorphism.

3.6. Classification of Cartan subgroups. We now suppose that G is a unitary group in 3 variables. Let $H = H_0 \times U(1)$ where H_0 is a unitary group in 2 variables. It is convenient to classify the Cartan subgroups of G according to four types. If K/F is a quadratic extension distinct from E, let $L' = KE$ and let α' be the automorphism of L' of the second kind which fixes K. Denote $T_{(L',\alpha')}$ by T_K. If L/F is a cubic extension, let $L' = LE$, let α' be the automorphism of the second kind of L' whose restriction to L is trivial, and denote $T_{(L',\alpha')}$ by T_L. By the results of §3.4, every Cartan subgroup of G or H is isomorphic to one of the following types of tori:

Type (0): $(\mathbf{G}_m)_{E/F} \times \underline{E}^1$

Type (1): $\underline{E}^1 \times \underline{E}^1 \times \underline{E}^1$

Type (2): $T_K \times \underline{E}^1$, where K is a quadratic extension of F distinct from E

Type (3): T_L where L/F is a cubic extension.

By Proposition 3.4.1(a), the stable conjugacy class of a Cartan subgroup of type (2) (resp., type (3)) is determined by the extension K/F (resp., L/F), and there is a unique stable conjugacy class of Cartan subgroups of type (0) or (1). We will say that a Cartan subgroup T_0 of H_0 is of type (j) if $T_0 \times U(1)$ is of type (j). According to Proposition 3.5.2(d), $\mathfrak{K}(T/F)$, is isomorphic to $\mathbf{Z}/2)^2$ (resp., $(\mathbf{Z}/2)$) if T is of type (1) (resp., type (2)), and $\mathfrak{K}(T/F)$ is trivial if T is of type (3).

LEMMA 3.6.1: *Let T be an anisotropic Cartan subgroup of H. Then natural map $\mathcal{E}_H(T/F) \longrightarrow \mathcal{E}_G(T/F)$ is injective and its image has index two.*

Proof: T is of the form $T_0 \times E^1$ where T_0 is a Cartan subgroup of H_0. The lemma follows easily from Proposition 3.5.2.

3.7. Weyl groups. We denote the Weyl group of a Cartan subgroup $\underline{T}$ of a group $\underline{G}$ will be denoted by $\Omega_G(T)$. Let $\Omega_F(T, G)$ be the subgroup of $\Omega_G(T)$ consisting of elements whose action is defined over F. The Galois group Γ acts on $\Omega_G(T)$ and $\Omega_F(T,G)$ is the group of fixed points of this action. Let $\Omega(T,G)$ be the Weyl group of T in G. The symbol G will be omitted from the notation when this causes no ambiguity. For simplicity, we will not distinghish in the notation between an element of a Weyl group and a representative for that element.

Now let $G = U(3)$ and let $H = U(2) \times U(1)$. We identify H with the subgroup G_γ of G, where $\gamma = d(1,-1,1)$.

PROPOSITION 3.7.1:

(a) $\Omega_F(T,G) = S_3$ *if* T *is of type* (1).

(b) $\Omega_F(T,G) = \mathbf{Z}/2$ *if* T *is of type* (2).

(c) $T \subset H$, *then* $\Omega_F(T,H) = \mathbf{Z}/2$.

Proof: If T is of type (1), then $T = (E^1)^3$ and $\Omega(T,G)$ acts by permuting the factors. This action is defined over F and (a) follows. If T is of type (2), then $\Omega_F(T,G)$ has order at most 2 and hence (c) implies (b). Part of (c) follows from the corresponding fact for Cartan subgroups of the derived group, $\mathrm{SL}_2(F)$, of H.

Let:

$$\Phi' = \begin{pmatrix} 1 & & \\ & 1 & \\ & & \alpha \end{pmatrix}$$

where $\alpha \in F^*$ is chosen so that G is isomorphic to $U_{\Phi'}$. Let T' be the diagonal subgroup in $U_{\Phi'}$. It is of type (1) and the elements:

$$w = \begin{pmatrix} 0 & 1 & 0 \\ 1 & 0 & 0 \\ 0 & 0 & 1 \end{pmatrix} \quad , \quad w' = \begin{pmatrix} 1 & 0 & 0 \\ 0 & 0 & a \\ 0 & b & 0 \end{pmatrix}$$

(for any choice of $a, b \in E^*$) generate $\Omega_F(T',G')$. Furthermore, $w \in \Omega(T',G')$ and $w' \in \Omega(T',G')$ if and only if $N_{E/F}(a)^{-1} = N_{E/F}(b) = \alpha$. It follows that $\Omega(T',G') = S_3$ or $\mathbf{Z}/2$ according as $\alpha \in NE^*$ or $\alpha \notin NE^*$.

If F is p-adic, then all unitary groups in 3 variables are isomorphic and α is arbitrary. Hence there exist conjugacy classes $\{T_1\}$ and $\{T_2\}$ of Cartan subgroups of type (1) such that $\Omega(T_1,G) = S_3$ and $\Omega(T_2,G) = \mathbf{Z}/2$. If $\gamma \in T_1$ is regular, then $\mathcal{O}_{\mathrm{st}}(\gamma)$ contains 4 conjugacy classes since $\mathcal{E}_G(T_1/F) = (\mathbf{Z}/2)^2$. If F is real, then there is a unique conjugacy class $\{T\}$ of Cartan subgroups of type (1) and $\Omega(T,G) = \mathbf{Z}/2$.

PROPOSITION 3.7.2: *Let F be a local field. There exists a Cartan subgroup T of type (1) contained in H such that $\Omega(T,G) = \mathbf{Z}/2$.*

Proof: Let $T \subset H$ be a Cartan subgroup of type (1). If $\Omega(T,G) = S_3$, then F is p-adic and $\Omega(T,H) = \mathbf{Z}/2$. Let γ be a regular element in T. Since $\mathcal{D}_H(T/F) = \mathcal{E}_H(T/F)$ has order two and injects in $\mathcal{E}_G(T/F)$ by Lemma 3.6.1, there exists $\gamma' \in H$ which is stably conjugate but not conjugate to γ in G. The centralizer T' of γ' is not conjugate to T in G and hence $\Omega(T,G) = \mathbf{Z}/2$.

3.8. Singular semisimple elements. Let $G = U(3)$. For $\xi \in F^*$, let H'_ξ be the unitary group in two variables defined by the Hermitian form:

$$\begin{pmatrix} \xi & \\ & -1 \end{pmatrix}.$$

The isomorphism class of H'_ξ depends only on ξ modulo NE^* and we obtain a bijection between F^*/NE^* and the set of isomorphism classes of unitary groups in two variables over F with respect to E. The group H_1 is quasi-split and is isomorphic to $U(2)$. The subgroup of elements of determinant one in H'_ξ is isomorphic to the norm one subgroup of the unique quaternion algebra over F which is ramified precisely at the set of places v of F at which ξ is not a norm from E_v. Set $H_\xi = H'_\xi \times E^1$ and let $H = H_1$.

PROPOSITION 3.8.1: *Let $\gamma \in G$ be a semisimple element which is singular but not central.*

(a) *If $\gamma' \in \mathcal{O}_{\mathrm{st}}(\gamma)$, then $G_{\gamma'}$ is isomorphic to H_ξ for some $\xi \in F^*/NE^*$. The map sending γ' to ξ defines a bijection between the set of conjugacy classes within $\mathcal{O}_{\mathrm{st}}(\gamma)$ and F^*/NE^*. In particular, two elements in $\mathcal{O}_{\mathrm{st}}(\gamma)$ are conjugate if and only if their centralizers are isomorphic.*

Let G' be an inner twist of G defined by a pair (D,α).

(b) *$\mathcal{O}_{\mathrm{st}}(\gamma)$ transfers to G' if and only if $D = M_3(E)$.*

(c) *Suppose that $F = \mathbf{R}$. If G' is compact, then $\mathcal{O}_{\mathrm{st}}(\gamma)$ transfers to a single conjugacy class in G'.*

(d) *If F is global, then the set of conjugacy classes in G' which transfer to $\mathcal{O}_{\mathrm{st}}(\gamma)$ is parametrized by the set of $\xi \in F^*/NE^*$ such that ξ is negative at all real places of F at which G is compact.*

Proof: Two semisimple elements in $G(E) = \mathrm{GL}_3(E)$ are conjugate if and only if they the same set of eigenvalues. Thus $\mathcal{O}_{\mathrm{st}}(\gamma)$ is determined by the set of eigenvalues of γ, which is of the form $\{\alpha,\alpha,\beta\}$ where $\alpha \neq \beta$. Furthermore, $\Gamma(\overline{F}/E)$ acts on $\{\alpha,\alpha,\beta\}$ and the action must be trivial. Hence $\alpha,\beta \in E^*$, and in fact $\alpha,\beta \in E^1$ since set of eigenvalues is also stable under $x \longrightarrow \overline{x}^{-1}$. Up to stable conjugacy, we can assume that

$$\gamma = \begin{pmatrix} \alpha & & \\ & \beta & \\ & & \alpha \end{pmatrix}.$$

Then $G_\gamma = H_1$ and $G_\gamma^d = \{(g,\lambda) \in H_1 : \det(d)\lambda = 1\}$ is isomorphic to the quasi-split unitary group $U(2)$ defined with respect to the Hermitian form $\left(\begin{smallmatrix} 0 & 1 \\ 1 & 0 \end{smallmatrix}\right)$.

The cohomology sequence associated to the determinant exact sequence

$$1 \longrightarrow \mathrm{SL}(2) \longrightarrow U(2) \longrightarrow E^1 \longrightarrow 1,$$

defines a map:

$$H^1(F,U(2)) \longrightarrow H^1(F,E^1) = F^*/NE^*. \tag{3.8.1}$$

This map is injective since $H^1(F,\mathrm{SL}(2)) = \{0\}$.

The map $H^1(F,G_\gamma^d) \longrightarrow H^1(F,G_\gamma)$ is injective since $G_\gamma = G_\gamma^d \times E^1$ and from (3.8.1) we obtain an embedding of $\mathcal{D}(G_\gamma/F)$ in F^*/NE^*. Let $\xi \in F^*$ and let $a_\xi(\sigma) = \left(\begin{smallmatrix} 0 & \xi \\ -1 & 0 \end{smallmatrix}\right)$. Regard $a_\xi(\sigma)$ as an element of $G_\gamma^d(E)$, which we identify with $\mathrm{GL}_2(E)$. Then $a_\xi(\sigma)\sigma(a_\xi(\sigma)) = 1$ and $a_\xi(\sigma)$ defines a cocycle in $H^1(\gamma(E/F),G_\gamma^d(E))$. It corresponds to the class of ξ in F^*/NE^* since $\det(a_\xi(\sigma)) = \xi$. In particular, (3.8.1) is surjective. Suppose that the cocycle belongs to $\mathcal{D}(G_\gamma/F)$ and let γ' be a representative for the associated conjugacy class in $\mathcal{O}_{\mathrm{st}}(\gamma)$. Then $G_{\gamma'}$ is the inner form of G_γ obtained via twisting by the cocycle $a_\xi(\sigma)$ and hence $G_{\gamma'}$ is isomorphic to H_ξ. In particular, the conjugacy class of γ' is determined by the isomorphism class of $G_{\gamma'}$.

If ξ is negative at the real places v such that G'_v is compact, then, by the classification of unitary groups (§1.9), there exists $\mu \in F^*$ such that G' is isomorphic to the unitary group of the Hermitian form

$$\begin{pmatrix} \xi & & \\ & -1 & \\ & & \mu \end{pmatrix}$$

The centralizer of the diagonal element in G' with diagonal entries α,α,β is isomorphic to H_ξ. Since H_ξ is non-compact at the real places at which

ξ is positive, H_ξ occurs as a centralizer in G' if and only if ξ is negative at the real places v such that G'_v is compact. For $G' = G$ this shows that $\mathfrak{D}(G_\gamma/F) = H^1(F, G_\gamma^d) = F^*/NE^*$ and proves (a) and (d). To prove (b), observe that if G' is defined by a pair (D, α) such that $D \neq M_3(E)$, then all non-regular elements are scalar. Part (c) follows since stable conjugacy coincides with conjugacy in a compact unitary group.

Suppose that E/F is local. Let $\gamma \in M$ be a singular, non-central element. Then $G_\gamma = H$. By Proposition 3.7.2, there exists a Cartan subgroup T of type (1) contained in H such that $\Omega(T, G) = \mathbf{Z}/2$. Suppose that $\gamma' \in T$ has the same eigenvalues as γ. Then γ' is stably conjugate to γ in G, but since $\Omega_F(T, G) = S_3$, we can choose γ' so that it is not conjugate to γ in G. In particular, γ' is regular as an element of H. Let δ_H be the non-trivial element in $\mathfrak{D}_H(T/F)$ and set $\gamma'' = (\gamma')^{\delta_H}$. Then γ' and γ'' are not conjugate in G since $\mathfrak{D}_H(T/F)$ injects in $\mathfrak{D}_G(T/F)$ by Lemma 3.6.1. This proves the next proposition.

PROPOSITION 3.8.2: *Assume that E/F is local. Let γ be a singular non-central element in M. Then there exist γ', $\gamma'' \in H$ which are regular as elements of H such that $\{\gamma', \gamma''\}$ is a set of representatives for the conjugacy classes within $\mathcal{O}_{\mathrm{st}}(\gamma)$.*

3.9. Non-semisimple classes. A unitary group in 3 variables contains non-semisimple elements if and only if it is quasi-split. Let $G = U(3)$. Every unipotent element in G is conjugate to an element of N, and an element $u(x, z)$ is regular if and only if $x \neq 0$. A unipotent element $u \in G$ will be called singular if it is not regular and $u \neq 1$.

PROPOSITION 3.9.1: *The set of regular unipotent elements in $U(3)$ consists of a single conjugacy class.*

Proof: The regular unipotent classes form a single G_{ad}-conjugacy class. The assertion follows from Lemma 3.5.3(a).

All singular unipotent elements are conjugate to an element of the form $n(t)$ for some $t \in E^\circ$. If $u = n(t)$, then $G_u = S \cdot N$ where $S = \{m \in M : \alpha_3(m) = 1\}$. The conjugacy class of u is determined by t mod NE^*.

Suppose that $\gamma \in G$ has non-trivial unipotent part and non-central semi-simple part. Then the unipotent part of γ is conjugate to $n(t)$ for some

$t \in F^*$ and there exist $\alpha, \beta \in E^1$ such that, up to conjugacy,

$$\gamma = \begin{pmatrix} \alpha & & \\ & \beta & \\ & & \alpha \end{pmatrix} \begin{pmatrix} 1 & 0 & t \\ 0 & 1 & 0 \\ 0 & 0 & 1 \end{pmatrix}$$

and $G_\gamma = S \cdot \{n(t) : t \in E^\circ\}$. The conjugacy class of γ is determined by t mod NE^* and α, β.

3.10. The norm map. Let $\underline{G}$ be a connected, reductive group over F such that $\underline{G}_{\text{der}}$ is simply connected and let $\underline{\widetilde{G}} = \text{Res}_{E/F}(\underline{G})$, where E/F is a cyclic extension of degree ℓ. Let σ be a generator of $\Gamma(E/F)$. Over E, $\underline{\widetilde{G}}$ is isomorphic to the product of ℓ copies of $\underline{G}$ in such a way that σ acts by

$$(x_1, x_2, \ldots, x_\ell) \longrightarrow (\sigma(x_\ell), \sigma(x_1), \ldots, \sigma(x_{\ell-1}))$$

and

$$\widetilde{G} = \{(x, \sigma(x), \ldots, \sigma^{\ell-1}(x)) : x \in G(E)\} .$$

We identify $\widetilde{G}$ with $G(E)$ by projection on the first factor. Let ε be the algebraic automorphism of $\underline{\widetilde{G}}$ consisting of a cyclic shift to the left. Then ε induces σ on $G(E)$ under the identification. The map $i : \underline{G} \to \underline{\widetilde{G}}$ defined by $x \to (x, x, \ldots, x)$ is an isomorphism of $\underline{G}$ with the subgroup of $\underline{\widetilde{G}}$ fixed by ε

Define the norm map $N : \widetilde{G} \longrightarrow \widetilde{G}$ by

$$N(\delta) = \delta\varepsilon(\delta)\varepsilon^2(\delta) \cdots \varepsilon^{\ell-1}(\delta) .$$

There is a bijection between the set of ε-conjugacy classes in $\widetilde{G}(\overline{F})$ and the set of conjugacy classes in $G(\overline{F})$, defined by sending an ε-conjugacy class $\mathcal{O}_\varepsilon(\delta)$ to $i^{-1}(\mathcal{O})$, where $\mathcal{O}$ is the intersection of $\{N(\delta') : \delta' \in \mathcal{O}_\varepsilon(\delta)\}$ with $i(G(\overline{F}))$. The relation

$$N(x^{-1}\delta\varepsilon(x)) = x^{-1}N(\delta)x$$

for $x \in \widetilde{G}(\overline{F})$ shows that $\mathcal{O}$ is a union of conjugacy class in $G(\overline{F})$. If $w_1, w_2 \in G(\overline{F})$ and $i(w_1)$ is conjugate to $i(w_2)$ in $G(\overline{F})$, then w_1 is conjugate to w_2 in $G(\overline{F})$ and hence $\mathcal{O}$ is a conjugacy class. If $\delta = (x_1, x_2, \ldots, x_\ell)$, then

$$N(\delta) = (\gamma, x_1^{-1}\gamma x_1, \ldots, (x_1 x_2, \ldots, x_{\ell-1})^{-1}\gamma(x_1 x_2, \ldots, x_{\ell-1}))$$

where $\gamma = x_1 x_2 \ldots x_\ell$, and $N(\delta)$ is conjugate in $\widetilde{G}(\overline{F})$ to $i(\gamma)$. This shows that the image of $\mathcal{O}_\varepsilon(\delta)$ is $\mathcal{O}(\gamma)$. It is easy to check that we obtain an injection. It is also surjective since every element of $G(\overline{F})$ is a norm. For example, $(\gamma, 1, \dots, 1)$ maps to γ.

Identify $\widetilde{G}$ with $G(E)$ and let $\delta \in \widetilde{G}$. The equation $\sigma(N(\delta)) = \delta^{-1}N(\delta)\delta$ shows that the conjugacy class of $N(\delta)$ in $G(E)$ is defined over F. If G is quasi-split and G_{der} is simply connected, then the stable conjugacy class of $N(\delta)$ intersects G in a stable conjugacy class which is determined by the stable ε-conjugacy class of δ (Theorem 3.2.1). This defines an injective map $\mathcal{N}$ from the set of stable ε-conjugacy classes in $\widetilde{G}$ to the set of stable conjugacy classes in G. For $\delta \in \widetilde{G}, \mathcal{N}(\delta)$ will denote the stable conjugacy class in G associated to $N(\delta)$. An element $\gamma \in G$ in $\mathcal{N}(\delta)$ will be called a norm of δ and we will also write $\mathcal{N}(\delta) = \gamma$. We call δ ε-regular (resp., ε-elliptic) if $\mathcal{N}(\delta)$ consists of regular (resp. elliptic) elements.

Let $\delta \in G(E)$ and set $\gamma = N(\delta)$. Then δ corresponds to $(\delta, \dots, \sigma^{\ell-1}(\delta))$ in $\widetilde{G}(E)$ and $\underline{\widetilde{G}}_{\delta\varepsilon}$ consists of elements of the form

$$(g, \delta^{-1}g\delta, \sigma(\delta)^{-1}\delta^{-1}g\delta\sigma(\delta), \dots) \tag{3.10.1}$$

where $g \in \underline{G}_\gamma$. Projection onto the first factor gives an isomorphism over E of $\underline{\widetilde{G}}_{\delta\varepsilon}$ with $\underline{G}_\gamma$. The element (3.10.1) belongs to $\widetilde{G}_{\delta\varepsilon}$ if and only if $g \in G_\gamma(E)$ and $g = \delta\sigma(g)\delta^{-1}$. Hence $\underline{\widetilde{G}}_{\delta\varepsilon}$ is the F-form of $\underline{G}_{\gamma/E}$ defined by the automorphism $g \to \delta\sigma(g)\delta^{-1}$. If $\gamma \in G$, then $\delta \in G_\gamma(E)$ since $\sigma(\gamma) = \delta^{-1}\gamma\delta$. In this case, $\underline{G}_{\delta\varepsilon}$ is an inner form of $\underline{G}_\gamma$. If δ lies in the center of $\underline{G}_{\delta\varepsilon}$ (in particular, if $\underline{G}_\gamma$ is abelian), then $\underline{G}_\gamma$ and $\underline{G}_{\delta\varepsilon}$ are isomorphic over F.

3.11. The norm map for unitary groups. In this section, let E/F be a quadratic extension and let $G = U(n)$. Set $\widetilde{G} = \text{Res}_{E/F}(U(n))$.

PROPOSITION 3.11.1: (a) *If T is a Cartan subgroup of G, then the norm map $T(E) \longrightarrow T(F)$ is surjective.*

(b) *Let $\gamma \in G$ be semisimple. There exists an element $\delta \in \widetilde{G}$ such that $N(\delta) = \gamma$, δ is central in $G_{\delta\varepsilon}$, and $G_{\delta\varepsilon} = G_\gamma$.*

(c) *The norm map defines a bijection between $\mathcal{O}_{\varepsilon-\text{st}}(\widetilde{G})$ and $\mathcal{O}_{\text{st}}(G)$.*

Proof: For (a), it suffices to show that the norm $N : T(E) \to T(F)$ is surjective for a torus of the form $T = T_{(L',\alpha')}$ such that (L', α) is irreducible. If L' is a field, then $T(F) = \{x \in (L')^* : N_{L'/L}(x) = 1\}$ and $T(E) = L'^*$. The norm is given by $N(y) = y\alpha(y)^{-1}$ and is surjective by theorem 90. If L' is not a field, then $L' = L'' \oplus L''$, where L'' is a field. The norm is then given by $(x,y) \to (x\alpha''(y)^{-1}, y\alpha''(x)^{-1})$ for some involution α'' of the second kind of L'' and surjectivity is clear in this case also. To prove (b), let $\gamma \in G$ be a semisimple element and let Z_γ be the center of G_γ. Then Z_γ is a torus and the subalgebra L of $M_n(E)$ spanned by Z_γ is semisimple. Since L is stable under the involution of the second kind α defining G, Z_γ is isomorphic to a torus of the form $T_{(L',\alpha')}$ and $\gamma = N(\delta)$ for some $\delta \in Z_\gamma(E)$ by (a). Since δ is central in $\underline{G}_\gamma$, $G_{\delta\varepsilon} = G_\gamma$ by the remark of §3.10. This proves (b). Part (c) follows from (a).

Let $\delta \in \widetilde{G}$ and assume that $\gamma = N(\delta)$ belongs to G. Assume also that δ lies in the center of $\widetilde{G}_{\delta\varepsilon}$. Then $\widetilde{G}_{\delta\varepsilon} = G_\gamma$ (when we identify $\widetilde{G}$ with $G(E)$). By Shapiro's lemma,

$$H^1(F,\widetilde{G}) = H^1(E,\mathrm{GL}_n) = 0$$

and it follows that there is a bijection between $H^1(F,G_\gamma)$ and the set of ε-conjugacy classes within the stable ε-class of δ.

Over E, G_γ is isomorphic to a product of GL_m's and hence $H^1(E,G_\gamma) = 0$. Let $\nu \in H^1(F,G_\gamma)$. By the inflation-restriction sequence, $H^1(F,G_\gamma) = H^1(\Gamma(E/F),G_\gamma(E))$ and ν has a representative cocycle of the form $a_\sigma = \alpha$ for some $\alpha \in G_\gamma(E)$ such that $\sigma(\alpha)\alpha = 1$. Under the identification of $\widetilde{G}(\overline{F})$ with $\mathrm{GL}_n(\overline{F}) \times \mathrm{GL}_n(\overline{F})$, δ corresponds to $(\delta,\sigma(\delta))$. Identify a_σ with (α,α) in $\widetilde{G}(E)$. Since $H^1(F,\widetilde{G}) = 0$, the image of ν splits in $\underline{\widetilde{G}}$, in fact, $a_\sigma = \sigma(g)g^{-1}$ where $g = (\alpha^{-1},1)$, and

$$g^{-1}(\delta,\sigma(\delta))\varepsilon(g) = (\alpha\delta,\sigma(\delta)\alpha^{-1}) = (\alpha\delta,\sigma(\alpha\delta)),$$

since $\alpha^{-1} = \sigma(\alpha)$ and δ commutes with α. A representative δ' for the ε-conjugacy class within $\mathcal{O}_{\varepsilon-\mathrm{st}}(\delta)$ corresponding to ν is therefore given by $\delta' = \alpha\delta$ in $\widetilde{G}$. In particular, $\det(\delta'\delta^{-1}) = \det(\alpha)$ lies in F^*, since $\det(\sigma(x)) = \overline{\det(x)}^{-1}$ for $x \in G(E)$, and coincides with the image of ν in $H^1(F,E^1) = F^*/NE^*$ under the determinant map.

If ν is the image of $z \in F^*$ under the map $H^1(F,Z_G) \to H^1(F,G_\gamma)$, then we can take $g = (z^{-1},1)$ and $\delta' = z\delta$. This shows that $z\delta$ is stably ε-conjugate to δ and that the image of a_σ under the map

$$H^1(F,Z_G) \to H^1(F,G_\gamma) \tag{3.11.1}$$

is $\mathrm{inv}(\delta,z\delta)$. Let $\mathcal{D}_\varepsilon(\delta/F)$ denote the cokernel of (3.11.1). Then $\mathcal{D}_\varepsilon(\delta/F)$ parametrizes the set of ε-conjugacy classes within $\mathcal{O}_{\mathrm{st}}(\delta)$, modulo multiplication by F^*. Let

$$\widetilde{G}(\delta\varepsilon)'' = \{g \in \widetilde{G} : g^{-1}\delta\varepsilon(g)\delta^{-1} \in F^*\} .$$

If γ is regular, we will often write $T = G_\gamma$ and $t_\nu = (\alpha, \alpha)$. Then t_ν belongs to the centralizer $\widetilde{T}$ of T in $\widetilde{G}$ and $\delta' = \delta t_\nu$. In addition, $G_{\delta'\varepsilon} = T$.

PROPOSITION 3.11.2.: *We retain the above notation. Assume that n is odd.*

(a) *$\mathfrak{R}(G_\gamma/F)$ is paired with $\mathfrak{D}_\varepsilon(\delta/F)$.*

(b) *$\mathfrak{D}_\varepsilon(\delta/F)$ is isomorphic to the image of $H^1(F, G_\gamma^d)$ in $H^1(F, G_\gamma)$. It contains $\mathfrak{D}(\gamma/F)$ as a subset.*

(c) *If $z \in F^*$, then δ is ε-conjugate to $z\delta$ if and only if $z \in NE^*$. Furthermore, $\widetilde{G}(\delta\varepsilon)'' = \widetilde{Z}\widetilde{G}(\delta\varepsilon)$.*

(d) *If F is p-adic or if $F = \mathbf{R}$ and γ is not central, then each ε-class in $\mathcal{O}_{\varepsilon-\mathrm{st}}(\delta)$ contains an element δ' such that $N(\delta') = \gamma'$ belongs to G and $\widetilde{G}_{\delta'\varepsilon} = G_\gamma$.*

Proof: Part (a) and (b) follow from (3.5.1). If $g^{-1}\delta\varepsilon(g) = z\delta$ where $z \in F^*$, then $z^n = N_{E/F}(\det(g))^{-1}$ and, since n is odd, $z \in NE^*$. On the other hand, if $z \in \widetilde{Z}$, then $z^{-1}\delta\varepsilon(z) = N_{E/F}(z)^{-1}\delta$. This proves (c). If γ is regular, (d) follows from the above discussion. If γ is central and F is p-adic, then $\mathcal{O}_{\varepsilon-\mathrm{st}}(\delta)$ consists of a single ε-class and we can choose $\delta \in \widetilde{Z}$. If γ is singular but not central, then $\mathfrak{D}_\varepsilon(\delta/F)$ and $\mathfrak{D}(\gamma/F)$ are both isomorphic to $H^1(F, G_\gamma^d)$ by Proposition 3.8.1. If $\{a_\tau\} \in H^1(F, G_\gamma^d)$, there exists $g \in G(\overline{F})$ such that $a_\tau = \tau(g)g^{-1}$ and we can take $\delta' = g^{-1}\delta\varepsilon(g)$, $\gamma' = g^{-1}\gamma g$.

3.12. ε-classes when $n = 2$. Let $G = U(2)$ and let T be an elliptic Cartan subgroup of $U(2)$.

PROPOSITION 3.12.1: *Let $\delta \in \widetilde{G}$ be ε-semisimple.*

(a) *If $\delta \in \widetilde{T}$ is ε-regular, there is a natural bijection between the set of ε-conjugacy classes in $\mathcal{O}_{\varepsilon-\mathrm{st}}(\delta)$ modulo F^* and $H^1(F, E^1)$.*

(b) *The sequence*

$$H^1(Z, F) \longrightarrow H^1(F, T) \xrightarrow{\det} H^1(F, E^1)$$

is exact. If we identify $H^1(F, E^1)$ with F^*/NE^*, then the determinant sends $\nu \in H^1(F, T)$ to the image of $\det(t_\nu)$ in F^*/NE^*.

(c) *δ is ε-conjugate to $z\delta$ for all $z \in F^*$ if and only if $\mathcal{N}(\delta)$ does not contain a regular element in a Cartan subgroup of type* (1).

(d) *If T is of type* (1), *then $H^1(Z,F)$ injects in $H^1(F,T)$. If T is of type* (2), *then the image of $H^1(Z,F)$ in $H^1(F,T)$ is trivial.*

(e) *Suppose that $N(\delta)$ is a scalar. Then $\mathcal{D}_\varepsilon(\delta/F)$ is isomorphic to $H^1(F,E^1)$. If $\delta' \in \mathcal{O}_{\varepsilon-\mathrm{st}}(\delta)$, then $\det(\delta'\delta^{-1}) \in F^*$ and the ε-class of δ' corresponds to the image of $\det(\delta'\delta^{-1})$ in F^*/NE^*.*

Proof: Let $T^d = T \cap G_{\mathrm{der}}$. The inclusion $T^d \to T$ yields an exact sequence

(3.12.1) $$H^1(F,T^d) \to H^1(F,T) \to H^1(F,E^1)$$

To prove (b), we show that the images of $H^1(Z,F)$ and $H^1(F,T^d)$ in $H^1(F,T)$ coincide. Since the determinant annihilates $H^1(F,Z)$, the image of $H^1(F,Z)$ in $H^1(F,T)$ is contained in the image of $H^1(F,T^d)$. If T is of type (1), then $H^1(F,T) = (F^*/NE)^2$. The groups $H^1(F,T^d)$ and $H(Z,F)$ embed in $H^1(F,T)$ as the subgroups of elements of the form (a,a^{-1}) and (a,a), respectively, and these coincide since $H^1(F,E^1)$ is killed by 2. Suppose that $T = T_K$ is of type (2). In this case, we show that the image of $H^1(F,T^d)$ in $H^1(F,T)$ is trivial. Suppose that F is global. By the Tate-Nakayama theorem, $H^1(F_v,T_v)$ (resp., $H^1(F_v,T_v^d)$) is isomorphic $\widehat{H}^{-1}(F_v,X_*(T))$ (resp., $\widehat{H}^{-1}(F_v,X_*(T_v^d))$), and by Proposition 3.5.2(e), the image of $\widehat{H}^{-1}(F_v,X_*(T^d))$ in $\widehat{H}^1(F_v,X_*(T))$ is trivial ($r = 1$ in this case). Hence the image of $H^1(F,T_v^d)$ in $H^1(F_v,T_v)$ is trivial for all v. Since $H^1(F,T) = K^*/N_{KE/K}((KE)^*)$, the map

$$H^1(F,T) \longrightarrow \oplus_v H^1(F_v,T_v)$$

is injective by the Hasse norm theorem and the image of $H^1(F,T^d)$ in $H^1(F,T)$ is therefore trivial. This, together with the remarks preceding Proposition 3.11.2 proves (b) and also (a). To prove (c), it remains to check that δ and $z\delta$ are ε-conjugate for all $z \in F^*$ if $\delta \in \widetilde{M}$ or if $N(\delta)$ is a scalar. If $\delta \in \widetilde{M}$, then $g^{-1}\delta\varepsilon(g) = z\delta$, where $g = d(z^{-1},1)$. Suppose that δ is a scalar. Then $G_{\delta\varepsilon} = G$ and the determinant defines an isomorphism between $H^1(F,G_{\delta\varepsilon})$ and $H^1(F,E^1)$ by (3.8.1). Since the determinant annihilates $H^1(F,Z)$, it follows that $\mathcal{D}_\varepsilon(\delta/F)$ is isomorphic to $H^1(F,E^1)$ and δ is ε-conjugate to $z\delta$ for $z \in F^*$. But then δ is ε-conjugate to $z\delta$ if δ is ε-stably conjugate to a scalar element, i.e., if $N(\delta)$ is a scalar. This completes the proof of (c) and (e) also follows from the remarks of §3.11.

3.13. ε-classes when $n = 3$. Let $G = U(3)$ and let $H = H_0 \times U(1)$, where $H_0 = U(2)$. Denote the determinant on the $U(2)$-factor of H by $\det_0$.

PROPOSITION 3.13.1: *Let T be a Cartan subgroup of H of type (1) or (2). Let $\delta \in \widetilde{T}$ and let $\nu \in \mathfrak{D}_\varepsilon(\delta/F)$. Then $\det_0(t_\nu) \in NE^*$ if and only if ν belongs to the image of $\mathfrak{D}_H(T/F)$.*

Proof: Let $T^d = \{(t, \det(t)^{-1}) \in H_0 \times U(1)\}$ and let $T_0^d = T \cap (H_0)_{\text{der}}$. Then $\mathfrak{D}_\varepsilon(\delta/F) = H^1(F, T^d)$ by Proposition 3.11.2(b), since $H^1(F, T^d)$ injects in $H^1(F, T)$. As with (3.12.1), $\det_0$ induces an exact sequence

$$H^1(F, T_0^d) \longrightarrow H^1(F, T^d) \longrightarrow H^1(F, E^1)\ .$$

Now $\mathfrak{D}_H(T/F)$ is contained in the image of $H^1(F, T_0^d)$ in $H^1(F, T^d)$. Since the determinant induces an isomorphism of $H^1(F, H_0)$ with $H^1(F, E^1)$, the image of $H^1(F, T_0^d)$ in $H^1(F, H_0)$ is trivial and $\mathfrak{D}_H(T/F)$ coincides with the image of $H^1(F, T_0^d)$ in $H^1(F, T^d)$.

PROPOSITION 3.13.2: *If $\delta \in \widetilde{G}$ is ε-semisimple but not ε-regular, then δ is stably ε-conjugate to an element of the form $d(\alpha, \beta, \alpha)$.*

(a) *If $N(\delta)$ is not central, $\mathfrak{D}_\varepsilon(\delta/F)$ is naturally isomorphic to F^*/NE^*.*

(b) *If $N(\delta)$ is central, $\mathfrak{D}_\varepsilon(\delta/F)$ is naturally isomorphic to $H^1(F, G_{\text{ad}})$.*

Proof: The first statement is clear since every non-regular diagonal element of G is of the form $N(d(\alpha, \beta, \alpha))$ and (a) follows from Proposition 3.11.2(b) and (3.8.1). If $N(\delta)$ is central, then δ is stably conjugate to a scalar element. We can assume that δ is scalar. Then the assertion follows from (3.5.2), since $H^1(F, G)$ maps onto $H^1(F, G_{\text{ad}})$ by Lemma 3.5.3.

The set $H^1(F, G_{\text{ad}})$ parametrizes the F-forms of G. If F is local, $H^1(F, G_{\text{ad}})$ is trivial unless $E/F = \mathbf{C}/\mathbf{R}$, and $H^1(\mathbf{R}, G_{\text{ad}}) = \mathbf{Z}/2$. For F global, $H^1(F, G_{\text{ad}}) = \Pi H^1(F_v, G_{\text{ad}})$ as observed in §3.5.

CHAPTER 4

Orbital integrals and endoscopic groups

In §1-6, we define endoscopic groups and the problem of transfer of orbital integrals. The endoscopic groups for unitary groups are determined in §7 and we describe the problems for unitary groups in two and three variables treated in this work from the point of view of functoriality in §8. The remaining sections deal with the transfer of orbital integrals for these cases.

4.1. κ-orbital integrals. For the general discussion, we assume that G_{der} is simply connected. Suppose that F is local and that ε is trivial. If $\gamma \in G$ is semisimple, an element $\kappa \in \mathcal{R}(G_\gamma/F)$ defines a function $\gamma' \to \kappa(\mathrm{inv}(\gamma,\gamma'))$ on $\mathcal{O}_{\mathrm{st}}(\gamma)$. If γ is regular, define the κ-orbital integral of $f \in C(G,\omega)$ by:

$$\Phi^\kappa(\gamma, f) = \sum_{\{\gamma'\}} \kappa(\mathrm{inv}(\gamma,\gamma'))\Phi(\gamma', f) \,. \tag{4.1.1.}$$

where $\{\gamma'\}$ is a set of representatives for conjugacy classes within $\mathcal{O}_{\mathrm{st}}(\gamma)$. If κ is trivial, $\Phi^\kappa(\gamma, f)$ is called a stable orbital integral and is denoted by $\Phi^{\mathrm{st}}(\gamma, f)$. The ordinary orbital integrals $\Phi(\gamma', f)$ determine the κ-orbital integrals and vice versa by the orthogonality relations for the finite abelian group $\mathcal{E}(G_\gamma/F)$.

A distribution T on $C(G,\omega)$ is called invariant if $T(f^g) = T(f)$ for all $g \in G$, where $f^g(x) = f(g^{-1}xg)$. If $\Phi^{\mathrm{st}}(\gamma, f) = 0$ for all regular semisimple γ, we will say that f is stably equivalent to zero. An invariant distribution will be called stable, or stably invariant, if $T(f) = 0$ for all f which are stably equivalent to zero.

If γ is not regular semisimple, then the sum (4.1.1) with κ trivial need not define a stable distribution. It is necessary to modify the definition (4.1.1) by inserting appropriate coefficients. The coefficients are determined by the asymptotic expansions of orbital integrals around semisimple elements. For any connected reductive group G, let $q(G)$ denote one-half the (real) dimension of the symmetric space attached to G if F is archimedean and let $q(G)$ be the F-rank of G_{der} if F is p-adic. Following [Kt$_5$], let $e(G) = (-1)^{q(G)-q(G')}$, where G' is the quasi-split form of G. If γ is semisimple, then G_γ is reductive and connected (since G_{der} is simply connected). Define

$$\Phi^\kappa(\gamma, f) = \sum_{\{\gamma'\}} e(\gamma')\kappa(\mathrm{inv}(\gamma,\gamma'))\Phi(\gamma', f). \tag{4.1.2}$$

where $e(\gamma')$. If γ is regular, then G_γ is a torus and the signs $e(\gamma')$ are trivial, so (4.1.2) is compatible with (4.1.1).

4.2. Endoscopic groups. To each pair (G, ε) is associated a set of auxiliary groups called endoscopic groups. These were first defined in [L_3] in the ordinary (ε trivial) case. The introduction of endoscopic groups in the twisted case is due to Shelstad. In this section, we follow [KS].

An element $s \in \widehat{G}$ will be called ε-semisimple if the endomorphism $\mathrm{ad}(s)\circ \varepsilon$ of $\widehat{G}$ fixes a Borel pair $(\widehat{B}, \widehat{T})$ (recall: $\widehat{B}$ is a Borel subgroup of $\widehat{G}$ and $\widehat{T}$ is a maximal torus in $\widehat{B}$). If s is ε-semisimple, then the connected component $\widehat{G}(s\varepsilon)^\circ$ of the ε-centralizer $\widehat{G}(s\varepsilon) = \{g \in \widehat{G} : g^{-1}s\varepsilon(g) = s\}$ is a connected reductive subgroup of $\widehat{G}$ by [St], Corollary 9.4.

An endoscopic triple is a triple (H, s, η), consisting of a quasi-split group H, an ε-semisimple element s in $\widehat{G}$ and an L-map $\eta : {}^LH \to {}^LG$, (we take the Weil forms of the L-groups) which satisfies the following two conditions.

(I): η restricts to an isomorphism of complex groups from $\widehat{H}$ to $\widehat{G}(s\varepsilon)^\circ$. Define $\lambda(w) = s\varepsilon(\eta(w))s^{-1}\eta(w)^{-1}$ for $w \in W_F$.

(II): λ takes values in $Z(\widehat{G})$ (in which case λ defines a cocycle with values in $Z(\widehat{G})$) and the class of λ in $H^1(W_F, Z(\widehat{G}))$ is locally trivial (resp. trivial) if F is global (resp. local).

The endoscopic group is the quasi-split connected reductive group H. The triple, or just H itself, is called elliptic if $\eta((Z(\widehat{H})^\Gamma)^\circ) \subset Z(\widehat{G})^\Gamma$. Note that if ε is trivial, then $(G, 1, id.)$ is an elliptic endoscopic triple for G.

Let (H_i, s_i, η_i), $i = 1, 2$, be endoscopic triples for G. An isomorphism between them is a pair (α, β) of maps

$$\alpha : H_1 \to H_2$$
$$\beta : {}^LH_2 \to {}^LH_1$$

where α is an F-isomorphism and β is an L-homomorphism satisfying:

(i) β induces an isomorphism of $\widehat{H}_2$ with $\widehat{H}_1$ dual to α.

(ii) There exists $g \in \widehat{G}$ such that $gs_1\varepsilon(g)^{-1}s_2^{-1} \in Z(\widehat{G})\mathrm{Cent}(\eta_2, \widehat{G})^\circ$ and $\eta_2 = \mathrm{ad}(g) \circ \eta_1 \circ \beta$.

In particular, if α is a 1-cocycle on W_F with values in $Z(\widehat{H})$, then $(H, s, \eta\alpha)$ is isomorphic to (H, s, η), where $\eta\alpha(h \times w) = \eta(\alpha(w)h \times w)$.

Every element of $\widehat{H}$ defines an automorphism of (H, s, η) by conjugation. Denote the automorphism group of (H, s, η) by $\mathrm{Aut}(H, s, \eta)$ and let $\Lambda(H, s, \eta) = \mathrm{Aut}(H, s, \eta)/\widehat{H}$. Then $\Lambda(H, s, \eta)$ is a finite group.

4.3. Transfer of orbital integrals. Recall that $\mathcal{O}_{st}(G)$ is the set of stable conjugacy classes of semisimple elements in G. Let T be a maximal F-torus in G. If $\gamma \in G(\overline{F})$ is semisimple, then the $G(\overline{F})$-conjugacy class of γ intersects $T(\overline{F})$ in an $\Omega_G(T)$-orbit. The orbit is fixed by Γ if and only if the $\overline{F}$-conjugacy class of γ is defined over F. Sending a stable conjugacy class to the image of its intersection with $T(\overline{F})$, we obtain a map

$$\mathcal{O}_{st}(G) \to [T(\overline{F})/\Omega_G(T)]^{\Gamma}.$$

Assume that G is quasi-split and G_{der} is simply connected. Then every F-rational $G(\overline{F})$-conjugacy class intersects G non-trivially by the Kottwitz-Steinberg theorem (Theorem 3.2.1) and in this case the map is bijection.

Assume further that ε is trivial. Let (H, s, η) be an endoscopic datum for G, and let $(\widehat{B}_H, \widehat{T}_H)$ and $(\widehat{B}, \widehat{T})$ be Borel pairs which are fixed by Γ in $\widehat{H}$ and $\widehat{G}$, respectively. There exists $y \in \widehat{G}$ such that $\mathrm{ad}(y) \circ \eta$ takes $(\widehat{B}_H, \widehat{T}_H)$ into $(\widehat{B}, \widehat{T})$. The set of roots of $\widehat{T}_H$ in $\widehat{B}_H$ maps to the set of roots α of $\widehat{T}$ in $\widehat{B}$ such that $\alpha(ysy^{-1}) = 1$. In particular, $\Omega_H(\widehat{T}_H)$ is mapped to a subgroup of $\Omega_G(\widehat{T})$ under $\mathrm{ad}(y) \circ \eta$. Define $\psi^{\wedge} : \widehat{T}_H \to \widehat{T}$ by $\psi^{\wedge}(t) = y\eta(t)y^{-1}$. For $\tau \in \Gamma$, there exists $x_\tau \in \widehat{G}$ such that

$$\eta(\tau(h)) = x_\tau \tau(\eta(h)) x_\tau^{-1} .$$

for $h \in \widehat{H}$ and consequently, $\psi^{\wedge}(\tau(h)) = n_\tau \tau(\psi^{\wedge}(h)) n_\tau^{-1}$, where $n_\tau = y x_\tau \tau(y^{-1})$ belongs to $\Omega_G(\widehat{T})$. The actions of Γ on $X_*(\widehat{T}_H)$ and $X_*(\widehat{T})$ therefore differ by a twisting with values in $\Omega_G(\widehat{T})$.

Let $(\underline{T}_H, \underline{B}_H)$ and $(\underline{T}, \underline{B})$ be Borel pairs in H and G, respectively. We have canonical identifications of $X^*(T_H)$ and $X^*(T)$ with $X_*(\widehat{T}_H)$ and $X^*(\widehat{T})$, respectively, under which $\Omega_G(\widehat{T})$ and $\Omega_H(\widehat{T}_H)$ are identified with $\Omega_G(T)$ and $\Omega_H(T_H)$, respectively. The map $\psi^{\wedge}$ defines an isomorphism $X^*(T_H) \to X^*(T)$ which carries $\Omega_H(T_H)$ into a subgroup of $\Omega_G(T)$. This gives rise to an $\overline{F}$-isomorphism $\psi : \underline{T}_H \to \underline{T}$, the inverse of the map naturally associated to $X^*(T_H) \to X^*(T)$.

Assume that $\underline{T}_H$ and $\underline{T}$ are defined over F. Then the Γ- actions on $X_*(\widehat{T})$ and $X^*(T)$ differ by a twisting with values in $\Omega_G(T)$. In fact, if $\underline{B}$ is defined over F, then the Γ-actions coincide. On the other hand, if $\underline{T}'$ is another Cartan subgroup of $\underline{G}$ defined over F, then $T' = gTg^{-1}$ for some $g \in G(\overline{F})$. The 1-cocycle $\{g^{-1}\tau(g)\}$ takes values in $\Omega_G(T)$ and $\mathrm{ad}(g)$ intertwines the Γ-action on $X^*(T')$ with the twist of the Γ-action on $X^*(T)$ by $\{g^{-1}\tau(g)\}$. The analogous statements hold for $X_*(\widehat{T}_H)$ and $X^*(T_H)$, and hence there is a 1-cocycle $\{w_\tau : \tau \in \Gamma\}$ with values in $\Omega_G(T)$ such that

$$\psi(\tau(t)) = \mathrm{ad}(w_\tau)(\tau(\psi(t))) .$$

In particular, ψ induces a Γ-equivariant map

$$T_H(\overline{F})/\Omega_H(T_H) \to T(\overline{F})/\Omega_G(T)$$

and hence a map:

$$A_{G/H} : \mathcal{O}_{\mathrm{st}}(H) \to \mathcal{O}_{\mathrm{st}}(G) .$$

The map $A_{G/H}$ is independent of all choices. Note also that ψ^{-1} restricts to an embedding of Z_G in Z_H which is defined over F.

Observe that T can be chosen so that ψ is defined over F. In fact, let $\gamma \in T_H$ be an element such that $\psi(\gamma)$ is regular in $G(\overline{F})$. The conjugacy class containing $\psi(\gamma)$ is defined over F and, by the Kottwitz-Steinberg theorem (Theorem 3.2.1), there exists $g \in G(\overline{F})$ such that $g\psi(\gamma)g^{-1}$ is F-rational. Let $\psi'(t) = g\psi(t)g^{-1}$. Then

$$\psi'(\gamma) = (gw_\tau\tau(g)^{-1})^{-1}\psi'(\gamma)(gw_\tau\tau(g)^{-1})$$

and hence $gw_\tau\tau(g)^{-1} \in gT(\overline{F})g^{-1}$ since $\psi'(\gamma)$ is regular. It follows that ψ' is defined over F and we can replace ψ and T by ψ' and gTg^{-1}.

A semisimple element $\gamma' \in H$ is called G-regular if $A_{G/H}(\gamma')$ is a regular class in G. Suppose that $\gamma' \in T_H$. Then γ' is called (G, H)-regular if $\alpha(\psi(\gamma')) \neq 1$ for each root α of T which is not the image of a root of T_H in H.

Let γ' be a (G, H)-regular element of T_H. Suppose that ψ is defined over F (this entails no loss of generality since the choice of T is arbitrary) and let $\gamma = \psi(\gamma')$. Then ψ defines an F-isomorphism between the root data of $\underline{T}_H$ in $\underline{H}_\gamma$, and that of $\underline{T}$ in $\underline{G}_\gamma$. It follows that ψ extends to an isomorphism of $\underline{H}_{\gamma'}$ with $\underline{G}_\gamma$ which is an inner twisting over F. In particular, we can identify $Z(\widehat{H}_{\gamma'})$ with $Z(\widehat{G}_\gamma)$ and, if F is local, we can choose compatible measures on $H_{\gamma'}$ and G_γ.

Let T be a maximal F-torus in $H_{\gamma'}$. Then $Z(\widehat{H})$ and $Z(\widehat{H}_{\gamma'})$ can both be viewed as subgroups of $\widehat{T}$ and as such, $Z(\widehat{H}) \subset Z(\widehat{H}_{\gamma'})$, where this inclusion is Γ-equivariant and independent of the choice of T. Since $Z(\widehat{H}_{\gamma'}) = Z(\widehat{G}_\gamma)$, we obtain a canonical Γ-equivariant inclusion of $Z(\widehat{H})$ in $Z(\widehat{G_\gamma})$. Now s defines an element s' of $Z(\widehat{G}_\gamma)$. By Condition (II) of §4.2, the image of s' in $Z(\widehat{G}_\gamma)/Z(\widehat{G})$ is Γ-invariant and the image of s in $\pi_0([Z(\widehat{G}_\gamma)/Z(\widehat{G})]^\Gamma)$ defines an element κ of $\mathcal{R}(G_\gamma/F)$. We also obtain an element κ_v of $\mathcal{R}(G_\gamma/F_v)$ for all v.

Assume now that F is local. According to a conjecture of Langlands, there is a function $\Delta_{G/H}(\gamma_H, \gamma)$ on pairs consisting of a G-regular element $\gamma_H \in H$ and $\gamma \in A_{G/H}(\mathcal{O}_{st}(\gamma_H))$ with the following property: for all $f \in C(G,\omega)$, there exists function $f^H \in C(H,\omega)$ whose orbital integrals match with those of f in the following sense:

$$\Phi^{st}(\gamma_H, f^H) = \Delta_{G/H}(\gamma_H, \gamma)\Phi^\kappa(\gamma, f). \tag{4.3.1}$$

for G-regular γ_H, where the orbital integrals are defined using compatible measures on $H_{\gamma'}$ and G_γ. We write $f \to f^H$ if f and f^H correspond via (4.3.1). The correspondence depends on the choice of measures on G and H, and f^H is determined only up to the addition of a function on H all of whose stable orbital integrals vanish. As suggested in [Kt$_4$], (4.3.1) should hold if γ_H is only assumed to be (G,H)-regular (cf. Proposition 8.1.3).

The function $\Delta_{G/H}(\gamma_H, \gamma)$ is called a transfer factor. It depends only on the stable conjugacy class of γ_H in H and the conjugacy class of γ in G. If γ' is stably conjugate to γ in G, then consistency implies that

$$\Delta_{G/H}(\gamma_H, \gamma') = \Delta_{G/H}(\gamma_H, \gamma)\kappa(\text{inv}(\gamma, \gamma'))\ . \tag{4.3.2}$$

A general definition of the transfer factor has been proposed by Langlands and Shelstad ([LS]). The definition specifies $\Delta_{G/H}$ up to a non-zero constant that depends on certain choices. Globally, there exists a compatible collection $\{\Delta_{G_v/H_v}\}$ of local transfer factors such that for all G-regular semisimple $\gamma_H \in H$ and all $\overline{\gamma} = (\overline{\gamma}_v) \in \mathbf{G}$ such that $\overline{\gamma}_v \in A_{G_v/H_v}(\mathcal{O}_{st}(\gamma_H))$ for all v, the product

$$\Delta_{G/H}(\gamma_H, \overline{\gamma}) = \prod_v \Delta_{G_v/H_v}(\gamma_H, \overline{\gamma}_v)$$

exists ($\Delta_{G_v/H_v}(\gamma_H, \overline{\gamma}_v) = 1$ for almost all v) and the value $\Delta_{G/H}(\gamma_H, \overline{\gamma})$ is independent of the choice of compatible collection. Furthermore, $\Delta_{G/H}$ satisfies the key property:

$$\Delta_{G/H}(\gamma_H, \overline{\gamma}) = \kappa(\text{obs}(\overline{\gamma})) \tag{4.3.3}$$

where $\mathrm{obs}(\overline{\gamma}) \in A(G_{\gamma}^{d})$ as in §3.3.

Suppose that $\gamma \in A_{G/H}(\mathcal{O}_{\mathrm{st}}(\gamma_H))$ and let $f = \Pi_v f_v \in C(G,\omega)$. For almost all v, $\Phi^{\kappa_v}(\gamma, f_v) = \Phi(\gamma, f_v)$, i.e., $\Phi(\gamma', f_v) = 0$ if γ' is stably conjugate but not conjugate to γ. This is the case if f_v is the unit in $\mathcal{H}_v$ and $(1-\alpha(\gamma))$ is a unit or zero for all roots of G ([Kt4], §7.3). We may therefore set $\Phi^{\kappa}(\gamma, f) = \Pi_v \Phi^{\kappa_v}(\gamma, f_v)$. By (4.3.2) and (4.3.3),

$$\Phi^{\kappa}(\gamma, f) = \sum_{\{\gamma'\}} \kappa(\mathrm{obs}(\gamma'))\Phi(\gamma', f)$$

where $\{\gamma'\}$ is a set of representatives for the $\mathbf{G}$-conjugacy classes in $\mathcal{O}_{\mathrm{st}}(\gamma/\mathbf{A})$. The sum is finite by the above remark. By (4.3.3), if γ_H is G-regular, then $\Delta_{G/H}(\gamma_H, \gamma) = 1$ since $\mathrm{obs}(\gamma)$ is trivial, and hence $\Phi^{\kappa}(\gamma, f) = \Phi^{\mathrm{st}}(\gamma_H, f^H)$, where $f^H = \Pi_v f_v^H$.

Functions f and f^H related by a relation of the type (4.3.1) will be considered in several different situations below. Whenever an equality is written between orbital integrals of functions defined on different groups, it will be tacitly assumed that they are related by the appropriate transfer.

4.4. Functoriality. The transfer between κ-orbital integrals on G and stable orbital integrals on H (in the cases where it is known), defines a dual map from stably invariant distributions on H to invariant distributions on G. The map on distributions should lead to character identities between L-packets on H and L-packets on G which are associated via η by the principle of functoriality. This program has been carried in various cases, (for example, in the real case [S1]) but is not yet known in general. It, as well as its global counterpart, is carried out in §13 for the cases described in §4.8 below.

4.5. The fundamental lemma. Suppose that F is p-adic. A connected reductive group G is said to be unramified over F if it is quasi-split and splits over an unramified extension of F. In this case, the action of W_F on $\widehat{G}$ factors through the projection of W_F onto $\Gamma(F^{\mathrm{un}}/F)$, where F^{un} is the maximal unramified extension of F. Assume that G is unramified and let T be a maximal torus contained in a Borel subgroup B of G. Let K be a hyperspecial maximal compact subgroup of G. An irreducible representation of G is called unramified if it contains a non-zero K-invariant vector. Let $E^{u}(G)$ be the set of unramified representations.

A character χ of T is said to be unramified if it is trivial on the maximal compact subgroup T_c. Let $\Pi^u(T)$ be the set of unramified characters of T. The space of K-fixed vectors in the principal series representation $i_G(\chi)$ is one-dimensional, by virtue of the Iwasawa decomposition $G = BK$, and $i_G(\chi)$ contains a unique irreducible unramified constituent π_χ. Two representations π_χ and $\pi_{\chi'}$ are equivalent if and only if χ and χ' lie in the same $\Omega(T,G)$-orbit and every element of $E^u(G)$ is of the form π_χ for some $\chi \in \Pi^u(T)$. Fix an element $w_F \in W_F$ whose projection to $\Gamma(F^{un}/F)$ is the Frobenius element. Recall ([Bo]) that there is a canonical bijection between $\Pi^u(T)/\Omega(T,G)$ (and hence $E^u(G)$) and the set of semisimple $\widehat{G}$- conjugacy classes in ${}^LG$ of the form $\{g \times w_F\}$. The conjugacy class $\{g(\pi)\}$ in ${}^LG$ associated to a representation $\pi \in E^u(G)$ is called the Langlands class of π. We can choose a representative $g \times w_F$ with $g \in \widehat{T}$.

Let (H, s, η) be an endoscopic datum for G. We will say that η is unramified if H is unramified and if the restriction of η to W_F is of the form $\eta(w) = \psi(w) \times w$ where ψ factors through the projection of W_F onto $\Gamma(F^{un}/F)$. Let T_H be a maximal torus contained in a Borel subgroup of H. Assume that η is unramified. We obtain a map $\Pi^u(T_H)/\Omega(T,H) \to \Pi^u(T_G)/\Omega(T,G)$ corresponding to the map sending $\{t \times w_F\}$ to $\{\eta(t \times w_F)\}$, and hence a map

$$E^u(H) \to E^u(G). \tag{4.5.1}$$

Suppose that ω is an unramified character of a torus Z contained in the center of G and let $\Pi^u(T,\omega)$ be the subgroup of $\Pi^u(T)$ consisting of characters whose restriction to Z is ω. Then $\Pi^u(T,\omega)$ has the structure of an algebraic variety isomorphic to $(\mathbf{C}^*)^r$ where r is the split rank of $Z\backslash G$ over F. The Satake transform of a function $f \in \mathcal{H}(G,\omega)$ is the function $f^\wedge(\chi)$ on $\Pi^u(T,\omega)$ defined by $f^\wedge(\chi) = \mathrm{Tr}(\pi_\chi(f))$. By the result of Satake ([Bo]), $f^\wedge(\chi)$ is an $\Omega(T,G)$-invariant polynomial function on $\Pi^u(T,\omega)$ (i.e., a Laurent polynomial in r variables) and the Satake transform defines an isomorphism between $\mathcal{H}(G,\omega)$ and the algebra of $\Omega(T,G)$-invariant polynomial functions on $\Pi^u(T,\omega)$.

The torus Z is canonically embedded in Z_H. There is a character μ of Z such that $\chi|Z = \chi'\mu|Z$ if $\pi_{\chi'}$ maps to π_χ under (4.5.1). The embedding η gives rise to an algebra homomorphism

$$\hat{\eta} : \mathcal{H}(G,\omega) \to \mathcal{H}(H,\omega\mu^{-1})$$

which sends $f \in \mathcal{H}(G,\omega)$ to the element $\hat{\eta}(f)$ such that $\mathrm{Tr}(\pi_{\chi'}(\hat{\eta}(f))) = \mathrm{Tr}(\pi_\chi(f))$ if $\pi_{\chi'}$ maps to π_χ.

The fundamental lemma is the assertion that there exists a choice of transfer factor $\Delta_{G/H}$ (recall that $\Delta_{G/H}$ is only defined up to a non-zero multiple) such that (4.3.1)) holds with $f^H = \hat{\eta}(f)$. It is not yet known in general.

4.6. Endoscopic groups associated to unitary groups. Let G be a unitary group in n variables. In this section we determine the isomorphism classes of elliptic endoscopic triples (H, s, η) for G (here ε is trivial).

PROPOSITION 4.6.1. *Let (H, s, η) be an elliptic endoscopic triple for G. Then H is isomorphic to $U(a) \times U(b)$ where a and b are positive integers such that $a + b = n$. The triple is determined by $\{a, b\}$ up to isomorphism. Furthermore, $\Lambda(H, s, \eta)$ has order 2 if $a = b$ and has order 1 otherwise.*

Proof: The centralizer of a semisimple element s in $\mathrm{GL}_n(\mathbf{C})$ is isomorphic to $\mathrm{GL}_{n_1}(\mathbf{C}) \times \ldots \times \mathrm{GL}_{n_m}(\mathbf{C})$, where $(n_1, \ldots, n_m)$ is a partition of n. By Condition (II) of §4.2, $s\eta(w)s^{-1}\eta(w)^{-1} = \lambda(w)$ where λ belongs to $\ker^1(W_F, Z(\widehat{G}))$. By Proposition 3.5.1(b), λ is a trivial cocycle. Since W_E acts trivially on $\widehat{G}$, $\lambda(w) = 1$ for $w \in W_E$.

We identify $\widehat{H}$ with the centralizer of s via η, keeping in mind that the Γ-action on $\widehat{H}$ is not the restriction of the Γ-action on $\widehat{G}$. If $\tau \in \Gamma_E$, then τ acts trivially on $\widehat{G}$ and τ acts on $\widehat{H}$ via $\mathrm{ad}(g)$ for some $g \in \widehat{G}$ which commutes with s. Hence $g \in \widehat{H}$ and since τ preserves a splitting, it acts trivially. This shows that the action of Γ on $\widehat{H}$ factors through $\Gamma(E/F)$. Replacing s by rs for some $r \in Z(\widehat{G})$ if necessary, we can assume that $s \in Z(\widehat{H})^\Gamma$.

Now σ acts by an automorphism of order 2 such that $(Z(\widehat{H})^\sigma)^\circ \subset Z(\widehat{G})^\sigma = \{\pm 1\}$, since (H, s, η) is elliptic. It is easy to see that if $(Z(\widehat{H})^\sigma)^\circ$ is finite, then σ must leave the blocks of $\widehat{H}$ stable and must act on $Z(\widehat{H})$ by $x \to x^{-1}$. The diagonal entries of s are therefore ± 1 and, if $\widehat{H} \neq \widehat{G}$, then $m = 2$.

Suppose that $\widehat{H} = \mathrm{GL}_a(\mathbf{C}) \times \mathrm{GL}_b(\mathbf{C})$ and that $\eta(w) = \psi(w) \times w$, for $w \in W_F$. Let w_σ be a fixed element of $W_{E/F}$ whose projection to $\Gamma(E/F)$ is σ. Then $W_F = W_E \cup W_E w_\sigma$. Up to equivalence, $\mathrm{ad}(\eta(\sigma))$ acts on $\widehat{H}$ by $g \to \Phi_{a,b}\, {}^t g^{-1} \Phi_{a,b}^{-1}$, where $\Phi_{a,b} = [\Phi_a, \Phi_b] \in \widehat{H}$, and $[g, h]$ denotes the block diagonal element in $\widehat{G}$ with g and h along the diagonal (Φ_j is defined in

§1.9). Hence $\eta(w_\sigma) = z\Phi_{a,b}\Phi^{-1} \times w_\sigma$ for some $z \in Z(\widehat{H})$ and it is easy to see that up to equivalence, we can assume that $\eta(w_\sigma) = \Phi_{a,b}\Phi^{-1} \times w_\sigma$. It follows that $H = U(a) \times U(b)$.

The action of W_E on $\widehat{H}$ is trivial. Hence $\psi(w) \in Z(\widehat{H})$ for $w \in W_E$, and there exist characters μ_1, μ_2 of W_E such that

$$\eta(z) = [\mu_1(z)|_a, \mu_2(z)|_b] \times z$$

where $|_m$ denotes the $m \times m$ identity matrix. We regard μ_1 and μ_2 as characters of C_E. If the image of w in C_E is z, then the image of $w_\sigma w w_\sigma^{-1}$ is $\overline{z}$ and since,

$$\psi(w_\sigma w w_\sigma^{-1}) = \mathrm{ad}(\eta(w_\sigma))(\psi(w)) = \psi(w)^{-1},$$

the μ_j are trivial on $N_{E/F}(C_E)$. The image of w_σ^2 in C_E belongs to $C_F - N_{E/F}(C_E)$. Since the restriction of μ_j to C_F has order two, $\mu_j|C_F = \omega_{E/F}$ if $\mu_j(w_\sigma^2) = -1$ and $\mu_j|C_F$ is trivial if $\mu_j(w_\sigma^2) = 1$. Since ${}^t\Phi_j^{-1} = \Phi_j$ and $\Phi_j^2 = (-1)^{j-1}$, we obtain

$$\eta(w_\sigma)^2 = [(-1)^{a+n}, (-1)^{b+n}] \times w_\sigma^2.$$

This shows that restriction of μ_1 to C_F is $\omega_{E/F}$ if $a \not\equiv n \pmod 2$ and is trivial if $a \equiv n \pmod 2$. Similarly, the restriction of μ_2 to C_F is $\omega_{E/F}$ if $b \not\equiv n \pmod 2$ and is trivial if $b \equiv n \pmod 2$. Furthermore, any choice of μ_j satisfying these conditions defines an embedding of ${}^LH$ into ${}^LG$.

Let α be a character of C_E with values in $Z(\widehat{H})$, which is trivial on C_F. Regard α as a character of W_E and extend it to a function on W_F by setting $\alpha(ww_\sigma) = \alpha(w)$ for $w \in W_E$. Then α defines a 1-cocycle on W_F with values in $Z(\widehat{H})$ and the map ρ defined by $\rho(h \times w) = h\alpha(w) \times w$ is an automorphism of ${}^LH$. It follows easily that the isomorphism class of (H, s, η) depends only on a and b, and in particular, is independent of the choice of μ_j. It is immediate that $\Lambda(H, s, \eta)$ has order 2 if $a = b$ and has order 1 otherwise.

4.7. Base change for unitary groups. Let $G = U(n)$ and set $\widetilde{G} = \mathrm{Res}_{E/F}(G)$. Let ε be the algebraic automorphism of $\widetilde{G}$ induced by the non-trivial element σ of $\Gamma(E/F)$. The dual group of $\widetilde{G}$ is $\widehat{G} \times \widehat{G}$. We also denote by ε the automorphism of $\widetilde{G}$ which interchanges the two copies of $\widehat{G}$. The Galois group Γ acts on $\widehat{G} \times \widehat{G}$ through its projection onto $\Gamma(E/F)$ and σ acts by $\sigma(x, y) = (\sigma(y), \sigma(x))$, where $\sigma(x) = \Phi_n\, {}^tx^{-1}\Phi_n^{-1}$.

There is a natural bijection between (equivalence classes of) L-parameters $\rho : W_E \to {}^L G_{/E}$ and $\rho' : W_F \to {}^L\widetilde{G}$ given by associating to ρ the map ρ' defined by

$$\rho'(w) = (\rho(w), \sigma(\rho(w_\sigma^{-1} w w_\sigma))) \times w, \quad \rho'(w_\sigma) = (\rho(w_\sigma^2), 1) \times w_\sigma ,$$

where $w \in W_E$. This corresponds to the fact that the representations of $\widetilde{G}(F)$ and $G(E)$ coincide since the two groups are canonically isomorphic. If E/F is unramified, then a Langlands class $\{(t_1, t_2) \times w_\sigma\}$ in ${}^L\widetilde{G}$ corresponds to the class $\{t_1\sigma(t_2) \times w_\sigma^2\}$ in ${}^L G_{/E}$

We now determine the elliptic endoscopic triples (H, s_1, η) associated to $(\widetilde{G}, \varepsilon)$. Every ε-semisimple element of $\widehat{G} \times \widehat{G}$ is ε-conjugate to an element of the form $(s, 1)$ where $s \in \widehat{G}$ is semisimple. Assume that $s_1 = (s, 1)$. Then $((\widehat{G} \times \widehat{G})(s_1\varepsilon))^\circ$ is the image of $\widehat{G}(s)$ under the diagonal embedding. Furthermore, let $\eta(w) = (\xi_1(w), \xi_2(w)) \times w$. By Condition (II), the cocycle

$$\lambda'(w) = \begin{cases} (s\xi_2(w)s^{-1}\xi_1(w)^{-1}, \xi_1(w)\xi_2(w)^{-1}) & \text{if } w \in W_E \\ (s\xi_2(w_\sigma)\xi_1(w_\sigma)^{-1}, \xi_1(w_\sigma)\sigma(s)^{-1}\xi_2(w_\sigma)^{-1}) & \text{if } w = w_\sigma \end{cases}$$

is locally trivial cocycle and takes values in $Z(\widehat{G} \times \widehat{G})$. The action of W_E on $\widehat{G}$ is trivial, hence $\lambda'(w) = 1$, $\xi_1(w) = \xi_2(w)$, and arguing as in the non-twisted case, we see that $\xi_j(w)$ belongs to $Z(\widehat{H})$ for $w \in W_E$. Furthermore, λ' defines a trivial cocycle of $\Gamma(E/F)$ and hence

$$(s\xi_2(w_\sigma)\xi_1(w_\sigma)^{-1}, \xi_1(w_\sigma)\sigma(s)^{-1}\xi_2(w_\sigma)^{-1}) = (\lambda, \lambda)$$

for some $\lambda \in Z(\widehat{G})$. It follows that $\xi_2(w_\sigma) = s^{-1}\lambda\xi_1(w_\sigma)$. Let $\xi = \xi_1(w_\sigma)$.

Now $\widehat{H}$ is isomorphic to $\mathrm{GL}_{n_1}(\mathbf{C}) \times \cdots \times \mathrm{GL}_{n_m}(\mathbf{C})$, where $(n_1, \ldots, n_m)$ is a partition of n, and $\mathrm{ad}(\eta(w_\sigma))$ induces the automorphism $h \to \xi\sigma(h)\xi^{-1}$ on $\widehat{H}$. Since (H, s_1, η) is elliptic, $\mathrm{ad}(\eta(w_\sigma))$ must preserve the factors of $\widehat{H}$. It induces $x \to x^{-1}$ on $Z(\widehat{H})$. Up to equivalence, we can assume that $\xi = \Phi^*\Phi^{-1}$, where Φ^* is the block-diagonal element $[\Phi_{n_1}, \ldots, \Phi_{n_m}]$ of $\widehat{H}$. Set $\overline{w} = w_\sigma w w_\sigma^{-1}$. From the relation,

$$(\xi_1(\overline{w}), \xi_1(\overline{w})) \times \overline{w} = \mathrm{ad}(\eta(w_\sigma))(\xi_1(w), \xi_1(w)),$$

we obtain $\xi_1(w\overline{w}) = 1$. It follows that $\xi_1(w_\sigma^2) = \pm 1$ and consideration of $\eta(w_\sigma)^2$ shows that $s^{-1}\lambda\xi\sigma(\xi) = \pm 1$. However $\sigma(\xi)\xi$ belongs to $Z(\widehat{H})$ and has eigenvalues ± 1, hence the eigenvalues of $s^{-1}\lambda$ are ± 1. Since we are free to replace s by αs where $\alpha \in Z(\widehat{G})$, we may assume that $s = \pm[|_a, -|_b]$ and $\lambda = \pm 1$. Then $\widehat{H} = \mathrm{GL}_a(\mathbf{C}) \times \mathrm{GL}_b(\mathbf{C})$ and $H = U(a) \times U(b)$. We can assume that the restriction of η to $\widehat{H}$ is the diagonal embedding and that η is defined on W_F by

$$\eta(w) = (\xi_1(w), \xi_1(w)) \times w$$
$$\eta(w_\sigma) = (\xi, \lambda s\xi) \times w_\sigma$$

where ξ_1 is a homomorphism of W_E in to $Z(\widehat{H})$ and $\xi = \Phi_{a,b}\Phi^{-1}$. We have

$$\eta(w_\sigma)^2 = (\lambda\xi\sigma(s\xi), \lambda s\xi\sigma(\xi)) \times w_\sigma^2 \,,$$

The equivalence class of (H, s, η) depends only on λ.

We now consider two cases. In the first case, we assume that $a \not\equiv b$ (mod 2). In this case, we can take $s = [(-1)^{a+n}|_a, (-1)^{b+n}|_b]$. Then $\sigma(s\xi) = \xi^{-1}$ and $\eta(w_\sigma)^2 = (\lambda, \lambda) \times w_\sigma^2$ Fix a character μ of C_F whose restriction to C_E is $\omega_{E/F}$. We define two embeddings $\eta_j : {}^LH \to {}^L\widehat{G}$ corresponding to the two possible values of λ by

$$\eta_1(w) = \begin{cases} 1 \times w & \text{for } w \in W_E \\ (\xi, s\xi) \times w_\sigma & \text{if } w = w_\sigma \,. \end{cases}$$

and

$$\eta_2(w) = \begin{cases} (\mu(w), \mu(w)) \times w & \text{for } w \in W_E \\ (\xi, -s\xi) \times w_\sigma & \text{if } w = w_\sigma \end{cases} .$$

The triples (H, s_1, η_1) and (H, s_1, η_2) are non-isomorphic.

Now assume that $a \equiv b$ (mod 2). We may take $s = [-(-1)^{a+n}, (-1)^{b+n}]$. Then $\sigma(\xi) = \xi^{-1}r$, where $r = (-1)^{a+n}$, and $\eta(w_\sigma)^2 = (\lambda rs, \lambda rs) \times w_\sigma^2$, since $\xi\sigma(s)\xi^{-1} = s$. Define two embeddings η_j by

$$\eta_1(w) = \begin{cases} ([\mu(w)|_a, |_b], [\mu(w)|_a, |_b]) \times w & \text{if } w \in W_E \\ (\xi, s\xi) \times w_\sigma & \text{if } w = w_\sigma \,. \end{cases}$$

and

$$\eta_2(w) = \begin{cases} ([|_a, \mu(w)|_b], [|_a, \mu(w)|_b]) \times w & \text{if } w \in W_E \\ (\xi, -s\xi) \times w_\sigma & \text{if } w = w_\sigma . \end{cases}$$

It is clear that η_1 and η_2 are inequivalent if $a \neq b$. If $a = b$, then $\eta_2 = \mathrm{ad}((\xi, \xi)) \circ \eta_1$, as can be checked directly. The next lemma follows.

LEMMA 4.7.1: *The endoscopic triples (H, s_1, η_1) and (H, s_1, η_2) are inequivalent, unless $a = b$. Furthermore, every endoscopic triple for $(\widetilde{G}, \varepsilon)$ is, up to equivalence, of the form (H, s_1, η_1) or (H, s_1, η_2) for some endoscopic triple (H, s, η) for G.*

The case $s = 1$ gives rise to two non-equivalent base change embeddings. Let $\psi_G : {}^LG \to {}^L\widetilde{G}$ denote the standard base change embedding

$$\psi_G(g \times w) = (g, g) \times w.$$

The second "primed" embedding ψ'_G is defined by $\psi'_G(g \times w) = \alpha(w)\psi_G(g \times w)$ where $\alpha(w)$ is the 1-cocycle defined by:

$$\alpha(w) = \begin{cases} (\mu(w), \mu(w)) & \text{if } w \in W_E \\ (\mu(w_0), -\mu(w_0)) & \text{if } w = w_0 w_\sigma, w_0 \in W_E \ . \end{cases}$$

Recall that μ is a fixed character of C_E whose restriction to C_F is $\omega_{E/F}$. By the Langlands correspondence, the cocycle α corresponds to the ε-invariant character χ_μ of $\widetilde{G}$ defined by $\chi_\mu(g) = \mu(\det(g))$. The base change transfers with respect to ψ_G and ψ'_G of an L-packet Π on G should therefore be related by $\psi_G(\Pi) = \psi'_G(\Pi) \otimes \chi_\mu$.

Observe that if π is a representation of $\widetilde{G}$ such that $\varepsilon(\pi)$ is equivalent to π, then the central character ω_π of π is trivial on the subgroup NE^* of $Z(\widetilde{G})$ since $\varepsilon(z) = \overline{z}^{-1}$ for $z \in Z(\widetilde{G})$. Here we identify $Z(\widetilde{G})$ with E^*. The restriction of ω_π to F^* is therefore either trivial or $\omega_{E/F}$. On the other hand, suppose that a base change transfer ψ_G is defined taking L-packets on G to representations of $\widetilde{G}$. If $\pi = \psi_G(\Pi)$ for some L-packet Π, then $\omega_\pi = \omega_\Pi \circ N$, where ω_Π is the central character of Π and N is the norm map from $Z(\widetilde{G})$ to $Z(G)$. The restriction of ω_π to F^* is trivial in this case. The restriction of χ_μ to F^* is either trivial or equal to $\omega_{E/F}$, according as n is even or odd. It follows that for n odd, the restriction of ω_π to F^* is trivial if π is in the image of the base change transfer via ψ_G and is equal to $\omega_{E/F}$ if π is in the image of the base change transfer via ψ'_G. On the other hand, if n is even, then the restriction of ω_π to F^* is trivial if π is in the image of either ψ_G or ψ'_G. This leads to the expectation in the global case (see Theorem 11.5.2 for the case $n = 2$) that if π is an ε-invariant cuspidal representation of $\mathrm{GL}(n)_{/E}$ for n even, then the restriction of ω_π to I_F must be trivial. This is consistent with the "Galois side" of the picture. Let $\tau : W_E \to \mathrm{GL}_n(\mathbf{C})$ be an irreducible representation, and let $\varepsilon(\tau)$ denote the representation $\varepsilon(\tau)(w) = \tau^*(w_\sigma w w_\sigma^{-1})$, where τ^* is the contragredient of τ. If n is even and $\varepsilon(\tau)$ is equivalent to τ, then the restriction of $\det(\tau)$ to I_F is trivial by Lemma 15.5.2(c).

Let $H = U(a) \times U(b)$ be an endoscopic group for $\widetilde{G}$. Assuming again that L-packets $\Pi = \Pi_1 \times \Pi_2$ on H are defined and that the transfers with respect to the endoscopic embeddings η_j exist, let $\pi^j = \eta_j(\Pi)$. Let P be the parabolic subgroup of $\widetilde{G}$ of type (a, b). If n is odd, then $a \not\equiv b \pmod 2$ and $\pi^1 = i_P(\pi_1 \otimes \pi_2)$, where π_j is the standard base change of Π_j. In particular, the restriction of ω_{π^1} to F^* is trivial and the restriction of ω_{π^2} to F^* is $\omega_{E/F}$. If n is even, then $a \equiv b \pmod 2$ and $\pi^1 = i_P(\pi_1 \otimes \pi_2)$, where π_1 is the primed base change of Π_1 to $\mathrm{GL}(a)_{/E}$ and π_2 is the standard base change of Π_2 to $\mathrm{GL}(b)_{/E}$. In this case, for $j = 1, 2$ the restriction of ω_{π^j} to F^* is $\omega_{E/F}$ if a and b are both odd, and is trivial if a and b are both even.

4.8. Endoscopy problems to be considered. This book is concerned mainly with the groups $U(3)$ and $\mathrm{Res}_{E/F}(U(3))$. However, to analyze this case completely, we must consider endoscopy and base change for $U(2)$ and the related group $U(2) \times U(1)$. The theory for $U(2) \times U(1)$ follows directly from the theory for $U(2)$. Let $\mathfrak{U}$ be the set of four pairs (G, ε) consisting of the $U(3)$, $U(2)$ with trivial ε and the associated twisted pairs $(\mathrm{Res}_{E/F}(U(3)), \varepsilon)$, $(\mathrm{Res}_{E/F}(U(2)), \varepsilon)$. For $(G, \varepsilon) \in \mathfrak{U}$, we now define the set $\mathcal{E}(G)$ of elliptic endoscopic triples (H, s, η) which will play a role in succeeding chapters. We omit s from the notation and also omit η if there is no ambiguity.

1. ordinary cases. In this case, each group has one proper elliptic endoscopic group H and $\mathcal{E}(G) = \{G, H\}$.

(a) $G = U(3), H = U(2) \times U(1)$;

(b) $G = U(2), H = U(1) \times U(1)$

We now fix L-embeddings for these cases.

Case (a): Let $G = U(3)$ and $H = U(2) \times U(1)$. We will identify H (resp. $\widehat{H}$) with the subgroup of G (resp. $\widehat{G}$) of matrices of the form

$$\begin{pmatrix} * & 0 & * \\ 0 & * & 0 \\ * & 0 & * \end{pmatrix}.$$

This also determines an identification of $\widetilde{H}$ with a subgroup of $\widetilde{G}$. As before, let μ be a fixed character of C_E whose restriction to C_F is $\omega_{E/F}$. Fix an element $w_\sigma \in W_{E/F}$ whose projection to $\Gamma(E/F)$ is σ. Let

$$\xi_H : {}^L H \to {}^L G$$

be the extension of the inclusion $\widehat{H} \subset \widehat{G}$ defined on W_F by:

$$\xi_H(w) = \begin{pmatrix} \mu(w) & 0 & 0 \\ 0 & 1 & 0 \\ 0 & 0 & \mu(w) \end{pmatrix} \times w \qquad \text{if} \qquad w \in W_E$$

and

$$\xi_H(w_\sigma) = \begin{pmatrix} 1 & 0 & 0 \\ 0 & 1 & 0 \\ 0 & 0 & -1 \end{pmatrix} \times w_\sigma \, .$$

Case (b): If $G = U(2)$ and $H = U(1) \times U(1)$, let $\xi_H : {}^L H \to {}^L G$ denote the embedding which identifies $\widehat{H}$ with the diagonal subgroup of $\widehat{G}$ and on W_F is given by:

$$\xi_H(w) = \begin{pmatrix} \mu(w)^{-1} & 0 \\ 0 & \mu(w)^{-1} \end{pmatrix} \times w \qquad \text{if} \qquad w \in W_E$$

$$\xi_H(w_\sigma) = \begin{pmatrix} 0 & -1 \\ 1 & 0 \end{pmatrix} \times w_\sigma .$$

For the endoscopic group $H = U(2) \times U(1)$ of $U(3)$, we must also consider an embedding $\xi_C : {}^L C \to {}^L H$ where $C = U(1) \times U(1) \times U(1)$. For $(a, b, c) \in \widehat{C} = (\mathrm{GL}_1(\mathbf{C}))^3$, let $\xi_C((a, b, c))$ be the diagonal element of $\widehat{H}$ with entries a, b, c. Define ξ_C by

$$\xi_C(w) = \begin{pmatrix} \mu(w)^{-1} & 0 & 0 \\ 0 & 1 & 0 \\ 0 & 0 & \mu(w)^{-1} \end{pmatrix} \times w \qquad \text{if} \qquad w \in W_E$$

$$\xi_C(w_\sigma) = \begin{pmatrix} 0 & 0 & -1 \\ 0 & 1 & 0 \\ 1 & 0 & 0 \end{pmatrix} \times w_\sigma \, .$$

2. Twisted cases. Let $\widetilde{G} = \mathrm{Res}_{E/F}(G)$.

(a) $G = U(3)$. Then $\mathcal{E}(\widetilde{G}) = \{(G, \psi_G), (H, \eta_1)\}$ where $H = U(2) \times U(1)$. Here ψ_G is the standard base change embedding and η_1 is the embedding defined in §4.7

(b) $G = U(2)$. In this case $\mathcal{E}(\widetilde{G}) = \{(G, \eta_1), (G, \eta_2)\}$. In §11, we will also consider the twisted endoscopic group $H = U(1) \times U(1)$ with respect to the two (equivalent) embeddings η_1 and η_2.

Let $G = U(3)$ and $H = U(2) \times U(1)$. Let H_0 be the $U(2)$-factor of H. We now define two embeddings

$$\psi'_H : {}^L H \to {}^L \widetilde{H}$$
$$\psi_H : {}^L H \to {}^L \widetilde{H}.$$

Let ψ_H be the embedding whose restriction to both the $U(2)$ and the $U(1)$ factor is the standard base change embedding. Let ψ'_H be the embedding whose restriction to the H_0-factor is ψ'_{H_0} and whose restriction to the $U(1)$ factor is the standard base change embedding. Observe that $\widetilde{C}$ is isomorphic to the Levi factor of a Borel subgroup of $\widetilde{G}$. Let $\xi_{\widetilde{C}}$ be the embedding of ${}^L\widetilde{C}$ in ${}^L\widetilde{G}$. The composition $\psi'_H \circ \psi_C$ is equivalent to $\xi_{\widetilde{C}} \circ \psi_C$, where ψ_C is the standard base change map from ${}^LC$ to ${}^L\widetilde{C}$.

Let $\xi_{\widetilde{H}}$ be the embedding of ${}^L\widetilde{H}$ in ${}^L\widetilde{G}$ whose restriction to $\widehat{H} \times \widehat{H} \times W_E$ is the natural inclusion and such that

$$\xi_{\widetilde{H}}(w_\sigma) = \left(\begin{pmatrix} 1 & & \\ & 1 & \\ & & -1 \end{pmatrix}, \begin{pmatrix} -1 & & \\ & 1 & \\ & & 1 \end{pmatrix} \right) \times w_\sigma \,.$$

The following diagram of L-maps:

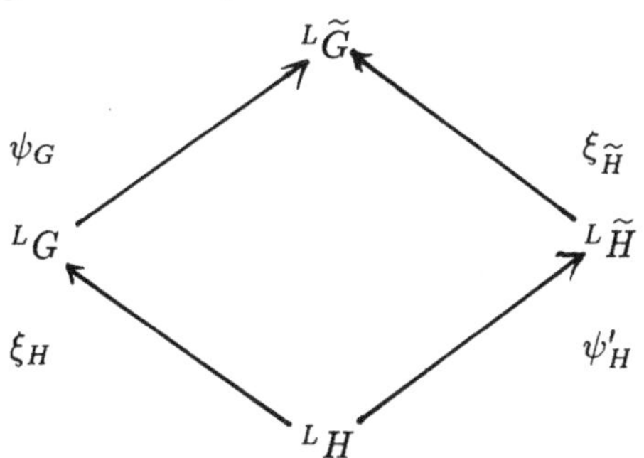

is commutative. The transfer of representations from $\widetilde{H}(F) = H(E)$ to $\widetilde{G}(F) = G(E)$ corresponding to the L-map $\xi_{\widetilde{H}}$ is the one obtained by regarding $H(E)$ as the Levi factor of a parabolic subgroup P of $G(E)$. If ρ is a representation of $H(E)$, then $\xi_{\widetilde{H}}(\rho) = i_P(\rho)$. Observe that the composite $\psi_G \circ \xi_H$ is not an endoscopic embedding of ${}^LH$ into ${}^L\widetilde{G}$. The transfer of an L-packet ρ on H to $\widetilde{G}$ via $\psi_G \circ \xi_H$ is $i_P(\psi'_H(\rho))$, where $\psi'_H(\rho)$ is the base change of ρ with respect to ψ'_H. On the other hand, the endoscopic embedding $\eta_1 : {}^LH \to {}^L\widetilde{G}$ is equivalent to $\xi_{\widetilde{H}} \circ \psi_H$ and the transfer of an L-packet ρ on H to $\widetilde{G}$ via η_1 is $i_P(\psi_H(\rho))$. The existence of the local and global base change transfer with respect to ψ_H and ψ'_H is established in §11.

We now fix some notation that will be used in the remainder of the book. In the context of $U(3)$, Z will denote the center of $G = U(3)$ and $\widetilde{Z}$ will denote the center of $\widetilde{G}$. We often identify Z and $\widetilde{Z}$ with E^1 and E^*, respectively, and view F^* as a subgroup of $\widetilde{Z}$. We will also regard Z and $\widetilde{Z}$ as subgroups of the centers of H and $\widetilde{H}$, respectively, viewing H as a subgroup of G as described above. In the context of $U(2)$, Z will denote the center of $U(2)$. The symbol ω will denote a fixed character of Z and $\tilde{\omega}$ will denote the character of $\widetilde{Z}$ defined by $\tilde{\omega}(z) = \omega(z/\overline{z})$.

4.9 Transfer factors. We retain the notation of the previous section. Assume that F is local. The map $A_{G/H} : \mathcal{O}_{st}(H) \to \mathcal{O}_{st}(G)$ is obtained from the inclusion $H \subset G$. For convenience, we regard elements H as elements of G and drop $A_{G/H}$ from the notation. Similarly, we fix an embedding of C in H and regard C as a subgroup of H.

Let T be a maximal torus in G and for $\gamma \in T$, define

$$D_G(\gamma) = |\Pi(1 - \alpha(\gamma))|^{1/2}$$

where the product, which lies in F, is over all roots of T in G. Define D_H similarly and if γ is a (G, H)-regular element of H, set

$$D_{G/H}(\gamma) = D_G(\gamma)/D_H(\gamma)\ .$$

Observe that $D_{G/H}(\gamma)$ is well-defined.

In the case $G = U(3)$, the transfer $f \to f^H$ defined in Proposition 4.9.1 below associates to $f \in C(G, \omega)$ an element in $C(H, \omega\mu^{-1})$, where μ is regarded as a character of $Z = E^1$ via its restriction from E^* to E^1. The change in central character is required by the form of the embedding ξ_H and is reflected in the factor $\Delta_{G/H}(\gamma)$.

Let $\gamma \in H$ and let $\gamma_1, \gamma_2, \gamma_3$ be the eigenvalues of γ, labelled so that γ is of the form

$$\begin{pmatrix} * & 0 & * \\ 0 & \gamma_2 & 0 \\ * & 0 & * \end{pmatrix}.$$

Set

$$\tau(\gamma) = \mu(\gamma_2)\mu^{-1}((\gamma_2\gamma_1^{-1} - 1)(1 - \gamma_2\gamma_3^{-1}))$$

and define the transfer factor

$$\Delta_{G/H}(\gamma) = \tau(\gamma) D_{G/H}(\gamma)$$

Note that $\Delta_{G/H}(z\gamma) = \mu(z)\Delta_{G/H}(\gamma)$ for $z \in Z$. Because we have fixed an embedding of H into G (which, in general, does not exist), we can identify γ_H with γ with γ and $\Delta_{G/H}(\gamma_H, \gamma) = \Delta_{G/H}(\gamma)$, in the notation of §4.3. It is easy to check that this definition of $\Delta_{G/H}$ arises from the construction of [LS].

PROPOSITION 4.9.1: *Let* $G = U(3)$, $H = U(2) \times U(1)$.

(a) *For all* $f \in C(G, \omega)$, *there exists* $f^H \in C(H, \omega\mu^{-1})$ *such that*

$$\Delta_{G/H}(\gamma)\Phi^{\kappa}(\gamma, f) = \Phi^{st}(\gamma, f^H) \tag{4.9.1}$$

for all G-regular elements γ *in* H. *Here* $\kappa \in \mathcal{R}(G_\gamma/F)$ *is the element corresponding to* H.

(b) *If* F *is p-adic,* E/F *is unramified, and the characters* μ *and* ω *are unramified, then* (4.9.1) *holds with* $f^H = \hat{\xi}_H(f)$ *if* $f \in \mathcal{H}(G, \omega)$.

Proof: In the archimedean case, the existence of f^H in the Schwartz space is a special case of the results of Shelstad ([S$_1$]). To obtain f^H of compact support, we use the results of Clozel-Delorme and argue as in [AC], §1.7. The existence of f^H in the p-adic case is contained in [LS$_2$]. In Proposition 8.3.1, we show that (4.9.1) holds for all (G, H)-regular γ.

Part (b) is the "fundamental lemma" for $U(3)$. It is proved in [BR$_1$]. To describe $\hat{\xi}_H$ explicitly, let χ_z denote the character $\chi_z(d(a, b, \overline{a}^{-1}) = z^{\mathrm{val}(a)}$, where val($\cdot$) is the valuation of E. The Satake transforms of f and f^H in their respective Hecke algebras can be regarded as functions of $z \in \mathbf{C}^*$, where $f^\wedge(z) = \mathrm{Tr}(i_G(\chi_z)(f))$ and $(f^H)^\wedge(z) = \mathrm{Tr}(i_H(\chi_z)(f^H))$. If E/F and μ are unramified, then $\mu = \chi_{-1}$ and (4.5.1) sends the unramified constituent in $i_H(\chi_z)$ to the unramified constituent in $i_G(\chi_{-z})$. Hence $\hat{\xi}_H(f)$ is the function whose Satake transform is $\hat{\xi}_H(f)^\wedge(z) = f(-z)$.

We regard μ as a character of M by setting

$$\mu\left(\begin{pmatrix} a & & \\ & b & \\ & & \overline{a}^{-1} \end{pmatrix}\right) = \mu(a) .$$

By functoriality, the embedding ξ_H carries $i_H(\chi)$ to $i_G(\chi\mu)$ when both are irreducible. The next lemma shows that the transfer $f \to f^H$ is compatible with this functoriality.

LEMMA 4.9.2: *Let χ be a character of M whose restriction to Z is ω and let $f \in C(G,\omega)$. Then*

$$Tr(i_G(\chi)(f)) = Tr(i_H(\chi\mu^{-1})(f^H))$$

if $f \to f^H$.

Proof: By the character formula for principal series representations

$$\mathrm{Tr}(i_G(\chi)(f)) = \int_{Z\backslash M} D_G(\gamma)\Phi(\gamma, f)\chi(\gamma)d\gamma \ .$$

The eigenvalues of $\gamma \in M$ satisfy $\overline{\gamma}_1 = \gamma_3^{-1}$ and $\overline{\gamma}_2 = \gamma_2^{-1}$, from which it follows that $\tau(\gamma) = \mu(\gamma_1)$. This gives

$$D_H(\gamma)\Phi(\gamma, f^H) = \mu(\gamma)D_G(\gamma)\Phi(\gamma, f)$$

and

$$\mathrm{Tr}(i_G(\chi)(f)) = \int_{Z\backslash M} D_H(\gamma)\Phi(\gamma, f^H)\chi\mu^{-1}(\gamma)d\gamma \ ,$$

and the lemma follows from the character formula for H.

For the case $G = U(2)$ and $H = U(1) \times U(1)$, we fix an embedding of H in G and define

$$\Delta_{G/H}(\gamma) = \mu^{-1}(\gamma_1 - \gamma_2)D_G(\gamma) \ ,$$

for $\gamma = (\gamma_1, \gamma_3) \in H$. Similarly, for the case $H = U(2) \times U(1)$ and $C = U(1) \times U(1) \times U(1)$, define

$$\Delta_{H/C}(\gamma) = \mu^{-1}(\gamma_1 - \gamma_3)D_H(\gamma)$$

for $\gamma = (\gamma_1, \gamma_2, \gamma_3) \in C$.

LEMMA 4.9.3: *Let $H = U(2) \times U(1)$. For all $f \in C(H, \omega\mu^{-1})$, there exists $f^C \in C(C, \omega)$ such that*

$$\Delta_{H/C}(\gamma)\Phi^\kappa(\gamma, f) = \Phi(\gamma, f^C) \tag{4.9.2}$$

for all H-regular γ in C, where $\kappa \in \mathfrak{R}(H_\gamma/F)$ is the element corresponding to C. If F is p-adic, E/F is unramified, and the characters μ and ω are unramified, then (4.9.2) holds with $f^C = \xi_C(f)$ if $f \in \mathcal{H}(H, \omega\mu^{-1})$.

The techniques of [LL] used in the case of SL(2) can be applied in a straightforward way to establish Lemma 4.9.2. We omit the details. A statement analogous to Proposition 4.9.2 for the pair $G = U(2), H = U(1)\times U(1)$ also holds.

4.10 The twisted case. In this section, let $G = U(3)$. We consider the orbital integral transfers needed to compare the ordinary trace formula for G with the twisted trace formula for $\widetilde{G}$. For $\delta \in \widetilde{G}$ and $\nu \in \mathfrak{D}_\varepsilon(\delta/F)$, let δ^ν denote a representative of the ε-conjugacy class within $\mathcal{O}_{\varepsilon-\mathrm{st}}(\delta)$ associated to ν. If δ is ε-semisimple, set $e(\delta) = e(G_{\delta\varepsilon})$. Assume that $\gamma = N(\delta)$ is semisimple and that $\widetilde{G}_{\delta\varepsilon} = G_\gamma$. Then $\mathfrak{R}(G_\gamma/F)$ and $\mathfrak{D}_\varepsilon(\delta/F)$ are paired by Proposition 3.11.2(a). For $\kappa \in \mathfrak{R}(G_\gamma/F)$, define

$$\Phi_\varepsilon^\kappa(\delta,\phi) = \sum_{\nu\in\mathfrak{D}_\varepsilon(\delta/F)} \kappa(\nu)e(\delta^\nu)\Phi_\varepsilon(\delta^\nu,\phi) \tag{4.10.1}$$

where

$$\Phi_\varepsilon(\delta,\phi) = \int_{\widetilde{G}(\delta\varepsilon)''\backslash\widetilde{G}} \phi(g^{-1}\delta\varepsilon(g))dg\ ,$$

Recall that $\widetilde{G}(\delta\varepsilon)'' = \{g \in \widetilde{G} : g^{-1}\delta\varepsilon(g) \in F^*\}$ (§3.11). If κ is trivial, denote $\Phi_\varepsilon^\kappa(\delta,\phi)$ by $\Phi_\varepsilon^{\mathrm{st}}(\delta,\phi)$. Note that $\Phi_\varepsilon^\kappa(\delta^\nu,\phi) = \kappa(\nu)^{-1}\Phi_\varepsilon^\kappa(\delta,\phi)$ if δ is ε-regular.

Let $H = U(2)\times U(1)$. Denote the first factor of H by H_0. The embedding of H as a subgroup of G of §4.9 defines an embedding of $\widetilde{H}$ in $\widetilde{G}$. Let $\delta \in \widetilde{H}$ and let γ be a norm of δ in H. If γ is (G,H)-regular, set $\tilde{\tau}(\delta) = \mu(\det_0(\delta))^{-1}\tau(\gamma)$ where $\det_0(\delta)$ denotes the determinant of the H_0-factor of δ. Note that $\mu(\det_0(g^{-1}\delta\varepsilon(g))) = \mu(\det_0(\delta))$ for $g \in \widetilde{H}$ since μ is trivial on $N_{E/F}(E^*)$ and also that $\tilde{\tau}(z\delta) = \tilde{\tau}(\delta)$ for $z \in \widetilde{Z}$. We now define the transfer factor with respect to the map $\eta_1 : {}^LH \to {}^L\widetilde{G}$ by

$$\widetilde{\Delta}(\delta) = \tilde{\tau}(\delta)D_{G/H}(\gamma) = \mu(\det_0(\delta))^{-1}\Delta_{G/H}(\gamma)$$

Let κ be the element of $\mathfrak{R}(G_\gamma/F)$ corresponding to the triple (H,η_1,s). By Proposition 3.13.1,

$$\kappa(\nu) = \omega_{E/F}(\det(t_\nu)) = \mu(\det_0(t_\nu))^{-1}\ .$$

and since $\tilde{\tau}(\delta t_\nu) = \mu(\det_0(t_\nu))^{-1}\tilde{\tau}(\delta)$, it follows that $\widetilde{\Delta}(\delta)\Phi_\varepsilon^\kappa(\delta,\phi)$ depends only on the stable ε-conjugacy class of δ in $\widetilde{H}$.

As in the ordinary case, a distribution T on $C(\widetilde{G},\tilde{\omega})$ is called ε-invariant if $T(\phi^g) = T(\phi)$ for all $g \in \widetilde{G}$, where $\phi^g(x) = \phi(g^{-1}x\varepsilon(g))$. If $\Phi_\varepsilon^{\mathrm{st}}(\gamma,\phi) = 0$ for all regular semisimple γ, we will say that ϕ is ε-stably equivalent at zero. An ε-invariant distribution will be called stable, or stably invariant, if $T(\phi) = 0$ for all ϕ which are ε-stably equivalent to zero.

If E/F is unramified, then the embeddings ψ_G and η_1 are unramified and define homomorphisms

$$\widehat{\psi_G} : \mathcal{H}(\widetilde{G}, \tilde{\omega}) \to \mathcal{H}(G, \omega)$$
$$\widehat{\eta_1} : \mathcal{H}(\widetilde{G}, \tilde{\omega}) \to \mathcal{H}(H, \omega) .$$

We also denote $\mathcal{H}(\widetilde{G}, \tilde{\omega})$ by $\widetilde{\mathcal{H}}$

PROPOSITION 4.10.1: (a) *Let $\phi \in C(\widetilde{G}, \tilde{\omega})$. Then there exists a function $f \in C(G, \omega)$ such that*

$$\Phi_{\varepsilon}^{\mathrm{st}}(\delta, \phi) = \Phi^{\mathrm{st}}(\gamma, f) \tag{4.10.2}$$

for all ε-regular semisimple $\delta \in \widetilde{G}$ and $\gamma \in \mathcal{N}(\delta)$. If E/F is unramified and $\phi \in \widetilde{\mathcal{H}}$, then (4.10.2) holds with $f = \widehat{\psi_G}(\phi)$. Converseley, if $f \in C(G, \omega)$, there exists $\phi \in C(\widetilde{G}, \tilde{\omega})$, such that (4.10.2) holds for ε-regular semisimple $\delta \in \widetilde{G}$ and $\gamma \in \mathcal{N}(\delta)$,

(b) *If $\phi \in C(\widetilde{G}, \tilde{\omega})$, then there exists $\phi^H \in C(H, \omega)$ with the following property. Let $\delta \in \widetilde{H}$ and let γ be a norm of δ in H. If γ is regular in G, then*

$$\widetilde{\Delta}(\delta)\Phi_{\varepsilon}^{\kappa}(\delta, \phi) = \Phi^{\mathrm{st}}(\gamma, \phi^H) . \tag{4.10.3}$$

If E/F is unramified and $\phi \in \widetilde{\mathcal{H}}$ then (4.10.3) holds with $\phi^H = \widehat{\eta_1}(\phi)$.

For proofs of the fundamental lemma for $\widehat{\eta_1}$ and $\widehat{\psi_G}$, we refer to [BR$_1$]. The existence of the transfers for a general function ϕ is given in §4.12.

Let χ be an ε-invariant character of $\widetilde{M}$ and let $\pi = i_{\widetilde{G}}(\chi)$. There is a natural choice of action of ε on π. Realize $i_G(\chi)$ on the space of smooth functions $\varphi(g)$ on $\widetilde{G}$ such that $\varphi(bg) = \delta^{1/2}\chi(b)\varphi(g)$ for $b \in \widetilde{B}$ and define $\pi(\varepsilon)(\varphi)(g) = \varphi(\varepsilon^{-1}(g))$. We use this choice to define the twisted trace in the next proposition.

PROPOSITION 4.10.2: *Let χ be a character of M whose restriction to Z is ω. Set $\tilde{\pi} = i_{\widetilde{G}}(\chi \circ N)$. For all $\phi \in C(\widetilde{G}, \tilde{\omega})$:*

$$\mathrm{Tr}(\tilde{\pi}(\phi)\tilde{\pi}(\varepsilon)) = \mathrm{Tr}(i_G(\chi)(f))$$
$$\mathrm{Tr}(\tilde{\pi}(\phi)\tilde{\pi}(\varepsilon)) = \mathrm{Tr}(i_H(\chi)(\phi^H)) .$$

Proof: The first equality follows from the standard calculation of the twisted character for principal series prepresentations:

$$\mathrm{Tr}(\pi(\phi)\pi(\varepsilon)) = \int_{\widetilde{Z}\widetilde{M}^{1-\varepsilon}\backslash\widetilde{M}} D_G(N(\delta))\Phi_\varepsilon(\delta,\phi)\chi(N(\delta))d\gamma$$
$$= \int_{Z\backslash M} D_G(\gamma)\Phi(\gamma,f)\chi(\gamma)d\gamma$$

(cf. $[L_1]$, §7). If $\delta = d(x,y,z) \in \widetilde{M}$, then $\gamma = d(x/\overline{z}, y/\overline{y}, z/\overline{x})$,

$$\tilde{\tau}(\delta) = \mu(xz)^{-1}\tau(\gamma) = \mu(xz)^{-1}\mu(x/\overline{z}) = 1.$$

and the second equality follows similarly.

4.11 Twisted transfer for $U(2)$. For the group $\widetilde{G} = \mathrm{Res}_{E/F}(U(2))$, define $\Phi_\varepsilon^\kappa(\delta,\phi)$, for ε-semisimple δ as in (4.10.1). If $N(\delta)$ is elliptic regular or scalar, then $\mathcal{D}_\varepsilon(\delta/F)$ is isomorphic to F^*/NE^* by Proposition 3.12.1. In this section, κ will denote the non-trivial character of $\mathcal{D}_\varepsilon(\delta/F)$ if $\mathcal{D}_\varepsilon(\delta/F)$ is non-trivial and the trivial character otherwise. Explicitly,

$$\kappa(\nu) = \begin{cases} \mu(\det(t_\nu)) & \text{if } N(\delta) \text{ is elliptic regular} \\ \kappa(\nu) = \mu(\det(\delta^\nu\delta^{-1})) & \text{is } N(\delta) \text{ is scalar} \end{cases}$$

(cf. §3.12). It follows from the relation $\Phi_\varepsilon^\kappa(\delta^\nu,\phi) = \kappa(\nu)^{-1}\Phi_\varepsilon^\kappa(\delta,\phi)$ that $\mu(\det(\delta))\Phi_\varepsilon^\kappa(\delta,\phi)$ depends only on $N(\delta)$. Note also that $\mu(\det(z\delta)) = \mu(z/\overline{z})\mu(\det(\delta))$. The embeddings η_j define homomorphisms

$$\widehat{\eta_1} : \mathcal{H}(\widetilde{G},\tilde{\omega}) \to \mathcal{H}(G,\omega)$$
$$\widehat{\eta_2} : \mathcal{H}(\widetilde{G},\tilde{\omega}) \to \mathcal{H}(G,\omega\mu^{-1}).$$

We define $\Phi_\varepsilon(\delta,\phi)$ as in §4.10, as an integral over $\widetilde{G}''(\delta\varepsilon)\backslash\widetilde{G}$.

PROPOSITION 4.11.1: *Let* $\phi \in C(\widetilde{G},\tilde{\omega})$.

(a) *There exist functions* $f_1 \in C(G,\omega)$ *and* $f_2 \in C(G,\omega\mu^{-1})$ *such that*

$$\Phi_\varepsilon^{st}(\delta,\phi) = \Phi^{st}(\gamma,f_1),\ \mu(\det(\delta))\Phi_\varepsilon^\kappa(\delta,\phi) = \Phi^{st}(\gamma,f_2) \tag{4.11.2}$$

for all ε*-regular semisimple* $\delta \in \widetilde{G}$ *and* $\gamma \in \mathcal{N}(\delta)$. *Conversely, given* $f_1 \in C(G,\omega)$ *and* $f_2 \in C(G,\omega\mu^{-1})$ *such that* $\Phi(\gamma,f_2) = \mu(\gamma)\Phi(\gamma,f_1)$ *for regular* $\gamma \in M$, *there exists* ϕ *such that* (4.11.2) *holds.*

(b) *If* $\phi \in \mathcal{H}(\widetilde{G},\tilde{\omega})$, *then* (4.11.2) *holds with* $f_j = \widehat{\eta_j}(\phi)$.

(c) *Let χ be a character of M whose restriction to Z is ω. Set $\tilde{\pi} = i_{\widetilde{G}}(\chi \circ N)$. Then*

$$\mathrm{Tr}(\tilde{\pi}(\phi)\tilde{\pi}(\varepsilon)) = \mathrm{Tr}(i_G(\chi)(f_1))$$
$$\mathrm{Tr}(\tilde{\pi}(\phi)\tilde{\pi}(\varepsilon)) = \mathrm{Tr}(i_G(\chi\mu^{-1})(f_2))$$

In the next section, we prove the existence of the transfers $\phi \to f$ and $\phi \to \phi^H$ described in Proposition 4.10.1. for F p-adic. We follow the method, based on a reduction argument of Harish-Chandra, used to prove similar results in [L$_2$] and [AC]. In the case $F = \mathbf{R}$, the existence of f and ϕ^H belonging to the Schwartz space is due to Shelstad ([S$_2$]). We appeal to the argument of [AC], §1.7, based on the Paley-Wiener theorem of Clozel-Delorme, to obtain f and ϕ^H which are K-finite and of compact support. It goes through without difficulty, provided one takes into account that ϕ specifies only the traces of f and ϕ^H on L-packets and not on individual representations.

4.12. Assume that F is p-adic. It will suffice to prove the existence of a transfer $\phi \to f$ satisfying (4.10.2) for functions ϕ such that $\phi(zg) = \phi(g)$ for all $z \in F^*$ and compactly supported f. Recall that $\widetilde{G}(\delta\varepsilon)'' = \widetilde{Z}\widetilde{G}(\delta\varepsilon)$, by Proposition 3.11.2. If we set:

$$\phi'(g) = \int_{F^*\backslash\widetilde{Z}} \phi(zg)\tilde{\omega}(z)dz$$

$$f'(g) = \int_{Z} f(zg)\omega(z)\, dz,$$

then

$$\int_{\widetilde{Z}G_{\delta\varepsilon}\backslash\widetilde{G}} \phi'(g^{-1}\delta\varepsilon(g))dg = \int_{\widetilde{Z}G_{\delta\varepsilon}\backslash\widetilde{G}} \int_{F^*\backslash\widetilde{Z}} \phi(g^{-1}z\delta\varepsilon(g))\omega(z/\overline{z})\, dz\, dg$$
$$= \int_{G_\gamma\backslash G} \int_{F^*\backslash\widetilde{Z}} f(g^{-1}(z/\overline{z})\gamma g)\omega(z/\overline{z})\, dz\, dg = \int_{G_\gamma\backslash G} f'(g^{-1}\gamma g)\, dz\, dg$$

and ϕ' and f' correspond. Furthermore, every function in $C(\widetilde{G},\tilde{\omega})$ and $C(G,\omega)$ is of the form ϕ' and f', respectively.

A partitition of unity argument shows that it suffices to construct f for ϕ with support modulo F^* in a small neighborhood of an ε-semisimple element $\delta_0 \in \widetilde{G}$. Let $M = \widetilde{G}(\delta_\circ\varepsilon)$. The first step is to show that there exists a smooth compactly supported function φ on M such that $\Phi_\varepsilon(\delta\delta_0, \phi) = \Phi_M(\delta, \varphi)$ for all $\delta \in M$ near 1. If δ_0 is ε-regular, this is easy since $\Phi_\varepsilon(\delta, \phi)$ is constant near δ_0. We therefore assume that δ_0 is not ε-regular.

LEMMA 4.12.1: *There is a neighborhood V of 1 in M with the following property: for any set $\Omega \subset \widetilde{G}$ which is compact modulo F^*, the image of the set*

$$\omega = \{g \in \widetilde{G} : g^{-1} V \delta_0 \varepsilon(g) \cap \Omega \neq \emptyset\}$$

in $\widetilde{Z}M\backslash\widetilde{G}$ is compact.

We refer to [AC], page 20, where a similar result is proved in the case of base change for GL(n).

Let Ω be the support of ϕ and let ω and V be as in the lemma. Choose a smooth compactly-supported function α on $\widetilde{Z}\backslash\widetilde{G}$ such that

$$\overline{\alpha}(g) = \int\limits_{\widetilde{Z}\backslash\widetilde{Z}M} \alpha(xg)\, dx$$

is equal to 1 if $g \in \omega$, and define

$$\varphi(m) = \int\limits_{\widetilde{Z}\backslash\widetilde{G}} \alpha(g)\phi(g^{-1} m \delta_0 \varepsilon(g))\, dg\ .$$

for $m \in M$. Then φ is smooth since $\alpha(g)$ has compact support, and to check that φ has compact support, observe that if $\varphi(m) \neq 0$, then $m\delta_0 \in \widetilde{Z}\Omega_1\Omega_2\varepsilon(\Omega_1)^{-1} \cap M$, where Ω_1 and Ω_2 are compact subsets of $\widetilde{G}$ such that $\mathrm{Supp}(\alpha) \subset \widetilde{Z}\Omega_1$ and $\mathrm{Supp}(\phi) \subset \widetilde{Z}\Omega_2$. It follows that m belongs to a compact set since $\widetilde{Z} \cap M$ is compact. For x close to $1, M_x = \widetilde{G}_{x\delta_\circ\varepsilon}$ (this can be checked by passing to an extension field over which $\widetilde{G}$ is isomorphic to $G \times G$ and $N(\delta_0)$ is diagonalizable). Therefore

$$\begin{aligned}\Phi_\varepsilon(x\delta_0, \phi) &= \int\limits_{\widetilde{Z}M\backslash\widetilde{G}} \int\limits_{M_x\backslash M} \overline{\alpha}(g)\phi(g^{-1}m^{-1}x\delta_0\varepsilon(mg))\, dm\, dg \\ &= \int\limits_{\widetilde{Z}\backslash\widetilde{G}} \int\limits_{M_x\backslash M} \alpha(g)\phi(g^{-1}(m^{-1}xm)\delta_0\varepsilon(g))\, dm\, dg \\ &= \Phi_M(x, \varphi).\end{aligned}$$

We can assume that φ has support in a small neighborhood of 1.

Let $\gamma_0 = N(\delta_0)$. By Proposition 3.11.2(d), we may assume that γ_0 lies in G and that $\widetilde{G}(\delta_0\varepsilon) = G(\gamma_0)$.

We next construct a function f on G such that $\Phi_G(N(x\delta_0), f) = \Phi_\varepsilon(x\delta_0, \phi)$ for x near 1 in M. Observe that $N(x\delta_0) = x^2\gamma_0$ for $x \in M$ and the map $x \to x^2$ is an isomorphism which preserves conjugacy for x near 1. Since γ_0 is central in M, there exists a function φ_1 on M with support in a small neighborhood of γ_0 such that $\Phi_M(N(x\delta_0), \varphi_1) = \Phi_M(x, \varphi)$ for x near 1 in M.

Let s be an analytic section of the map $G \to M\backslash G$ defined in a small neighborhood V_1 of the identity coset. Let V_2 be a small neighborhood of γ_0 in M. Then the map $(w, \gamma) \to s(w)^{-1}\gamma s(w)$ defines an isomorphism of $V_1 \times V_2$ with a neighborhood V_3 of γ_0 in G. Let β be a smooth function on $M\backslash G$ with support in V_1 such that $\int_{M\backslash G} \beta(g)dg = 1$. Suppose that $\mathrm{supp}(\varphi_1) \subset V_2$ and set

$$f(s(w)^{-1}\gamma s(w)) = \beta(w)\varphi_1(\gamma).$$

Then f is a smooth function with support in V_3. If $\gamma \in V_2$, then $G_\gamma = M_\gamma$ and

$$\begin{aligned}\int_{G_\gamma\backslash G} f(g^{-1}\gamma g)dg &= \int_{M\backslash G}\int_{M_\gamma\backslash M} f(g^{-1}m^{-1}\gamma m g)dm\, dg \\ &= \int_{M_\gamma\backslash M} \varphi_1(m^{-1}\gamma m)dm\ .\end{aligned}$$

Hence $\Phi_G(\gamma, f) = \Phi_M(\gamma, \varphi_1)$ for $\gamma \in V_2$.

LEMMA 4.12.2: *There exists a neighborhood U_M of 1 in M such that the map $x\delta_0 \to N(x\delta_0) = x^2\gamma_0$ for $x \in U_M$ induces a bijection between the set of ε-semisimple conjugacy classes in $\widetilde{G}$ which pass near δ_0 and the set of semisimple conjugacy classes in G which pass near γ_0.*

Proof: The maps

$$\psi_1 : \widetilde{G} \times M \to \widetilde{G}$$
$$\psi_2 : G \times M \to G$$

defined by $\psi_1(g, x) = g^{-1}x\delta_0\varepsilon(g)$ and $\psi_2(g, x) = g^{-1}x\gamma_0 g$ are submersive ([C2], §3.4) for x in a small neighborhood U_M of 1 in M and g near 1 in

G. Hence, every conjugacy class $\{\gamma\}$ (resp., ε-conjugacy class $\{\delta\}$) passing near γ_0 (resp., δ_0) has a representative of the form $\gamma = x\gamma_0$ (resp., $x\delta_0$) where $x \in U_M$. Suppose that $x, y \in M$ are both close to 1. It follows (e.g., by passing to an extension field over which γ_0 is diagonalizable) that if $g^{-1}x\gamma_0 g = y\gamma_0$ for $g \in G$, then $g^{-1}\gamma_0 g = \gamma_0$ and $g^{-1}xg = y$, and hence $g \in M$. Choose U_M sufficiently small so that for $x, y \in U_M$, $x\gamma_0$ and $y\gamma_0$ are conjugate in G if and only if x and y are conjugate in M. Similarly, we can assume that if $x, y \in U_M$, then $x\delta_0$ and $y\delta_0$ are ε-conjugate in $\widetilde{G}$ if and only if x and y are conjugate in M. The lemma follows since the map $x \to x^2$ is an isomorphism near the identity.

We can assume that ϕ is supported on the ε-conjugacy classes which, modulo F^*, intersect $U_M\delta_0$, where U_M is as in the previous lemma. Let $\delta \in U_M\delta_0$ be an ε-regular element and set $\gamma = N(\delta)$. Let $\{\delta_i\}$ be a set of representatives in $U_M\delta_0$ for the ε-classes within $\mathcal{O}_{\varepsilon-\mathrm{st}}(\delta)$ which intersect $U_M\delta_0$ and set $\gamma_i = N(\delta_i)$. Then $\Phi_\varepsilon(\delta_i, \phi) = \Phi(\gamma_i, f)$ and for any coefficients $\{a_i\}$ we have

$$\sum a_i\Phi_\varepsilon(\delta_i, \phi) = \sum a_i\Phi(\gamma_i, f) \,. \tag{4.12.1}$$

Taking $a_i = 1$ for all i, we obtain (4.10.2).

To construct ϕ^H, suppose that $\gamma_0, \gamma \in H$ and that $\delta_0, \delta \in \widetilde{H}$. Then $\delta = y\delta_0$ and $\gamma = y^2\gamma_0$ for some unique $y \in U_M \cap \widetilde{H}$. Let κ be the character defined by the endoscopic group H and set $a_i = \kappa(\mathrm{inv}(\gamma, \gamma_i))$. Then (4.12.1) gives $\Phi_\varepsilon^\kappa(\delta, \phi) = \Phi^\kappa(\gamma, f)$. Let f^H be the function corresponding to f by Proposition 4.9.1. We can assume that f^H is supported in a small neighborhood of γ_0. Define ϕ^H by $\phi^H(\gamma) = \mu(\det_0(\delta))^{-1}f^H(\gamma)$. As observed in §4.10, if δ is ε-semisimple, $\mu(\det_0(\delta))$ depends only on the ε-conjugacy class of δ in $\widetilde{H}$. It follows from Proposition 3.13.1 that it depends in fact only on the stable ε-conjugacy class of δ, in $\widetilde{H}$, and hence depends only on the stable conjugacy class of γ in H. Therefore:

$$\widetilde{\Delta}(\delta)\Phi_\varepsilon^\kappa(\delta, \phi) = \widetilde{\Delta}(\delta)\Phi^\kappa(\gamma, f) = \mu(\det_0(\delta))^{-1}\Phi^{\mathrm{st}}(\gamma, f^H) = \Phi^{\mathrm{st}}(\gamma, \phi^H).$$

To prove the existence of ϕ given f, note that we can use the above construction to obtain a smooth compactly-supported function φ on M such that $\Phi(N(x\delta_0), f) = \Phi_M(x, \varphi)$ for $x \in M$ near 1. Let $\alpha(g)$ be a smooth compactly supported function on $\widetilde{Z}M\backslash\widetilde{G}$ such that $\int_{\widetilde{Z}M\backslash\widetilde{G}} \alpha(g)dg = 1$ and let s_1 be an analytic section of the map $\widetilde{G} \to \widetilde{Z}M\backslash\widetilde{G}$. Define ϕ by

$\phi(s_1(w)^{-1}zx\delta_0\varepsilon(s_1(w))) = \alpha(w)\varphi(x)$ for $x \in M$ near 1 and $z \in F^*$. Then $\Phi_\varepsilon(x\delta_0, \phi) = \Phi_M(x, \varphi)$ and (4.10.2) holds as above.

4.13 The case of split primes: In this section, we assume that E/F is global and that v is a place of F which splits in E. Then $E_v = E_w \times E_{w'}$ where w, w' are the places of E above v. If $G = U(n)$, then

$$\widetilde{G}_v = \mathrm{GL}_n(E_w) \times \mathrm{GL}_n(E_{w'})$$

The action of ε sends (g, h) to $(\varepsilon_0(h), \varepsilon_0(g))$ where $\varepsilon_0(g) = \Phi^t\overline{g}^{-1}\Phi^{-1}$ and G_v is the subgroup fixed by ε. Projection onto the first component yields an isomorphism of G_v with $\mathrm{GL}_n(E_w)$. If (π, v) is an irreducible representation of $\mathrm{GL}_n(E_w)$, let $\varepsilon(\pi)$ be the representation $\pi(\varepsilon_0^{-1}(g))$ of $\mathrm{GL}_n(E_{w'})$ on V. The base change transfer (with respect to the standard base change map) of π to $\widetilde{G}_v$ is the representation $\tilde{\pi} = \pi \otimes \varepsilon(\pi)$. It acts on $V \otimes V$ and we let ε act by $\tilde{\pi}(\varepsilon)(v \otimes w) = w \otimes v$.

The transfers of functions are easily described explicitly. In this case, there is no difference between ordinary, stable, and κ-orbital integrals. For example $\Phi(\gamma, f)$, $\Phi^\kappa(\gamma, f)$, $\Phi^{st}(\gamma, f)$ coincide for $f \in C(G_v, \omega_v)$. We distinguish between them in the notation to make the notation in the global case uniform. In the case $n = 3$, we identify H_v with $\mathrm{GL}_2(E_w) \times \mathrm{GL}_1(E_w)$ and regard it as the Levi factor of a parabolic subgroup P of G_v of type (2,1). The transfer of a representation ρ of H_v to $\widetilde{G}_v$ via η_1 is $i_{G_w}(\rho) \times i_{G_{w'}}(\varepsilon(\rho))$.

LEMMA 4.13.1: (a) *Let $G = U(3)$ and $H = U(2) \times U(1)$. For all $f \in C(G_v, \omega_v)$, there exists a function $f^H \in C(H_v, \omega_v\mu_v^{-1})$ such that*

$$\Delta_{G/H}(\gamma)\Phi^\kappa(\gamma, f) = \Phi^{st}(\gamma, f^H)$$

for all (G, H)-regular $\gamma \in H_v$.

(b) *Let ρ be a representation of H_v whose restriction to Z_v is $\omega_v\mu_v^{-1}$ and let $\rho' = \rho \otimes \mu_w \circ \det_0$. Then* $\mathrm{Tr}(i_G(\rho')(f)) = \mathrm{Tr}(\rho(f^H))$.

Proof: Let $\gamma = (\gamma', \gamma'')$ be a semisimple element of H_v. Let γ_1, γ_2, γ_3 be the eigenvalues of γ', labelled as in §4.9. Then the eigenvalues of γ'' are $\overline{\gamma}_1^{-1}$, $\overline{\gamma}_2^{-1}$, $\overline{\gamma}_3^{-1}$ and

$$\tau(\gamma) = \mu_w(\gamma_2)\mu_w^{-1}((\gamma_2\gamma_1^{-1} - 1)(1 - \gamma_2\gamma_3^{-1}))\mu_{w'}(\overline{\gamma}_2^{-1})\mu_{w'}^{-1}((\overline{\gamma}_1\overline{\gamma}_2^{-1} - 1)(1 - \overline{\gamma}_3\overline{\gamma}_2^{-1})).$$

This is equal to $\mu_w(\gamma_1\gamma_3)$, and hence $\tau(\gamma) = \mu_w(\det_0(\gamma))$.

Identify G_v with $\mathrm{GL}_3(E_w)$ and let P be the standard parabolic subgroup of G_v with H_v as Levi factor. Observe as a character of the center Z_v, μ_v

becomes the character μ_w^2 of E_w^*. Define

$$f^H(h) = \mu_w(\det_0(h))\delta_P(h)^{1/2} \int_{\mathbf{N}_P} f^K(hn)\, dn\ ,$$

where $f^K(g) = \int_K f(k^{-1}gk)dk$. Then $f^H \in C(G_v, \omega_v\mu_v^{-1})$ if $f \in C(G_v, \omega_v)$. If γ is (G, H)-regular, then $H_{v\gamma} = G_{v\gamma}$ and

$$|(1-\gamma_2\gamma_1^{-1})(1-\gamma_2\gamma_3^{-1})|_w \Phi(\gamma, f) = \int_{H_{v\gamma}\backslash H_v} \int_{\mathbf{N}_P} f^K(h^{-1}\gamma hn)\, dn\, dh$$

by a change of variables. We have

$$\delta_P(\gamma)^{1/2} = |\gamma_1\gamma_3\gamma_2^{-2}|_w^{1/2}\ ,$$

$$D_{G/H}(\gamma) = \delta_P(\gamma)^{1/2}|(1-\gamma_2\gamma_1^{-1})(1-\gamma_2\gamma_3^{-1})|_w$$

and hence $\Delta_{G/H}(\gamma)\Phi(\gamma, f) = \Phi(\gamma, f^H)$. If μ is unramified and $f \in \mathcal{H}$, then $f^H = \hat{\xi}(f)$, as is well-known, and (b) is also standard.

PROPOSITION 4.13.2: *Let $G = U(3)$.*

(a) *Let $\phi = \varphi \times \varphi' \in C(\widetilde{G}_v, \tilde{\omega}_v)$, where φ and φ' are functions on* $\mathrm{GL}_3(E_w)$ *and* $\mathrm{GL}_3(E_{w'})$, *respectively. Then there exists a function $f \in C(G_v, \omega_v)$ such that $\Phi_\varepsilon^{\mathrm{st}}(\delta, \phi) = \Phi^{\mathrm{st}}(\gamma, f)$ for all $\delta \in \widetilde{G}_v$ and $\gamma \in \mathcal{N}(\delta)$. Conversely, if F is p-adic, then for all $f \in C(G_v, \omega_v)$, there exists $\phi = \varphi\times\varphi'$ such that $\Phi_\varepsilon^{\mathrm{st}}(\delta, \phi) = \Phi^{\mathrm{st}}(\gamma, f)$ for all $\delta \in \widetilde{G}_v$ and $\gamma \in \mathcal{N}(\delta)$. If F is archimedean, the same statement holds for all smooth f, but with ϕ equal to a finite linear combination of functions of the form $\varphi \times \varphi'$, where for prescribed $N > 0$, φ and φ' can be chosen to have smooth derivatives of order $\leq N$. If π is a representation of G, then $Tr(\tilde{\pi}(\phi)\tilde{\pi}(\varepsilon)) = Tr(\pi(f))$, where $\tilde{\pi} = \pi \otimes \varepsilon(\pi)$.*

(b) *For all $\phi \in C(\widetilde{G}_v, \tilde{\omega}_v)$ there exists a function $\phi^H \in C(H_v, \omega_v)$ with the following property. Let $\delta \in \widetilde{G}_v$ and $\gamma = N(\delta)$. If γ lies in H_v and is (G_v, H_v)-regular, then $\widetilde{\Delta}(\delta)\Phi_\varepsilon^\kappa(\delta, \phi) = \Phi^{\mathrm{st}}(\gamma, \phi^H)$. Let ρ be a representation of H_v and let $\tilde{\pi} = i_{G_w}(\rho) \times i_{G_{w'}}(\varepsilon(\rho))$. Then* $\mathrm{Tr}(\tilde{\pi}(\phi)\tilde{\pi}(\varepsilon)) = \mathrm{Tr}(\rho(\phi^H))$.

Proof: Let $f = \varphi * \varphi'$, where

$$\varphi * \varphi'(g) = \int_{Z_w\backslash \mathrm{GL}_3(E_w)} \varphi(gg'^{-1})\varphi'(\varepsilon_0(g'))\, dg'\ . \tag{4.13.1}$$

Let $\delta \in \widetilde{G}_v$. Up to ε-conjugacy, we can assume that $\delta = (\gamma, 1)$ where $\gamma \in \mathrm{GL}_3(E_w)$. Then

$$\begin{aligned}\Phi_\varepsilon(\phi, \delta) &= \int\limits_{\widetilde{Z}_v \widetilde{G}_{v\delta\varepsilon} \backslash \widetilde{G}_v} \varphi(g^{-1}\gamma\varepsilon(g'))\varphi'(g'^{-1}\varepsilon_0(g))dg' \\ &= \int\limits_{\mathrm{GL}_3(E_w)_\gamma \backslash \mathrm{GL}_3(E_w)} \int\limits_{Z_w \backslash \mathrm{GL}_3(E_w)} \varphi(g^{-1}\gamma g g'^{-1})\varphi'(\varepsilon_0(g'))\, dg\, dg' \\ &= \Phi(f, \gamma)\,.\end{aligned}$$

If v is finite, then every f is of the form $\varphi * \varphi'$. If v is infinite, then every smooth function f is equal to a finite linear combination of functions $\varphi * \varphi'$, where the φ, φ' can be chosen to have smooth derivatives of order $\leq N$ for fixed N, by a result of Duflo-Labesse ([A$_3$], Cor. 4.2).

To prove the identity $\mathrm{Tr}(\tilde{\pi}(\phi)\tilde{\pi}(\varepsilon)) = \mathrm{Tr}(\pi(f))$, let $\{v_j\}$ be a basis of V. Then $v_{ij} = v_i \otimes v_j$ is a basis of $V \times V$ and the $v_i \otimes v_j$-component of $\tilde{\pi}(\phi)\tilde{\pi}(\varepsilon)(v_i \otimes v_j)$ is $\pi(\varphi)_{ji}\varepsilon(\pi)(\varphi')_{ij}$ where the subscripts denote the matrix entries with respect to this basis. Hence

$$\mathrm{Tr}(\tilde{\pi}(\phi)\tilde{\pi}(\varepsilon)) = \sum \pi_{ji}(\varphi)\varepsilon(\pi)(\varphi')_{ij} = \mathrm{Tr}(\pi(\varphi)\varepsilon(\pi)(\varphi'))$$

and this is equal to $\mathrm{Tr}(\pi(\varphi * \varphi'))$.

To prove (b), observe that $\tilde{\tau}(\delta) = 1$ since $\mu_v(\det_0(\delta))$ and $\tau(\gamma)$ are both equal to $\mu_w(\det_0(\gamma))$. Furthermore, $\Phi_\varepsilon^\kappa(\delta, \phi) = \Phi_\varepsilon(\delta, \phi)$ when v splits. Define

$$\phi^H(h) = \delta_P(h)^{1/2} \int\limits_{N_P} (\varphi * \varphi')^K(hn)\, dn.$$

It follows, as in the proof of Proposition 4.13.1, that $\phi \to \phi^H$. The character relation follows by combining the argument of (a) and Proposition 4.13.1.

As observed in [L$_1$], §8, for the case of GL(2), the twisted trace formula is valid for functions whose archimedean component is sufficiently differentiable. In §13, we will apply the trace formula to functions whose component at split archimedean primes is of this type.

PROPOSITION 4.13.3: *Let $G = U(2)$.*

(a) *For all* $\phi \in C(\widetilde{G}_v, \tilde{\omega}_v)$, *there exists functions* $f_1 \in C(G_v, \omega_v)$ *and* $f_2 \in C(G_v, \omega_v \mu_v^{-1})$ *such that*

$$\Phi_\varepsilon^{\mathrm{st}}(\delta, \phi) = \Phi^{\mathrm{st}}(\gamma, f_1), \ \ \mu_v(\det(\delta))\Phi_\varepsilon^\kappa(\delta, \phi) = \Phi^{\mathrm{st}}(\gamma, f_2)$$

for all $\delta \in \widetilde{G}_v$ *and* $\gamma \in \mathcal{N}(\delta)$. *Conversely, for all* $f \in C(G_v, \omega_v)$, *there exists* $\phi \in C(\widetilde{G}_v, \tilde{\omega}_v)$, *such that* $\Phi_\varepsilon^{\mathrm{st}}(\delta, \phi) = \Phi^{\mathrm{st}}(\gamma, f)$ *for all* $\delta \in \widetilde{G}_v$ *and* $\gamma \in \mathcal{N}(\delta)$, *with the same qualification as in Proposition* 4.13.2(a) *if* F *is archimedean.*

(b) *Let* π *be a representation of* G_v *and set* $\tilde{\pi} = \pi \otimes \varepsilon(\pi)$, $\tilde{\pi}' = \tilde{\pi} \otimes \mu_v \circ \det$. *Then*

$$\mathrm{Tr}(\tilde{\pi}(\phi)\tilde{\pi}(\varepsilon)) = \mathrm{Tr}(\pi(f_1)) \ , \ \mathrm{Tr}(\tilde{\pi}'(\phi)\tilde{\pi}'(\varepsilon)) = \mathrm{Tr}(\pi(f_2)) \ .$$

Proof: As above, if $\phi = \varphi \times \varphi'$, where φ and φ' are functions on $\mathrm{GL}_2(E_w)$ and $\mathrm{GL}_2(E_{w'})$, respectively, we define $f_1(g)$ by the formula (4.13.1) with the integration over $\mathrm{GL}_2(E_w)$ instead of $\mathrm{GL}_3(E_w)$. Since $\mu_v(\det(\delta)) = \mu_w(\det(\gamma))$, we can set $f_2(g) = \mu_w(\det(g)) f_1(g)$. Parts (a) and (b) follow as above.

CHAPTER 5

Stabilization

5.1. The endoscopic groups serve to express ε-invariant distributions on G in terms of ε-stably invariant distributions on the endoscopic groups. In the global case, the goal is to obtain an expression for θ_G of the type

$$\theta_G(f) = \sum i(G,H) S\theta_H(f^H) \tag{5.1.1}$$

where the sum is over the set $\mathcal{E}(G)$ of elliptic endoscopic groups H of G and $S\theta_H$ is a certain ε-stably invariant distributions on H. The coefficient $i(G,H)$ is defined below (§5.3).

5.2. A plan for establishing (5.1.1) by induction on $\dim(G)$ was outlined by Langlands in [A]. Assume that G is quasi-split and that ε is trivial. If H is an elliptic endoscopic group for G such $H \neq G$, then $\dim(H) < \dim(G)$ and we can assume by induction that $S\theta_H$ is defined and is a stable distribution on H. Then $f \to S\theta_H(f^H)$ is well-defined as a distribution on G, since the stable orbital integrals of f^H are determined by f. We define $S\theta_G(f)$ by equation (5.1.1). The problem is then to prove that $S\theta_G(f)$ is a stable distribution. In contrast, if ε is non-trivial, then G itself does not appear on the right-hand side of (5.1.1) and the problem in this case is to prove (5.1.1). This is carried out in §10 (Theorem 10.3.1) for the cases described in §4.8, namely, for $U(n)$ and the corresponding base change problem on $\mathrm{GL}(n)$, for $n = 2, 3$. The consequences of (5.1.1) are worked out in §11 and §13.

Denote the distribution $\theta_{G,M}$ defined in §2.5 by θ_M. In §7.6, we define, for the groups we study, distributions J_M on G such that the following equality holds:

$$\sum_{M \in \mathcal{L}(G)} \theta_M(f) = \sum_{M \in \mathcal{L}(G)} J_M(f) \, . \tag{5.2.1}$$

Each side gives an expression for $T_G(f)$. The decomposition in terms of the J_M's is called the "fine $\mathcal{O}$-expansion". It exists in general ([A$_2$]). A key point is that $J_G(f)$ is an invariant distribution which is independent of T

and is supported on the elliptic conjugacy classes. It is equal to a sum of terms

$$J_G(f) = \sum J_G(\mathcal{O}_{\text{st}}, f)$$

indexed by the elliptic stable conjugacy classes $\mathcal{O}_{\text{st}}$ modulo Z in G. If the class is regular and $\mathcal{O}$ is a conjugacy class contained in $\mathcal{O}_{\text{st}}$, then $J_{\mathcal{O}}^T(f)$ is simply an orbital integral and $J_G(\mathcal{O}_{\text{st}}, f)$ is a sum of the $J_{\mathcal{O}}^T(f)$ (cf. §5.4). However, if $\mathcal{O}_{\text{st}}$ is singular, then $J_{\mathcal{O}}^T(f)$ is neither invariant nor independent of T and $J_G(\mathcal{O}_{\text{st}}, f)$ is more complicated. It is built out of invariant pieces of $J_{\mathcal{O}}^T(f)$ which are supported on conjugacy classes whose closure contains a class contained in $\mathcal{O}_{\text{st}}$ (cf. §7.6).

A Levi subgroup M_H of H determines a Levi subgroup M_G of G as follows. Let A_H be the maximal split torus in the center of M_H. Then A_H is contained in a Cartan subgroup T_H of H and an embedding $j : T_H \to G$ is determined up to stable conjugacy. The centralizer M_G of $j(A_H)$ is the Levi factor of a parabolic subgroup of G. Conjugating j if necessary, we may assume that M_G is a Levi subgroup, i.e., that it contains M_0. The conjugacy class of M_G is uniquely determined by M_H. We write $M_H \to M_G$ if M_H and M_G are related in this way.

Now define distributions $S\theta_M$ and SJ_M on G, assuming by induction that they have been defined for the proper elliptic endoscopic groups of G, by the equalities:

$$\theta_M(f) = \sum_{(H,M_H)} i(G,H) S\theta_{M_H}(f^H),\ J_M(f)$$

$$= \sum_{(H,M_H)} i(G,H)\, SJ_{M_H}(f^H). \tag{5.2.2}$$

The sums range over pairs (H, M_H) consisting of an elliptic endoscopic group H and a Levi subgroup M_H, given up to conjugacy, such that $M_H \to M$. The term

$$J_G(f) = \sum i(G,H) SJ_H(f^H)$$

is the elliptic part of the stable trace formula.

The proof that $S\theta_G$ is a stable distribution proceeds in two steps. The equality

$$\sum_{M_H \in \mathcal{L}(H)} S\theta_{M_H}(f^H) = \sum_{M_H \in \mathcal{L}(H)} SJ_{M_H}(f^H).$$

follows from (5.2.1), (5.2.2), and induction on $\dim(H)$. For G itself we write this as

$$S\theta_G(f) - SJ_G(f) = \sum_{M \neq G} SJ_M(f) - \sum_{M \neq G} S\theta_M(f). \tag{5.2.3}$$

We want to show that $S\theta_G(f - f^*) = 0$ whenever f^* is a function whose stable orbital integrals at regular semisimple elements are the same as those of f. The first step is to show directly that $SJ_G(f)$ is a stable distribution. Then $SJ_G(f - f^*) = 0$ and $S\theta_G(f - f^*)$ is equal to the right hand side of (5.2.3) applied to $f - f^*$. The second step is based on an idea used in §11 of $[L_1]$ for the case of base change. Let S be a finite set of places of F containing the infinite places and let w be an auxiliary finite place of F. Assume that f_v and f_v^* are equal to the unit in the Hecke algebra for all places $v \notin S \cup \{w\}$ and suppose further that $f_w = f_w^*$. Fix f_v for $v \neq w$ and regard f_w as a varying element of $\mathcal{H}(G_v, w_v)$. Observe that $S\theta_G$ is a countable sum of traces of the form $\mathrm{Tr}(\pi(f^H))$ for $H \in \mathcal{E}(G)$. By the fundamental lemma (cf. §4.5), if π_w is unramified, then $\mathrm{Tr}(\pi_w(f_w^H)) = \mathrm{Tr}(\pi'_w(f_w))$ where π'_w is the unramified representation of G_w corresponding to π_w. Therefore

$$S\theta_G(f - f^*) = \sum c(\pi_w)\mathrm{Tr}(\pi(f_w))\ ,$$

where the sum is over a countable set of unitary unramified representations of G_w and $c(\pi_w) \in \mathbf{C}$. However, the right hand side of (5.2.3) yields a distribution defined by a continuous measure on the unitary unramified spectrum of G_w. This implies, by the Riesz representation theorem, that both sides of (5.2.3) applied to $f - f^*$ must be zero. A variant of this argument is carried out with ε non-trivial in §10.3.

5.3. The constant $\imath(G, H)$ was specified in $[\mathrm{L}_2]$. The following formula is given in $[\mathrm{Kt}_3]$:

$$\imath(G, H) = \tau(G)\tau(H)^{-1}\lambda^{-1}$$

where $\lambda_H = |\Lambda(H, s, \eta)|$ and $\tau(G)$ denotes the Tamagawa number of G. Furthermore we have the formula

$$\tau(G) = |\pi_0(Z(\widehat{G})^\Gamma)| \cdot |\ker^1(F, Z(\widehat{G}))|^{-1}.$$

(This formula is proved in $[\mathrm{Kt}_3]$ under the assumption that $\tau(G_{sc}) = 1$, but this is known for the groups dealt with here. In fact it is known for all G such that G_{sc} has no E_8-factors by $[\mathrm{Kt}_6]$).

PROPOSITION 5.3.1: *Let G be a unitary group in n variables and let H be the endoscopic group $U(a) \times U(b)$. Then*

$$i(G,H) = \begin{cases} \frac{1}{2} & \text{if } \ a \neq b \\ \frac{1}{4} & \text{if } \ a = b\,. \end{cases}$$

Proof: By Lemma 3.5.2(b), $|\ker^1(F, Z(\widehat{G}))| = 1$, hence $\tau(G) = 2$ and $\tau(H) = 4$. By Proposition 4.6.1, $\lambda_H = 2$ if $a = b$ and is equal to 1 otherwise.

5.4. Elliptic regular terms. Let $\mathcal{O}$ be a semisimple conjugacy class containing an element γ_0. Let $G'_{\gamma_0} = \{g \in G : g^{-1}\gamma_0 g \gamma_0^{-1} \in Z\}$ and set $\varepsilon(\gamma_0) = [G'_{\gamma_0} : G_{\gamma_0}]$. If G_{γ_0} is anisotropic, then, by §2.2,

$$\begin{aligned} J_{\mathcal{O}}^T(f) &= \int\limits_{\mathbf{Z}G\backslash\mathbf{G}} \sum_{\gamma \in \underline{\mathcal{O}}'} f(g^{-1}\gamma g)\, dg. \\ &= m(\mathbf{Z}G'_{\gamma_0}\backslash\mathbf{G}'_{\gamma_0}) \int\limits_{\mathbf{G}'_{\gamma_0}\backslash\mathbf{G}} f(g^{-1}\gamma_0 g)\, dg\,. \\ &= \frac{m(\mathbf{Z}G'_{\gamma_0}\backslash\mathbf{G}'_{\gamma_0})}{[\mathbf{G}'_{\gamma_0} : \mathbf{G}_{\gamma_0}]} \int\limits_{\mathbf{G}_{\gamma_0}\backslash\mathbf{G}} f(g^{-1}\gamma_0 g)\, dg \end{aligned}$$

where $\mathbf{G}'_{\gamma_0} = \{g \in \mathbf{G} : g^{-1}\gamma_0 g \gamma_0^{-1} \in Z\}$. Now

$$\frac{m(\mathbf{Z}G'_{\gamma_0}\backslash\mathbf{G}'_{\gamma_0})}{[\mathbf{G}'_{\gamma_0} : \mathbf{G}_{\gamma_0}]} = \frac{m(\mathbf{Z}G_{\gamma_0}\backslash\mathbf{G}'_{\gamma_0})}{[\mathbf{Z}G'_{\gamma_0} : \mathbf{Z}G_{\gamma_0}][\mathbf{G}'_{\gamma_0} : \mathbf{G}_{\gamma_0}]} = \varepsilon(\gamma_0)^{-1} m(\mathbf{Z}G_{\gamma_0}\backslash\mathbf{G}_{\gamma_0})$$

and

$$J_{\mathcal{O}}^T(f) = \varepsilon(\gamma_0)^{-1} m(\mathbf{Z}G_{\gamma_0}\backslash\mathbf{G}_{\gamma_0})\Phi(\gamma_0, f)$$

Let $\mathcal{C}$ be a set of representatives for the conjugacy classes in $\mathcal{O}_{\mathbf{st}}(\gamma_0)$. Let $\varepsilon_{\mathbf{st}}(\gamma_0) = \varepsilon_{\mathbf{st}}(\gamma_0, G)$ be the number of $z \in Z$ such that γ_0 is stably conjugate to $z\gamma_0$. For each $\gamma \in \mathcal{C}$, the number of $\gamma' \in \mathcal{C}$ such that γ is stably conjugate to $z\gamma'$ for some $z \in Z$ is equal to $\varepsilon_{\mathbf{st}}(\gamma_0)\varepsilon(\gamma_0)^{-1}$.

Set

$$J_G(\mathcal{O}_{\mathbf{st}}, f) = \sum J_{\mathcal{O}}^T(f)$$

where the sum is over a set of representative for the conjugacy classes $\mathcal{O}$ containing a stable conjugate of γ_0 modulo Z. Assume that γ_0 is elliptic regular. Then G_γ is anisotropic for all $\gamma \in \mathcal{C}$, and

$$J_G(\mathcal{O}_{st}, f) = \varepsilon_{st}(\gamma_0)^{-1} m(\mathbf{Z}G_{\gamma_0}\backslash\mathbf{G}_{\gamma_0}) \sum_{\gamma\in\mathcal{C}} \Phi(\gamma, f) \,. \tag{5.4.1}$$

Let $\mathcal{C}_{\mathbf{A}}$ be a set of representatives for the $G(\mathbf{A})$-conjugacy classes within the $G(\overline{\mathbf{A}})$-conjugacy class of γ_0. The orbital integral $\Phi(\gamma, f)$ depends only on the image of γ in $\mathcal{C}_{\mathbf{A}}$ under the natural map $\mathcal{C} \to \mathcal{C}_{\mathbf{A}}$. Let

$$k(\gamma_0) = |\ker(\mathfrak{D}(\gamma_0/F) \to \mathfrak{D}(\gamma_0/\mathbf{A}))| \,.$$

It follows from Corollary 3.3.2 that (5.4.1) is equal to

$$\frac{k(\gamma_0)}{|\mathfrak{R}(G_{\gamma_0}/F)|} \varepsilon_{st}(\gamma_0)^{-1} m(\mathbf{Z}G_{\gamma_0}\backslash\mathbf{G}_{\gamma_0}) \sum_{\gamma\in\mathcal{C}_{\mathbf{A}}} \sum_{\kappa\in\mathfrak{R}(G_{\gamma_0}/F)} \kappa(\mathrm{obs}(\gamma))\Phi(\gamma, f) \,.$$

The following cohomological formula

$$\frac{k(\gamma_0)}{|\mathfrak{R}(G_{\gamma_0}/F)|} = \frac{\tau(G)}{\tau(G_{\gamma_0})} \tag{5.4.2}$$

holds by [Kt$_4$], §9, and gives

$$J_G(\mathcal{O}_{st}, f) = \varepsilon_{st}(\gamma_0)^{-1} z_G(\gamma_0)\tau(G) \sum_{\kappa\in\mathfrak{R}(G_{\gamma_0}/F)} \Phi^\kappa(\gamma_0, f) \tag{5.4.3}$$

where $z_G(\gamma_0) = \tau(G_{\gamma_0})^{-1} m(\mathbf{Z}G_{\gamma_0}\backslash\mathbf{G}_{\gamma_0})$ (cf. §4.3). Set

$$SJ_G(\mathcal{O}_{st}, f) = \varepsilon_{st}(\gamma_0)^{-1} z_G(\gamma)\tau(G)\Phi^{st}(\gamma, f) \,.$$

The following proposition, which yields a stabilization of the elliptic regular term, was carried out in general by Langlands ([L$_2$]). It provides a basic motivation for the stable trace formula. The formalism was recast and generalized to include the elliptic singular terms by Kottwitz ([Kt$_4$]).

PROPOSITION 5.4.1: *Let $G = U(3)$, $U(2) \times U(1)$, or $U(2)$, and let $\mathcal{O}_{st}$ be a stable elliptic regular class in G. Then*

$$J_G(\mathcal{O}_{st}, f) = SJ_G(\mathcal{O}_{st}, f) + i(G,H)\sum SJ_H(\mathcal{O}'_{st}, f^H) \tag{5.4.4}$$

where H is the unique proper elliptic endoscopic group for G and $\mathcal{O}'_{st}$ ranges over the stable elliptic classes modulo Z in H which transfer to $\mathcal{O}_{st}$.

Proof: By definition, $i(G,H)\tau(H) = \tau(G)\lambda_H^{-1}$. The assertion to be checked is that (5.4.3) is equal to the sum of $SJ_G(\mathcal{O}_{st}, f)$ and

$$\tau(G)\sum \varepsilon_{st}(\gamma_H, H)^{-1} z_H(\gamma_H)\lambda_H^{-1}\Phi^{st}(\gamma_H, f^H)$$

where the sum is over a set of representatives for the stable conjugacy classes modulo Z of elements $\gamma_H \in H$ which transfer to γ. This is clear if γ_0 is not an image of an element in H since $\mathfrak{K}(G_{\gamma_0}/F)$ is then trivial. If γ_0 is an image of $\gamma_H \in H$, then $z_G(\gamma_0) = z_H(\gamma_H)$ since $G_{\gamma_0} = H_{\gamma_H}$. In this case, G_{γ_0} is a Cartan subgroup of type (1) or (2).

Let $G = U(3)$. Then $H = U(2)\times U(1)$ and $\lambda_H = 1$. If γ_0 belongs to a Cartan subgroup T of type (1), then $|\mathfrak{K}(G_{\gamma_0}/F)| = 4$ (Prop. 3.5.2) and there are three stable conjugacy classes in H with representatives $\{\gamma_0, \gamma_1, \gamma_2\}$ in T which transfer to γ_0. If γ_0 belongs to a Cartan subgroup T of type (2), then $|\mathfrak{K}(G_{\gamma_0}/F)| = 2$ and the stable class $\{\gamma_H\}_{st}$ is unique. Now we have:

$$\sum_{\kappa \in \mathfrak{K}(G_{\gamma_0}/F)} \Phi^{\kappa}(\gamma_0, f)$$

(5.4.5)

$$= \begin{cases} \Phi^{st}(\gamma_0, f) + \sum_{0 \leq i \leq 2} \Phi^{st}(\gamma_i, f^H) & \text{for } T \text{ of type (1)} \\ \Phi^{st}(\gamma_0, f) + \Phi^{st}(\gamma_0, f^H) & \text{for } T \text{ of type (2)} \end{cases}$$

This is clear if T is of type (2). In this case, the unique non-trivial κ corresponds to H and $\Phi^{\kappa}(\gamma_0, f) = \Phi^{st}(\gamma_0, f^H)$ by the main property of the transfer (cf. §4.3). Suppose that T is of type (1) and that $\Omega(T, G) = \mathbf{Z}/2$ (Prop. 3.7.2). Let $s \in \Omega_F(T)$. The $\{\tau(s^{-1})s\}$ defines an element of $\mathfrak{D}(\gamma_0/F)$. If $\{g\}$ is a set of elements in $G(\overline{F}_v)$ such that $\{g^{-1}\gamma g\}$ is a set of representatives for the conjugacy classes within the stable class of γ in G_v, for regular $\gamma \in T$, then

$$\begin{aligned} \Phi^{\kappa_v}(s^{-1}\gamma_0 s, f_v) &= \sum \kappa_v(\{\tau(g)g^{-1}\})\Phi(g^{-1}s^{-1}\gamma_0 s g, f_v) \\ &= \sum \kappa_v(\{\tau(s^{-1}g)g^{-1}s\})\Phi(g^{-1}\gamma_0 g, f_v) \\ &= \kappa_v(\{\tau(s^{-1})s\})\Phi^{\kappa'_v}(\gamma_0, f_v) \end{aligned}$$

where κ' is defined by $\kappa'(x) = \kappa(s^{-1}xs)$. Globally, $\kappa(\{\tau(s^{-1})s\}) = 1$ and $\Phi^{\kappa}(s^{-1}\gamma s, f) = \Phi^{\kappa'}(\gamma, f)$. If s has order 3, then it permutes $\{\gamma_0, \gamma_1, \gamma_2\}$ and the action $\kappa \to \kappa'$ also permutes the three non-trivial elements of $\mathfrak{K}(G_{\gamma_0}/F)$. Therefore

$$\sum_{\substack{\kappa \in \mathfrak{K}(G_{\gamma_0}/F) \\ \kappa \neq 1}} \Phi^{\kappa}(\gamma_0, f) = \sum_{0 \leq i \leq 2} \Phi^{\kappa'}(\gamma_i, f)$$

where κ' is any non-trivial element in $R(G_{\gamma_0}/F)$. We may take κ' to be the character defined by H and (5.4.5) follows.

Observe that $\varepsilon_{\rm st}(\gamma, H) = 1$ for all $\gamma \in H$ because the $U(1)$-factor of stably conjugate elements in H coincide. The equation $g^{-1}\gamma_0 g = z\gamma_0$ implies that $z^3 = 1$ and hence $\varepsilon_{\rm st}(\gamma_0, G) = 1$ or 3. If $\varepsilon_{\rm st}(\gamma_0, G) = 1$, then the assertion follows from (5.4.5). If $\varepsilon_{\rm st}(\gamma_0, G) = 3$, then T is of type 1. In this case, the elements $\gamma_0, \gamma_1, \gamma_2$ are equivalent modulo $Z, \Phi^{\rm st}(\gamma_i, f^H)$ is independent of i, and only one of the $\Phi^{\rm st}(\gamma_i, f^H)$ occurs on the right-hand side of (5.4.4). We obtain

$$\varepsilon_{\rm st}(\gamma_0, G)^{-1} \sum_{\substack{\kappa \in \mathcal{R}(G_{\gamma_0}/F) \\ \kappa \neq 1}} \Phi^{\kappa}(\gamma_0, f) = \Phi^{\rm st}(\gamma_0, f^H)$$

as required.

If $G = U(2)$, then $H = U(1) \times U(1)$ and $\lambda_H = 2$. Let $\gamma_1, \gamma_2 \in H$ be distinct elements which are conjugate in G (recall that we have fixed an embedding of H in G). In this case, $|\mathcal{R}(G_{\gamma_1}/F)| = 2$ and we have to check that

$$\varepsilon_{\rm st}(\gamma_1, G)^{-1}\Phi^{\kappa}(\gamma_1, f) = \frac{1}{2}\left(f^H(\gamma_1) + f^H(\gamma_2)\right) ,$$

where the term $f^H(\gamma_2)$ on the right does not occur if $\gamma_1 = z\gamma_2$ for some $z \in Z$. Arguing as above, we obtain $f^H(\gamma_1) = f^H(\gamma_2)$. However $\varepsilon_{\rm st}(\gamma_1, G) = 1$ if γ_1 and γ_2 are distinct modulo Z and $\varepsilon_{\rm st}(\gamma_1, G) = 2$ otherwise. The assertion follows. The case $G = U(2) \times U(1)$ is essentially the same.

5.5. The twisted case. Let $G = U(2)$ or $U(3)$ and let $\widetilde{G} = \operatorname{Res}_{E/F}(G)$. Let $\delta_0 \in \mathcal{O}$, where $\mathcal{O}$ is an ε-semisimple conjugacy class in $\widetilde{G}$. Let

$$\begin{aligned} \widetilde{G}(\delta_0\varepsilon)' &= \{g \in \widetilde{G} : g^{-1}\delta_0\varepsilon(g)\delta_0^{-1} \in \widetilde{Z}\} \\ \widetilde{\mathbf{G}}(\delta_0\varepsilon)' &= \{g \in \widetilde{\mathbf{G}} : g^{-1}\delta_0\varepsilon(g)\delta_0^{-1} \in I_F\widetilde{Z}\} \\ \widetilde{\mathbf{G}}(\delta_0\varepsilon)'' &= \{g \in \widetilde{\mathbf{G}} : g^{-1}\delta_0\varepsilon(g)\delta_0^{-1} \in I_F\} \end{aligned}$$

and let $\varepsilon(\delta_0) = [\widetilde{G}(\delta_0\varepsilon)' : \widetilde{G}(\delta_0\varepsilon)'']$. As in the previous section, if $\widetilde{G}(\delta_0\varepsilon)$ is anisotropic, then

$$J_{\mathcal{O}}^T(\phi) = m(\widetilde{\mathbf{Z}}\widetilde{G}(\delta_0\varepsilon)'\backslash\widetilde{\mathbf{G}}(\delta_0\varepsilon)') \int_{\widetilde{\mathbf{G}}(\delta_0\varepsilon)'\backslash\widetilde{\mathbf{G}}} \phi(g^{-1}\delta_0\varepsilon(g))\, dg$$

$$= \frac{m(\widetilde{\mathbf{Z}}\widetilde{G}(\delta_0\varepsilon)'\backslash\widetilde{\mathbf{G}}(\delta_0\varepsilon)')}{[\mathbf{G}(\delta_0\varepsilon)' : \widetilde{\mathbf{G}}(\delta_0\varepsilon)'']}\Phi_\varepsilon(\delta_0,\phi)\ .$$

We have

$$\frac{m(\widetilde{\mathbf{Z}}\widetilde{G}(\delta_0\varepsilon)'\backslash\widetilde{\mathbf{G}}(\delta_0\varepsilon)')}{[\widetilde{\mathbf{G}}(\delta_0\varepsilon)' : \widetilde{\mathbf{G}}(\delta_0\varepsilon)'']} = \varepsilon(\delta_0)^{-1}\frac{m(\widetilde{\mathbf{Z}}\widetilde{G}(\delta_0\varepsilon)''\backslash\widetilde{\mathbf{G}}(\delta_0\varepsilon)')}{[\mathbf{G}(\delta_0\varepsilon)' : \widetilde{\mathbf{G}}(\delta_0\varepsilon)'']}$$
$$= \varepsilon(\delta_0)^{-1}m(\widetilde{\mathbf{Z}}\widetilde{G}(\delta_0\varepsilon)''\backslash\widetilde{\mathbf{G}}(\delta_0\varepsilon)'').$$

Let $a(\delta_0) = [\widetilde{\mathbf{G}}(\delta_0\varepsilon)'' : \widetilde{\mathbf{Z}}\widetilde{\mathbf{G}}(\delta_0\varepsilon)\widetilde{G}(\delta_0\varepsilon)'']$. Then

$$m(\widetilde{\mathbf{Z}}\widetilde{G}(\delta_0\varepsilon)''\backslash\widetilde{\mathbf{G}}(\delta_0\varepsilon)'') = a(\delta_0)m(\widetilde{\mathbf{Z}}\widetilde{G}(\delta_0\varepsilon)\backslash\widetilde{\mathbf{Z}}\widetilde{\mathbf{G}}(\delta_0\varepsilon))$$
$$= a(\delta_0)m(\mathbf{Z}\widetilde{G}(\delta_0\varepsilon)\backslash\widetilde{\mathbf{G}}(\delta_0\varepsilon)) = a(\delta_0)m(\mathbf{Z}G_{\gamma_0}\backslash\mathbf{G}_{\gamma_0})$$

if $\gamma_0 \in G$ is a norm of δ_0, since G_{γ_0} is then an inner form of $\widetilde{G}(\delta_0\varepsilon)$. We obtain

$$J_{\mathcal{O}}^T(f) = \varepsilon(\delta_0)^{-1}a(\delta_0)m(\mathbf{Z}G_{\gamma_0}\backslash\mathbf{G}_{\gamma_0})\Phi_\varepsilon(\delta_0,\phi).$$

Let $\varepsilon_{\mathrm{st}}(\delta_0)$ be the number of $z \in \widetilde{Z}$ modulo F^* such that δ_0 is stably ε-conjugate to $z\delta_0$.

LEMMA 5.5.1: (a) *If $G = U(3)$, then $a(\delta_0) = 1$. If $G = U(2)$, then $a(\delta_0) = 1$ unless γ_0 is central or G_{γ_0} is a Cartan subgroup of type* (0) *or* (2). *In these two cases, $a(\delta_0) = 2$.*

(b) $\varepsilon_{\mathrm{st}}(\delta_0) = \varepsilon_{\mathrm{st}}(\gamma_0)$.

Proof: If $G = U(3)$ or if $G = U(2)$ and G_{γ_0} is a Cartan subgroup of type (1), then $\widetilde{\mathbf{G}}(\delta_0\varepsilon)'' = \mathbf{Z}\widetilde{\mathbf{G}}(\delta_0\varepsilon)$ by Propositions 3.11.2(c) and 3.12.1(c), respectively. If $G = U(2)$ and γ_0 is central or G_{γ_0} is a Cartan subgroup of type (0) or (2), the map $g \to g^{-1}\delta_0\varepsilon(g)\delta^{-1}$ defines an isomorphism of $\widetilde{\mathbf{Z}}\widetilde{\mathbf{G}}(\delta_0\varepsilon)\widetilde{G}(\delta_0\varepsilon)''\backslash\widetilde{\mathbf{G}}(\delta_0\varepsilon)''$ with $F^*N_{E/F}(I_E)\backslash I_F$ by Proposition 3.12.1(c). Part (a) follows. To prove (b), we can assume that $N(\delta_0) = \gamma_0$ and that $G(\delta_0\varepsilon) = G_{\gamma_0}$. Identify $\widetilde{G}$ over $\overline{F}$ with $\mathrm{GL}_n \times \mathrm{GL}_n (n = 2$ or $3)$. Then δ_0 corresponds to $(\delta_0, \sigma(\delta_0))$, and if

$$(g^{-1}, h^{-1})(\delta_0, \sigma(\delta_0))(h, g) = (z\delta_0, \sigma(z\delta_0)) \tag{5.5.1}$$

for $(z,\sigma(z)) \in \widetilde{Z}$, then $g^{-1}\gamma_0 g = (z/\overline{z})\gamma_0$. Conversely, if $g \in \mathrm{GL}_n(\overline{F})$ and $g^{-1}\gamma_0 g = (z/\overline{z})\gamma_0$, then (5.5.1) holds with $h = \delta_0^{-1} g z \delta_0$.

Let $\widetilde{\mathcal{O}}_{\mathrm{st}}$ be the stable ε-elliptic regular conjugacy class containing δ_0 and set

$$J_{\widetilde{G}}(\widetilde{\mathcal{O}}_{\mathrm{st}}, \phi) = \sum J_{\mathcal{O}}^{T}(\phi)$$

where the sum is over a set of representatives modulo $\widetilde{Z}$ for the ε-conjugacy classes $\mathcal{O}$ contained in $\widetilde{\mathcal{O}}_{\mathrm{st}}$. Let $\mathcal{C}_\varepsilon$ be a set of representatives for the ε-conjugacy classes in $\widetilde{\mathcal{O}}_{\mathrm{st}}$ modulo F^*. For each $\delta \in \mathcal{C}$, the number of $\delta' \in \mathcal{C}$ such that δ is stably ε-conjugate to $z\delta'$ for some $z \in \widetilde{Z}$ is equal to $\varepsilon_{\mathrm{st}}(\gamma_0)\varepsilon(\gamma_0)^{-1}$ by Lemma 5.5.1(b). Hence

$$J_{\widetilde{G}}(\widetilde{\mathcal{O}}_{\mathrm{st}}, \phi) = \varepsilon_{\mathrm{st}}(\gamma_0)^{-1} a(\delta_0) m(\mathbf{Z}G_{\gamma_0}\backslash \mathbf{G}_{\gamma_0}) \sum_{\delta \in \mathcal{C}} \Phi_\varepsilon(\delta, \phi). \tag{5.5.2}$$

PROPOSITION 5.5.2: *Let $\widetilde{\mathcal{O}}_{\mathrm{st}}$ be a stable ε-elliptic regular conjugacy class in $\widetilde{G}$ and let $\mathcal{O}_{\mathrm{st}}$ be the stable conjugacy class in G associated to $\widetilde{\mathcal{O}}_{\mathrm{st}}$ via the norm.*

(a) *Let $G = U(3)$ and let $H = U(2) \times U(1)$. Then*

$$J_{\widetilde{G}}(\widetilde{\mathcal{O}}_{\mathrm{st}}, \phi) = SJ_G(\mathcal{O}_{\mathrm{st}}, f) + i(G,H) \sum SJ_H(\mathcal{O}'_{\mathrm{st}}, \phi^H)$$

where $\mathcal{O}'_{\mathrm{st}}$ ranges over the stable elliptic classes in H which transfer to $\mathcal{O}_{\mathrm{st}}$.

(b) *If $G = U(2)$, then*

$$J_{\widetilde{G}}(\widetilde{\mathcal{O}}_{\mathrm{st}}, \phi) = SJ_G(\mathcal{O}_{\mathrm{st}}, f_1) + SJ_G(\mathcal{O}_{\mathrm{st}}, f_2)\ .$$

Proof: Let $\delta_0 \in \widetilde{\mathcal{O}}_{\mathrm{st}}$. We may assume that $\gamma_0 = N(\delta_0) \in G$ and that $\widetilde{\mathbf{G}}_{\delta_0\varepsilon} = G_\gamma$ (Proposition 3.11.1). Let $T = G_{\gamma_0}$.

Assume first that $G = U(3)$. We have $H^1(F,T) = \mathcal{D}_\varepsilon(\delta_0/F) \times H^1(F,Z)$ and $\mathcal{D}_\varepsilon(\delta_0/F)$ is the image of $H^1(F,T^d)$ in $H^1(F,T)$ (Proposition 3.11.2(b)). The sequence

$$\mathcal{D}_\varepsilon(\delta_0/F) \to \bigoplus_v \mathcal{D}_\varepsilon(\delta_0/F_v) \to \mathcal{E}(T/F) \tag{5.5.3}$$

is exact, as follows from (3.3.1). Let $k(\delta_0)$ be the cardinality of the kernel of the first arrow. If $x \in \mathcal{D}_\varepsilon(\delta_0/F)$ is in the kernel, then the image of $\times$ in $H^1(F, G_d)$ is locally trivial, hence trivial by the Hasse principle for G_d, and x belongs to the subset $\mathcal{D}(\gamma_0/F)$ of $\mathcal{D}_\varepsilon(\delta_0/F)$. Hence $k(\delta_0) = k(\gamma_0)$.

Let $\mathcal{C}_{\mathbf{A}}$ be a set of representatives for the $\widetilde{\mathbf{G}}$-ε-conjugacy classes modulo I_F of elements $\delta = (\delta_v) \in \widetilde{\mathbf{G}}$ such that δ_v is stably ε-conjugate to δ_0 in $\widetilde{G}_v$ for all v and δ_v is ε-conjugate to δ_0 in $\widetilde{\mathbf{G}}_v$ for almost all v. An element $\delta \in \mathcal{C}_{\mathbf{A}}$ determines an element $\overline{\delta}$ in $\oplus_v \mathcal{D}_\varepsilon(\delta_0/F_v)$ and we let $\kappa(\delta)$ denote the pairing of κ with the image of $\overline{\delta}$ in $\varepsilon(T/F)$, for $\kappa \in \mathcal{R}(T/F)$.

As in the non-twisted case, the product $\Phi_\varepsilon^\kappa(\delta_0,\phi) = \Pi_v \Phi_\varepsilon^{\kappa_v}(\delta_0,\phi_v)$ is well-defined. For almost all v, $\Phi_\varepsilon^{\kappa_v}(\delta_0,\phi_v) = \Phi_\varepsilon(\delta_0,\phi_v)$ (this follows from the non-twisted case and the comparison of twisted and ordinary orbital integrals for units in Hecke algebras [Kt1]). We have

$$\Phi_\varepsilon^\kappa(\delta_0,\phi) = \sum_{\delta \in \mathcal{C}_{\mathbf{A}}} \kappa(\delta)\Phi(\delta,\phi)$$

(the sum is finite).

By Lemma 3.5.1(c), $\mathcal{R}(T/F)$ is dual to $\varepsilon(T/F)$. By Lemma 5.5.1(a), $a(\delta_0) = 1$. It follows, using (5.4.2), that

$$J_{\widetilde{G}}(\widetilde{\mathcal{O}}_{\mathrm{st}},\phi) = \varepsilon_{\mathrm{st}}(\gamma_0)^{-1} z_G(\gamma_0)\tau(G) \sum_{\kappa \in \mathcal{R}(T/F)} \Phi_\varepsilon^\kappa(\delta_0,\phi).$$

The proof of (a) can now be completed by an argument parallel to the one used to prove Proposition 5.4.1, in which $\Phi^\kappa(\gamma_0,f)$ is replaced by $\Phi_\varepsilon^\kappa(\delta_0,\phi)$, using, of course, that $\Phi_\varepsilon^{\mathrm{st}}(\delta_0,\phi) = \Phi^{\mathrm{st}}(\gamma_0,\phi)$ and $\Phi_\varepsilon^\kappa(\delta_0,\phi) = \Phi^{\mathrm{st}}(\gamma_0,\phi^H)$ if κ corresponds to H. Observe that if T is of type (1), then we may assume that $\Omega(T,G) = \mathbf{Z}/2$ and we have $\Phi^\kappa(s^{-1}\delta_0\varepsilon(s),\phi) = \Phi^{\kappa'}(\delta_0,\phi)$ for $s \in \Omega_F(T)$, in the notation of the proof of Proposition 5.4.1.

Now suppose that $G = U(2)$. By Proposition 3.12.1(a), there is a bijection between $\mathcal{C}$ and $H^1(F,E^1) = F^*/NE^*$. The sequence

$$1 \to F^*/NE^* \to \bigoplus_v (F_v^*/NE_v^*) \to I_F/F^*NI_E \to 1$$

is exact and, defining $\Phi_\varepsilon^\kappa(\delta_0,\phi)$ as in the previous case, we obtain

$$\sum_{\delta \in \mathcal{C}} \Phi_\varepsilon(\delta,\phi) = \frac{1}{2}\sum_\kappa \Phi_\varepsilon^\kappa(\delta_0,\phi)$$

where κ ranges over the two characters of I_F/F^*NI_E. This is equal to

$$\frac{1}{2}(\Phi^{\mathrm{st}}(\gamma_0,f_1) + \Phi^{\mathrm{st}}(\gamma_0,f_2))$$

by Propositions 4.11.1 and 4.13.3. By Lemma 5.5.(b),

$$\begin{aligned} J_{\widetilde{G}}(\widetilde{\mathcal{O}}_{\mathrm{st}}, f) &= \frac{1}{2}\varepsilon_{\mathrm{st}}(\gamma_\circ)^{-1}a(\delta_0)m(\mathbf{Z}G_{\gamma_0}\backslash\mathbf{G}_{\gamma_0})[\Phi^{\mathrm{st}}(\gamma_\circ, f_1) + \Phi^{\mathrm{st}}(\gamma_\circ, f_2)] \\ &= \frac{1}{2}\varepsilon_{\mathrm{st}}(\gamma_\circ)^{-1}a(\delta_0)\tau(G_{\gamma_0})z_G(\gamma)[\Phi^{\mathrm{st}}(\gamma_\circ, f_1) + \Phi^{\mathrm{st}}(\gamma_\circ, f_2)]. \end{aligned}$$

To complete the proof, it suffices to show that

$$\frac{1}{2}a(\delta_0)\tau(G_{\gamma_0}) = \tau(G). \tag{5.5.2}$$

If G_{γ_0} is a Cartan subgroup of type (1) (resp., type (2)), then $|\mathfrak{R}(\gamma_\circ/F)| = 2$ (resp., $|\mathfrak{R}(\gamma_\circ/F)| = 1$) by Proposition 3.5.2(c), and hence $\frac{1}{2}a(\delta_0) = |\mathfrak{R}(\gamma_\circ/F)|^{-1}$ by Lemma 5.5.1(a). Equality (5.5.2) will follow from (5.4.2) if we check that $k(\gamma_\circ) = 1$. By Proposition 3.5.2(a), $H^1(F, G_{\gamma_0})$ is isomorphic to $L^*/N_{L'/L}(L'^*)$ where L'/L is a quadratic extension if G_{γ_0} is of type (2) and is isomorphic to $F^*/N_{E/F}(E^*) \oplus F^*/N_{E/F}(E^*)$ if G_{γ_0} is of type (1). The assertion follows since an everywhere local norm is a global norm.

5.6 Let $i(\widetilde{G}, H) = i(G, H)$ if $G = U(3)$ and let $i(\widetilde{G}, U(2)) = 1$ if $G = U(2)$. We can summarize the results of Propositions 5.4.1 and 5.5.1 by the following formula, valid for all our pairs $(G, \varepsilon) \in \mathfrak{U}$. If $\mathcal{O}_{\mathrm{st}}$ is a stable ε-elliptic conjugacy class in G, then

$$J_G(\mathcal{O}_{\mathrm{st}}, f) = \sum i(G, H) \sum SJ_H(\mathcal{O}'_{\mathrm{st}} f^H)$$

where H runs over $\mathcal{E}(G)$ and $\mathcal{O}'_{\mathrm{st}}$ runs over the stable ε-elliptic classes in H which transfer to $\mathcal{O}_{\mathrm{st}}$.

CHAPTER 6

Weighted orbital integrals

In this chapter, the terms $J_{\mathcal{O}}^T(f)$ associated to ($\varepsilon-$) conjugacy classes containing a regular element in the quasi-split torus are calculated.

6.1. Weighted orbital integrals. We first deal with the non-twisted case. Let γ be a regular element in M and let $\mathcal{O} = \mathcal{O}(\gamma)$. Then $J_{\mathcal{O}}^T(f)$ can be expressed as a weighted orbital integral. It is equal to $\varepsilon(\gamma)^{-1}$ times

$$m(\mathbf{Z}M\backslash\mathbf{M}^1) \int_{\mathbf{M}\backslash\mathbf{G}} f(g^{-1}\gamma g)\mathcal{W}(g)\, dg \tag{6.1.1}$$

where $\mathcal{W}(g)$ is a weight factor which, according to [M], is equal to the volume of the convex hull of subset

$$\{s^{-1}T - s^{-1}H(w_s g) : s \in \Omega(M)\}$$

of $\mathcal{A}_M$, where w_s is a representative of s in G. Let $\|\cdot\|$ and $|\cdot|$ denote the absolute values of E and F, respectively. We identify $\mathcal{A}_M$ with $\mathbf{R}$, so that $H(d(\alpha,\beta,\overline{\alpha}^{-1})) = ln\|\alpha\|$ if $G = U(3)$ or $U(2)\times U(1)$ and $d(\alpha,\beta,\overline{\alpha}^{-1}) \in \mathbf{M}$, and $H(d(\alpha,\overline{\alpha}^{-1})) = ln\|\alpha\|$ if $G = U(2)$. If $w \in \Omega(M)$, then

$$\{s^{-1}T - s^{-1}H(swg) : s \in \Omega(M)\} = w\{s^{-1}T - s^{-1}H(sg) : s \in \Omega(M)\}\ ,$$

and since the action of $\Omega(M)$ preserves the volume, the integral depends only on the orbit of γ under $\Omega(M)$. The integral (6.1.1) blows up as γ approaches a singular element. Our goal (Proposition 6.3.1) is to define a continuous function $J_M^T(\gamma, f)$ on $\mathbf{M}$ such that

$$J_{\mathcal{O}}^T(f) = \varepsilon(\gamma)^{-1} m(\mathbf{ZM}\backslash\mathbf{M}^1) J_M^T(\gamma, f)$$

when γ is a regular element in M.

The first step is to apply a change of variables to $J_{\mathcal{O}}^T(f)$. The map $n \to \gamma^{-1}n^{-1}\gamma n$ is a bijection on $\mathbf{N}$ if γ is a regular element of $\mathbf{M}$. Its modulus is equal to $\Pi|1 - \alpha(\gamma^{-1})|$, where the product is over the roots of M in N, and for $\gamma \in M$, the modulus is 1. Define n_1 by $\gamma n = n_1^{-1}\gamma n_1$, and for $n \in \mathbf{N}$, let $\mathcal{W}(\gamma, n) = \mathcal{W}(n_1)$. By the Iwasawa decomposition, the integral in (6.1.1) is equal to

$$\int_{\mathbf{N}} f^K(n^{-1}\gamma n)\mathcal{W}(n)\, dn = \int_{\mathbf{N}} f^K(\gamma n)\mathcal{W}(\gamma, n)\, dn.$$

Let

$$w = \begin{pmatrix} 0 & 0 & 1 \\ 0 & 1 & 0 \\ 1 & 0 & 0 \end{pmatrix} \quad \text{or} \quad \begin{pmatrix} 0 & 1 \\ -1 & 0 \end{pmatrix}$$

be a representative for the unique non-trivial element in $\Omega(M)$ in the cases $G = U(3)$, $U(2) \times U(1)$ or $G = U(2)$, respectively. If $g = mnk$, then

$$\mathcal{W}(g) = (T - H(m)) - (-T + H(wmw^{-1}wn)) = 2T - H(wn)\ .$$

Set $\mathcal{V}(\gamma, n) = -H(wn_1)$ and let

$$J_M(\gamma, f) = \int_{\mathbf{N}} f^K(\gamma n)\mathcal{V}(\gamma, n)\, dn\ .$$

Note that $J_M(\gamma, f)$ is invariant under $\Omega(M)$. For $\gamma \in M_v$, set

$$\Phi^M(\gamma, f_v) = \int_{N_v} f_v^K(\gamma n)\, dn$$

By the Iwasawa decomposition and a change of variables for γ regular,

$$\Phi^M(\gamma, f_v) = |\Pi(1 - \alpha^{-1}(\gamma))|_v \Phi(\gamma, f_v),$$

where the product is over the roots of M in N. Set $\Phi^M(\gamma, f) = \Pi\Phi^M(\gamma, f_v)$. If $\gamma \in M$ is regular, then $\Phi^M(\gamma, f) = \Phi(\gamma, f)$. It is immediate from the definition that $\mathcal{V}(\gamma, n) = \sum \mathcal{V}(\gamma, n_v)$. This gives

$$\text{(6.1.2)} \qquad J_M(\gamma, f) = \sum_v \int_{\mathbf{N}_v} f_v^K(\gamma n_v)\mathcal{V}(\gamma, n_v)dn_v \prod_{w \neq v} \Phi^M(\gamma, f_w)$$

where the sum is over all places of F, and

$$\text{(6.1.3)} \qquad J_{\mathcal{O}}^T(f) = \varepsilon(\gamma)^{-1} m(\mathbf{Z}M\backslash\mathbf{M}^1)[2T\Phi(\gamma, f) - J_M(\gamma, f)]\ .$$

It will be clear from the calculations of the next section that the term corresponding to v in the sum defining $J_M(\gamma, f)$ is zero if $\gamma_v \in K_v$ and f_v is the unit in the Hecke algebra. The sum over v in (6.1.2) is therefore finite.

6.2. Weight functions. Let k be a local field with an absolute value $|\cdot|$. Define the norm of a vector $(x_1, \dots, x_n) \in k^n$ by:

$$|(x_1, \dots, x_n)| = \begin{cases} \max\{|x_1|, \dots, |x_n|\} & \text{if } k \text{ is non-archimedean} \\ (x_1^2 + \cdots + x_n^2)^{1/2} & \text{if } k = \mathbf{R} \\ x_1\overline{x}_1 + \cdots + x_n\overline{x}_n & \text{if } k = \mathbf{C}. \end{cases}$$

We will treat the two cases in which v splits or remains prime in E with a uniform notation. If v splits, then $E_v = E_w \times E_{w'}$, where w and w' are the places of E dividing v. Suppose that $n' = u(x,z)$, where $x, z \in E_v$ and $z + \overline{z} = x + \overline{x}$. We have $x = (x', \overline{y}')$ and $z = (z', \overline{z}'')$, where $x', y', z', z'' \in E_w$ and $z'' = x'y' - z'$. Under the identification of G_v with $\mathrm{GL}_3(E_w)$, n' corresponds to

$$\begin{pmatrix} 1 & x' & z' \\ 0 & 1 & y' \\ 0 & 0 & 1 \end{pmatrix} \tag{6.2.1}$$

We recall that $\lambda \in E_v$ is defined in §1.10. It is equal to 1 if v splits or v does not divide 2.

PROPOSITION 6.2.1: (a) $H(wu(x,z)) = -ln\|(\lambda, \lambda x, z)\|$.
(b) $H(wn(x)) = -2ln|(1,x)|_v$.

Proof: Suppose that v is finite and let w be a place of E dividing v. Let $g = (g_{ij}) = m'n'k'$ be an element of $\mathrm{GL}_3(E_w)$, where

$$m' = \begin{pmatrix} \alpha & & \\ & \alpha^{-1}\beta & \\ & & \beta^{-1} \end{pmatrix},$$

n' is unipotent, and $k \in \mathrm{GL}_3(\mathcal{O}_w)$. Then $|(0,0,1)g|_w = |\beta|_w^{-1}$ since k' preserves the norm. Let Λ^2 be the second exterior power of the standard representation. The norm with respect to the standard basis of the image of $(0,1,0) \wedge (0,0,1)$ under $\Lambda^2(g)$ is $|\alpha|_w^{-1}$. It follows that

$$|\beta|_w^{-1} = |(g_{31}, g_{32}, g_{33})|_w$$
$$|\alpha|_w^{-1} = |(g_{21}g_{32} - g_{22}g_{31},\ g_{21}g_{33} - g_{23}g_{31}, g_{22}g_{33} - g_{23}g_{32})|_w.$$

If $g = wn'$, where n' is the element (6.2.1), then

$$ln|\alpha\beta|_w = -ln(|(1, x', z')|_w \cdot |(1, y', z' - x'y')|_w) . \tag{6.2.2}$$

Suppose that v splits in E and let $n = u(x, z)$, where $x = (x', \overline{y}')$, $z = (z', \overline{z}'')$ as above. Then $H(wn)$ is equal to (6.2.2). On the other hand, $|(1, x, z)|_w = |(1, x', z')|_w$ and $|(1, x, z)|_{w'} = |(1, y', z'')|_w$ and (a) follows in this case since $z'' = x'y' - z'$.

Now suppose that v does not split in E. Let $g = mnk$, where $m = d(\alpha, \beta, \overline{\alpha}^{-1}) \in M_v$, $n \in N_v$, and $k \in K_v$. Let $\gamma = d(1, 1, \lambda)$. Then $\gamma k_v \gamma^{-1} \subset \mathrm{GL}_3(\mathcal{O}_w)$ and $\gamma g \gamma^{-1} = mn'k'$, where $n' \in N_v$ and $k' \in \mathrm{GL}_3(\mathcal{O}_w)$. We have $H(g) = ln|\alpha\overline{\alpha}|_v$ and by the above calculation.

$$ln|\alpha\overline{\alpha}|_v = -ln(\max\{\|\lambda g_{31}\|, \|\lambda g_{32}\|, \|g_{33}\|\}) \; .$$

If $g = wu(x, z)$, then $(g_{31}, g_{32}, g_{33}) = (1, x, z)$ and (a) follows. The cases of infinite v and (b) are similar.

6.3 The function $J_M^T(\gamma, f)$. To define $J_M^T(\gamma, f)$, we follow a lecture of Langlands ([M]). For $s \in \Omega(M)$, let $X_s(\gamma, n) = s^{-1}T - s^{-1}H(w_s n_1)$, so that $\mathcal{W}(\gamma, n)$ is the volume of the convex hull of $\{X_s(\gamma, n)\}$. The calculation below shows that as γ approaches a singular element, $\mathcal{W}(\gamma, n)$ goes to infinity. We therefore replace $\mathcal{W}(\gamma, n)$ by a new weight function defined as the volume of the convex hull of $\{Y_s(\gamma, n)\}$, where

$$Y_s(\gamma, n) = X_s(\gamma, n) - \sum_{\alpha} \sum_{v \in S_0} ln|1 - \alpha^{-1}(\gamma)|_v H_\alpha \; .$$

Here α ranges over the positive roots separating $s^{-1}(W)$ from W, where W is the positive Weyl chamber corresponding to N, S_0 is a set of places of F, and H_α is the co-root associated to α. If $\gamma \in M$, the product formula implies that this new weight function coincides with the old one provided that S_0 contains all places v such that $|1 - \alpha^{-1}(\gamma)|_v \neq 1$. If S_0 is sufficiently large, this leads to a definition of (6.1.1) which extends to a continuous function on $\mathbf{M}$.

Suppose that $G = U(3)$. Let $\gamma \in M_v$ be a regular element. For $j = 1, 2, 3$, let $\alpha_j = \alpha_j(\gamma)$ and set

$$A_j = A_j(\gamma) = 1 - \alpha_j(\gamma)^{-1}.$$

Let $n_1 = u(x_1, z_1) \in N_v$ and $n = \gamma^{-1} n_1^{-1} \gamma n_1$. Then

$$x_1 = A_1^{-1} x, \; \overline{x}_1 = A_2^{-1} \overline{x}, \; z_1 = A_3^{-1}(z + \alpha_1^{-1} A_1^{-1} \overline{x}),$$

and

$$\mathcal{V}(\gamma, n) = ln(\|(1, A_1^{-1} x, \lambda^{-1} A_3^{-1}(z + \alpha_1^{-1} A_1^{-1} x\overline{x})\|_v) + ln\|\lambda\|_v \; .$$

Here we see that $\mathcal{V}'(\gamma, n)$ goes to infinity and the term associated to v in (6.1.2) blows up as γ approaches a singular element. Define the following modified weight function:

$$\begin{aligned}\mathcal{V}(\gamma, n, v) &= \mathcal{V}'(\gamma, n) + \mathit{ln}|A_1 A_2 A_3^2|_v \\ &= \mathit{ln}(\max\{\|A_1 A_3\|_v, \|A_3 x\|_v, \|\lambda^{-1}(A_1 z + \alpha_1^{-1} x\overline{x})\|_v) + \mathit{ln}\|\lambda\|_v\end{aligned}$$

If $G = U(2)$ or $U(2) \times U(1)$, then $\mathcal{V}'(\gamma, n(x)) = 2\mathit{ln}|(1, A^{-1}x)|$ for $n = n(x) \in N_v$, where $A = A(\gamma) = 1 - \alpha^{-1}(\gamma)$. Set

$$\mathcal{V}(\gamma, n, v) = \mathcal{V}'(\gamma, n) + 2\mathit{ln}|A|_v = 2\mathit{ln}(\max\{|A|_v, |x|\})\ .$$

In both cases, define a modified local weighted orbital integral:

$$\mathcal{I}(\gamma, f_v) = \int\limits_{\mathbf{N}_v} f_v^K(\gamma n)\mathcal{V}(\gamma, n, v)\ dn\ .$$

It is clear that for all $\gamma \in M_v$, $\mathcal{V}(\gamma, n, v)$ is defined for almost all n and that the integral converges. The analytic properties of $\mathcal{I}(\gamma, f_v)$ are studied in §9.

For $\gamma \in M$, the product formula implies that $\sum_v \mathcal{V}(\gamma, n_v, v) = \sum_v \mathcal{V}'(\gamma, n_v)$ and the right hand side of (6.1.2) is unchanged if for all v, the weighted orbital integral of f_v is replaced by $\mathcal{I}(\gamma, f_v)$. This would produce a function on $\mathbf{M}$, but it does not give a definition which is usable in later arguments involving the trace formula. We therefore proceed as follows.

Let Ω be the set of $\gamma \in M$ such that $f(k^{-1}\gamma n k)$ is not identically zero as a function of k and n. For fixed f, Ω is finite modulo Z. There is a finite set of places S_0 of F, which includes the infinite places, such that:

(i) f_v is the unit in $\mathcal{H}_v$ for $v \notin S_0$

(ii) $|A_j(\gamma)|_w = 1$ for $j = 1, 2, 3$ and all places w of E dividing some $v \notin S_0$, and all $\gamma \in \Omega$ such that $A_j(\gamma) \neq 0$.

Assume also that S_0 contains the finitely many places v for which the constant λ (cf. §1.10) is not equal to 1. The set S_0 clearly depends on f and will be further enlarged as necessary in the sequel. Define

$$\mathcal{I}_M(\gamma, f, v) = \mathcal{I}(\gamma, f_v)\left(\prod_{w \neq v} \Phi^M(\gamma_w, f_w)\right)$$

$$\mathcal{I}_M(\gamma, f) = \sum_{v \in S_0} \mathcal{I}_M(\gamma, f, v)\ .$$

For convenience, we omit the dependence of $\mathcal{J}_M(\gamma, f)$ on the choice of S_0 in the notation. Define

$$J_M^T(\gamma, f) = 2T\Phi^M(\gamma, f) + \mathcal{J}_M(\gamma, f) \ .$$

Suppose that $\gamma \in M$ is singular. If $G = U(3)$ and $n = u(x, z)$, set:

$$\beta(\gamma, n) = A_1(\gamma)z + \alpha_1(\gamma)^{-1}x\overline{x}. \tag{6.3.1}$$

Then $\mathcal{V}(\gamma, n, v) = ln\|\beta(\gamma, n)\|_v$ since $A_3 = 0$. If $G = U(2) \times U(1)$ or (2), let

$$\beta(\gamma, n) = x \ . \tag{6.3.2}$$

Then $\mathcal{V}(\gamma, n, v) = ln\|\beta(\gamma, n)\|_v$.

PROPOSITION 6.3.1: *Let $G = U(3)$, $U(2) \times U(1)$ or $U(2)$. Let $\gamma \in M$ and let $\mathcal{O} = \mathcal{O}(\gamma)$.*

(a) *If γ is regular then*

$$J_{\mathcal{O}}^T(f) = \varepsilon(\gamma)^{-1}m(\mathbf{Z}M\backslash\mathbf{M}^1)J_M^T(\gamma, f) \ .$$

(b) *If γ is singular, then*

$$\mathcal{J}_M(\gamma, f, v) = \left(\prod_{w \neq v} \Phi^M(\gamma, f_w)\right)\left(\int_{N_v} f_v^K(\gamma n) ln\|\beta(\gamma, n)\|_v dn\right)$$

Proof: We treat the case $G = U(3)$; the other cases are similar. Part (b) is immediate from the preceding calculation. Suppose that γ is regular. Then $\mathcal{J}_M(\gamma, f)$ and $J_M(\gamma, f)$ both vanish unless $\gamma \in \Omega$. If $v \notin S_0$ and $f_v(k^{-1}\gamma n_v k)$ is non-zero, then γ and n_v belong to K_v. However, in this case, $\mathcal{V}'(\gamma, n_v) = 0$ by (ii). It follows that

$$f_v(k^{-1}\gamma n_v k)\mathcal{V}'(\gamma, n_v) = 0$$

if $v \notin S_0$. By (ii) and the product rule,

$$\sum_{v \in S_0} ln|A_1 A_2 A_3^2|_v = 0$$

and

$$\mathcal{V}'(\gamma, n) = \sum_{v \in S_0} \mathcal{V}(\gamma, n, v),$$

for all regular $\gamma \in \Omega \cap M$. Part (a) follows.

6.4 The twisted case. In this section, let $\widetilde{G} = \mathrm{Res}_{E/F}(G)$, where G is $U(3)$ or $U(2)$. Let $\delta \in \widetilde{M}$ be a ε-regular and let $\gamma = N(\delta)$. Let $\mathcal{O}$ be the ε-conjugacy class containing δ. As in the non-twisted case,

$$J_{\mathcal{O}}^T(\phi) = \varepsilon(\delta)^{-1} a(\delta) m(\mathbf{Z}M\backslash\mathbf{M}^1) \int_{\mathbf{M}''\backslash\widetilde{\mathbf{G}}} \phi(g^{-1}\delta\varepsilon(g))\mathcal{W}(g)dg$$

where $\mathbf{M}'' = \{g \in \widetilde{G} : g^{-1}\delta\varepsilon(g)\delta^{-1} \in I_F\}$ and $a(\delta) = [\mathbf{M}'' : \widetilde{\mathbf{Z}}\mathbf{M}M'']$. Recall that $a(\delta) = 1$ (resp., $a(\delta) = 2$) if $G = U(3)$ (resp., $U(2)$) by Lemma 5.5.1(a). For $g = mnk$, where $m \in \widetilde{\mathbf{M}}$, $n \in \widetilde{\mathbf{N}}$, and $k \in \widetilde{\mathbf{K}}$, define $H(g) = \frac{1}{2} ln\|\frac{a}{c}\|$, where $m = d(a, b, c)$. If $m \in \mathbf{M}$, this definition coincides with the definition in §6.1. The weight factor is given by the formula

$$\mathcal{W}(g) = 2T - H(g) - H(wg)$$

([M]). Observe that if $\delta' \in \widetilde{M}$ and $N(\delta) = N(\delta')$, then $\delta' = zm^{-1}\delta\varepsilon(m)$ for some $z \in \widetilde{Z}$ and $m \in \widetilde{M}$. Since $\mathcal{W}(mg) = \mathcal{W}(wg) = \mathcal{W}(g)$ it follows from a change of variables that $J_{\mathcal{O}}^T(\phi)$ depends only on the orbit of $N(\delta)$ under $\Omega(M)$.

We first re-write $J_{\mathcal{O}}^T(\phi)$ to obtain the analogue of (6.1.3). The map $n \to \delta^{-1}n^{-1}\delta\varepsilon(n)$ is a bijection on $\widetilde{\mathbf{N}}$ if δ is an ε-regular element of $\widetilde{\mathbf{M}}$. Its modulus is equal to the product of $|(1 - \alpha(\gamma^{-1}))|$ over the roots in N and thus depends only on $N(\delta)$. If $\delta \in \widetilde{M}$, the modulus is 1. If $g = mnk$ is an Iwasawa decomposition for g, then $\mathcal{W}(g) = 2T - H(wn)$. Set $\delta_m = m^{-1}\delta\varepsilon(m)$ and let $\mathcal{W}(\delta_m, n) = \mathcal{W}(n_1)$, where $n = \delta_m^{-1}n_1^{-1}\delta_m\varepsilon(n_1)$. By a change of variables,

$$\int_{\mathbf{M}''\backslash\widetilde{\mathbf{G}}} \phi(g^{-1}\delta\varepsilon(g))\mathcal{W}(g)\, dg = \int_{\mathbf{M}''\backslash\widetilde{\mathbf{M}}} \int_{\widetilde{\mathbf{N}}} \phi^K(\delta_m n)\mathcal{W}(\delta_m, n)\, dmdn.$$

Let $\mathcal{V}(\delta_m, n) = -H(wn_1)$, and define

$$J_{\widetilde{M}}(\delta, \phi) = \int_{\mathbf{M}''\backslash\widetilde{\mathbf{M}}} \int_{\widetilde{\mathbf{N}}} \phi^K(\delta_m n)\mathcal{V}(\delta_m n)\, dmdn$$

For $\delta \in \widetilde{M}_v$, set

$$\Phi_\varepsilon^M(\delta, \phi_v) = \int_{M_v''\backslash\widetilde{M}_v} \int_{\widetilde{N}_v} \phi_v^K(\delta_m n)\, dn\, dm$$

and for $\delta \in \widetilde{\mathbf{M}}$, let $\Phi_\varepsilon^M(\delta,\phi) = \Pi\Phi_\varepsilon^M(\delta,\phi_v)$. By a change of variables,

$$\Phi_\varepsilon^M(\delta,\phi_v) = |\Pi(1-\alpha^{-1}(\gamma))|_v\Phi_\varepsilon(\delta,\phi_v),$$

and $\Phi_\varepsilon^M(\delta,\phi) = \Phi_\varepsilon(\delta,\phi)$ for ε-regular $\delta \in \widetilde{M}$. Then

$$J_{\mathcal{O}}^T(\phi) = \varepsilon(\delta)^{-1}a(\delta)m(M\backslash\mathbf{M}^1)[2T\Phi_\varepsilon(\delta,\phi) - J_{\widetilde{M}}(\delta,\phi)]$$

Suppose that $G = U(3)$. Let

$$n_1 = \begin{pmatrix} 1 & x_1 & z_1 \\ 0 & 1 & y_1 \\ 0 & 0 & 1 \end{pmatrix} \quad , \quad n = \delta^{-1}n_1^{-1}\delta\varepsilon(n_1) = \begin{pmatrix} 1 & x & z \\ 0 & 1 & y \\ 0 & 0 & 1 \end{pmatrix} ,$$

and let $A_j = A_j(\gamma)$. Then

$$\varepsilon(n_1) = \begin{pmatrix} 1 & \overline{y}_1 & \overline{x}_1\overline{y}_1 - \overline{z}_1 \\ 0 & 1 & \overline{x}_1 \\ 0 & 0 & 1 \end{pmatrix}$$

and a calculation gives:

$$\begin{pmatrix} x \\ \overline{y} \end{pmatrix} = \frac{1}{A_1}\begin{pmatrix} \overline{\alpha}_2(\delta)^{-1} & 1 \\ 1 & \alpha_1(\delta)^{-1} \end{pmatrix}\begin{pmatrix} x_1 \\ \overline{y}_1 \end{pmatrix} \tag{6.4.1}$$

$$\begin{pmatrix} y \\ \overline{x} \end{pmatrix} = \frac{1}{A_2}\begin{pmatrix} \overline{\alpha}_1(\delta)^{-1} & 1 \\ 1 & \alpha_2(\delta)^{-1} \end{pmatrix}\begin{pmatrix} y_1 \\ \overline{x}_1 \end{pmatrix}$$

and we obtain

$$x_1 = A_1^{-1}(\overline{\alpha}_2(\delta)^{-1}x + \overline{y}), y_1 = A_2^{-1}(\overline{x} + \overline{\alpha}_1(\delta)^{-1}y)$$

$$z_1 = \frac{1}{A_3}[(\overline{z} - \overline{\alpha}_3(\delta)^{-1}z) - A_1^{-1}(\overline{\alpha}_2(\delta)^{-1}x + \overline{y})(\overline{x} + \overline{\alpha}_3(\delta)^{-1}\alpha_1(\delta)^{-1}y]$$

$$x_1y_1 - z_1 = \frac{-1}{A_3}[(\overline{z} - \overline{\alpha}_3(\delta)^{-1}z) + \overline{\alpha}_1(\delta)^{-1}A_2^{-1}(\overline{\alpha}_2(\delta)^{-1}x + \overline{y})(\alpha_2(\delta)^{-1}\overline{x} + y)].$$

Let

$$X(\delta,n) = (1,x_1,z_1), Y(\delta,n) = (1,y_1,z_1 - x_1y_1).$$

By the calculation for GL_3 in the proof of Proposition 6.2.1, for $n_v \in \widetilde{N}_v$ we have

$$\mathcal{V}(\delta,n) = \frac{1}{2}ln(\|X(\delta,n)\|_v \cdot \|Y(\delta,n)\|_v) .$$

If $G = U(2)$ and $n_v = n(x) \in \widetilde{N}_v$, then $n_1 = n(x_1)$ where

$$x_1 = A^{-1}(\overline{\alpha}_3(\delta)^{-1}x + \overline{x}) .$$

In this case, $\mathcal{V}'(\delta, n_v) = ln\|(1, x_1)\|_v$.

Define

$$\mathcal{V}(\delta, n, v) = \begin{cases} \mathcal{V}'(\delta, n) + ln\|A\|_v & \text{if } G = U(2) \\ \mathcal{V}'(\delta_v, n_v) + \frac{1}{2} ln\|A_1 A_2 A_3^2\|_v & \text{if } G = U(3) \end{cases}$$

and for ε-regular $\delta \in M_v$ set:

$$\mathcal{J}_\varepsilon(\delta, \phi_v) = \int\limits_{M_v'' \backslash \widetilde{M}_v} \int\limits_{\widetilde{N}_v} \phi_v^K(\delta_m n) \mathcal{V}(\delta_m, n, v) dn\, dm \; .$$

It is again immediate that for all $\delta \in \widetilde{M}_v$, $\mathcal{V}(\delta_m, n, v)$ is defined for almost all n, and the integral defining $\mathcal{J}_\varepsilon(\delta, \phi_v)$ converges. If $\delta \in \widetilde{M}$, then

$$\sum_v \mathcal{V}(\delta, n_v, v) = \sum_v \mathcal{V}'(\delta, n_v)$$

by the product formula and hence

$$J_{\widetilde{M}}(\delta, \phi) = \sum_v \mathcal{J}_\varepsilon(\delta, \phi_v) \prod_{w \neq v} \Phi_\varepsilon^M(\delta, \phi_w)$$

For $\delta \in \widetilde{\mathbf{M}}$, set:

$$\mathcal{J}_{\widetilde{M}}(\delta, \phi, v) = \mathcal{J}_\varepsilon(\delta_v, \phi_v) \left[\prod_{w \neq v} \Phi_\varepsilon^{\widetilde{M}}(\delta_w, \phi_w) \right]$$

The set Ω of $\gamma \in M$ such that $\phi(k^{-1}\delta_m n\varepsilon(k))\mathcal{V}'(\delta, n)$ is not identically zero in m, n, and k is finite modulo Z (Recall that $\gamma = N(\delta)$). Let S_0 be a finite set of places of F, which includes the infinite places, such that:

(i) ϕ_v is the unit in $\widetilde{\mathcal{H}}_v$ for $v \notin S_0$

(ii) $|A_j(\gamma)|_w = 1$ for $j = 1, 2, 3$ and all places w of E dividing some $v \notin S_0$ and for all $\gamma \in \Omega$ such that $A_j(\gamma) \neq 0$.

Set

$$\mathcal{J}_{\widetilde{M}}(\delta, \phi) = \sum_{v \in S_0} \mathcal{J}_{\widetilde{M}}(\delta, \phi, v) \; .$$

and for $\delta \in \mathbf{M}$, define

$$J_{\widetilde{M}}^T(\delta, \phi) = 2T\Phi_\varepsilon^{\widetilde{M}}(\delta, \phi) + \mathcal{J}_{\widetilde{M}}(\delta, \phi) \; .$$

Observe that $\Phi_\varepsilon^M(\delta, \phi)$ and $\mathcal{J}_{\widetilde{M}}(\delta, \phi)$ depend only on the $\Omega(M)$-orbit of $N(\delta)$ if $\delta \in M$.

Suppose that $N(\delta) = \gamma$ is a singular element in M. If $G = U(3)$, then $\alpha_3(\delta)\overline{\alpha_3(\delta)} = 1$. With n as above, set

$$\beta(\delta,n) = -A_1(\gamma)(\overline{z} - \overline{\alpha}_3(\delta)^{-1}z) + \alpha_2(\delta)(\overline{\alpha}_2(\delta)^{-1}x + \overline{y})(\alpha_2(\delta)^{-1}\overline{x} + y)\,. \tag{6.4.3}$$

Then

$$\overline{\beta(\delta,n)} = \overline{\alpha}_3(\delta)[A_2(\gamma)(\overline{z}-\overline{\alpha}_3(\delta)^{-1}z)+\overline{\alpha}_1(\delta)^{-1}(\overline{\alpha}_2(\delta)^{-1}x+\overline{y})(\alpha_2(\delta)^{-1}\overline{x}+y)].$$

Since $\|\alpha_3(\delta)\|_v = 1$, a short calculation gives

$$\mathcal{V}(\delta,n,v) = ln\|\beta(\delta,n)\|_v\,.$$

If $G = U(2)$ and $n = n(x)$, set

$$\beta(\delta,n) = (\overline{\alpha}(\delta)^{-1}x + \overline{x})\,. \tag{6.4.4}$$

Then

$$\mathcal{V}(\delta,n,v) = ln\|\beta(\delta,n)\|_v\,.$$

PROPOSITION 6.4.1: *Let $\widetilde{G} = \mathrm{Res}_{E/F}(G)$ where $G = U(3)$ or $U(2)$. Let $\delta \in \widetilde{M}$ and set $\gamma = N(\delta)$. Let $\mathcal{O}$ be the ε-conjugacy class containing δ.*

(a) *If γ is regular, then*

$$J_{\mathcal{O}}^T(\phi) = \varepsilon(\delta)^{-1}a(\delta)m(\mathbf{Z}M\backslash\mathbf{M}^1)J_{\widetilde{M}}^T(\delta,f)$$

(b) *If γ is singular, then*

$$\mathcal{J}_{\widetilde{M}}(\delta,\phi,v) = \left(\prod_{w\neq v}\Phi_\varepsilon^M(\delta,\phi_w)\right)\left(\int_{M_v''\backslash\widetilde{M}_v\widetilde{N}_v}\int\phi_v^K(\delta_m n)ln\|\beta(\delta_m,n)\|_v dndm\right)$$

Proof: Part (b) follows from the above calculation. For $v \notin S_0$ and regular $\gamma \in \Omega$, the function $\phi_v(k^{-1}\delta_m n\varepsilon(k))\mathcal{V}(\delta,n)$ is identically zero in k, n, and m, for if $\phi_v(k^{-1}\delta_m n\varepsilon(k)) \neq 0$, then δ_m and n lie in $\widetilde{Z}_v\widetilde{K}_v$. Then $\mathcal{V}(\delta_m,n) = 0$ by (ii). Part (a) follows as in the proof of Proposition 6.3.1.

CHAPTER 7

Elliptic singular terms

In this chapter, we calculate $J_G^T(\mathcal{O}_{st}, f)$ for singular elliptic stable classes.

7.1. Preliminary lemma. Fix an additive character $\alpha = \Pi\alpha_v$ of $\mathbf{A}_F$ and let $da = \prod da_v$, where da_v is the Haar measure on F_v self-dual with respect to α_v. Let $\zeta(s) = \Pi\zeta_v(s)$ be the zeta function of F and define Haar measures $d^*a_v = \zeta_v(1)|a|_v^{-1}da_v$ and $d^*a = \Pi d^*a_v$ on F_v^* and I_F, respectively. Let $\psi = \Pi\psi_v$ be a Schwartz-Bruhat function on $\mathbf{A}_F$ and let $\hat{\psi} = \Pi\hat{\psi}_v$ be its Fourier transform with respect to α and da. Let χ be a character of finite order of $F^*\backslash I_F$ and for $\mathrm{Re}(s) > 1$, define the Mellin transform

$$\theta(s,\chi) = \int_{I_F} \psi(a)\chi(a)|a|^s \ d^*a.$$

Let $\theta_v(s,\chi)$ be the corresponding local integral. For almost all v, $\theta_v(s,\chi) = L(s,\chi_v)$. Denote $\theta_v(s,\chi)$ by $\theta(s)$ if χ is trivial.

The Mellin transform extends to a meromorphic function of s. It is holomorphic if χ is non-trivial trivial, and if χ is trivial, then it has at most a simple pole at $s = 1$([We]). Let

$$\zeta(s) = \lambda_{-1}(s-1)^{-1} + \lambda_0 + \dots$$

be the Taylor expansion of $\zeta(s)$ about $s = 1$. Let τ denote the characteristic function of the positive real numbers.

LEMMA 7.1.1: *Let $T \in \mathbf{R}$. Then*

$$\int_{F^*\backslash I_F} \left[\sum_{t\in F^*} \psi(at) - |a|^{-1}\tau(ln|a|^{-1} - T)\hat{\psi}(0)\right] \chi(a)|a|d^*a$$

converges. Furthermore,

(a) *If χ is non-trivial, its value is $\theta(1,\chi)$.*

(b) *If χ is trivial, its value is*

$$\lambda_{-1}T\hat{\psi}(0)+\lambda_o\hat{\psi}(0)+\lambda_{-1}\sum_v\left(\prod_{w\neq v}\hat{\psi}_w(0)\right)\left(\frac{d}{ds}(\theta_v(s)\zeta_v(s)^{-1})\right)_{s=1}$$

Proof: Let $I_F^1=\{a\in I_F:|a|=1\}$ and identify I_F/I_F^1 with $\mathbf{R}_+^*$ by the modulus map. Let $m(F^*\backslash I_F^1)$ be the measure of $F^*\backslash I_F^1$ with respect to the quotient of d^*a by the measure $t^{-1}dt$ on $\mathbf{R}_+^*$. Then $m(F^*\backslash I_F^1)=\lambda_{-1}$ (cf. [We], VII, §6, Prop. 12). The integral

$$\int_{F^*\backslash I_F}\tau(ln|a|^{-1}-T)\chi(a)|a|^{s-1}d^*a$$

converges for $\mathrm{Re}(s)>1$ and is equal to 0 if χ is non-trivial. If χ is trivial, it is equal to:

$$m(F^*\backslash I_F^1)\int_0^{e^{-T}}t^{s-2}dt=m(F^*\backslash I_F^1)(s-1)^{-1}e^{-T(s-1)}$$

Consider the integral

$$\int_{F^*\backslash I_F}\left[\sum_{t\in F^*}\psi(at)-|a|^{-1}\tau(ln|a|^{-1}-T)\hat{\psi}(0)\right]\psi(a)|a|^sd^*a \tag{7.1.1}$$

The integrand decreases rapidly as $|a|$ tends to infinity. By the Poisson summation formula, we have

$$\sum_{t\in F^*}\psi(at)-|a|^{-1}\hat{\psi}(0)=|a|^{-1}\sum_{t\in F^*}\hat{\psi}(a^{-1}t)-\psi(0).$$

The right hand side shows that it remains bounded as $|a|$ tends to 0. Hence (7.1.1) is analytic for $\mathrm{Re}(s)>0$. If χ is non-trivial, then its value at $s=1$ is $\theta(1,\chi)$. If χ is trivial, then for $\mathrm{Re}(s)>1$, it is equal to:

$$\theta(s)-\lambda_{-1}\frac{e^{-T(s-1)}}{(s-1)}\hat{\psi}(0)$$

The Taylor expansion of $\theta(s)$ about $s=1$ is of the form

$$\lambda_{-1}\frac{\hat{\psi}(0)}{(s-1)}+c+\cdots$$

where c is the constant term $\theta(s)$, since

$$\frac{e^{-T(s-1)}}{(s-1)} = \frac{1}{(s-1)} - T + O(s-1),$$

and the integral is of the form

$$\lambda_{-1}T\hat{\psi}(0) + c + O((s-1)).$$

It remains to calculate c. Let $T(s) = \zeta(s)^{-1}\theta(s)$. Then $T(s)$ is holomorphic at $s=1$ and $T(1) = \hat{\psi}(0)$. Since $\theta(s) = \zeta(s)T(s)$, the constant term at $s=1$ is $c = \lambda_0 T(1) + \lambda_{-1}T'(1)$. For almost all finite v, ψ_v is the characteristic function of $\mathcal{O}_v$ and $\text{meas}(\mathcal{O}_v) = 1$. For such v, $\theta_v(s) = \zeta_v(s)$. Furthermore, $\theta_v(1)\zeta_v(1)^{-1} = \hat{\psi}_v(0)$ for all v. The lemma follows.

With the above choice of Tamagawa measures, we have

$$\frac{d}{ds}\left(\theta_v(s)\zeta_v(s)^{-1}\right)_{s=1} = \int_{F_v} \psi_v(t) ln|t| dt - \hat{\psi}_v(0)\zeta_v'(1)\zeta_v(1)^{-1}.$$

As observed, this is zero for almost all v and the sum in 7.1.1(b) is finite.

7.2 Singular non-central term. Let $(G,\varepsilon) \in \mathcal{U}$ (if ε is non-trivial, we temporarily denote $\widetilde{M}$ and $\widetilde{Z}$ by M and Z, respectively). Every non-regular stable ε-elliptic conjugacy class in G is the form $\mathcal{O}_{st} = \mathcal{O}_{st}(\gamma_0)$, where γ_0 is an ε-singular element of M. Let $\{\gamma\}$ be a set of representatives for the ε-conjugacy classes in $\mathcal{O}_{st}$ modulo Z. Observe that $G_{\gamma\varepsilon}$ is an inner form of either $U(3), U(2) \times U(1)$, or $U(2)$ and if γ is not ε-conjugate to $z\gamma_0$ for any $z \in Z$, then $G_{\gamma\varepsilon}$ is an anisotropic form of $G_{\gamma_0\varepsilon}$. Choose compatible measures on $\mathbf{G}_{\gamma_0\varepsilon}$ and $\mathbf{G}_{\gamma\varepsilon}$ (cf. §1.8). The measure $m(\mathbf{Z}G_{\gamma\varepsilon}\backslash\mathbf{G}_{\gamma\varepsilon})$ is independent of γ. If $G_{\gamma\varepsilon}$ is anisotropic, then

$$J_{\mathcal{O}}^T(f) = a(\gamma)m(\mathbf{Z}G_{\gamma\varepsilon}\backslash\mathbf{G}_{\gamma\varepsilon})\Phi_\varepsilon(\gamma,f)$$

by §5.4 and §5.5, since $\varepsilon(\gamma)=1$ when γ is stably conjugate to an ε-singular element of M. Here $a(\gamma)=2$ if $G = Res_{E/F}(U(2))$ and $a(\gamma)=1$ otherwise.

Let $\mathcal{O}'$ be a set of representatives for the set, modulo Z, of non-ε-semisimple elements whose ε-semisimple parts belong to $\mathcal{O}_{st}$. By §2.2, $J_G^T(\mathcal{O}_{st},f)$ is equal to the sum of

$$a(\gamma_0)\cdot m(\mathbf{Z}G_{\gamma_0\varepsilon}\backslash\mathbf{G}_{\gamma_0\varepsilon})\sum_{\{\gamma\}}\Phi_\varepsilon(\gamma,f) \tag{7.2.1}$$

and integral over $\mathbf{Z}G\backslash\mathbf{G}$ of:

$$\sum_{\gamma'\in\mathcal{O}'} f(g^{-1}\gamma'\varepsilon(g))$$

(7.2.2)

$$-\sum_{\delta\in B\backslash G}\sum_{\gamma''}\sum_{\eta}\tau(H(\delta g)-T)\int_{\mathbf{N}_{\gamma''\varepsilon}} f(g^{-1}\delta^{-1}\eta^{-1}n^{-1}\gamma''\varepsilon(\eta\delta g))dn$$

where γ'' ranges $\mathcal{O}_{\mathrm{st}}\cap M$ modulo Z, and η ranges over $N_{\gamma''\varepsilon}\backslash N$.

PROPOSITION 7.2.1: *The integral over* $\mathbf{Z}G\backslash\mathbf{G}$ *of* (7.2.2) *is equal to*

(7.2.3)

$$\int_{\mathbf{Z}B'_{\gamma\varepsilon}\backslash\mathbf{G}}\left[\sum_{\substack{n\in N_{\gamma\varepsilon}\\ n\neq 1}} f(g^{-1}n^{-1}\gamma\varepsilon(g))-\tau(H(g)-T)\int_{\mathbf{N}_{\gamma\varepsilon}} f(g^{-1}n^{-1}\gamma\varepsilon(g))dn\right]dg$$

where $B'_{\gamma\varepsilon}=\{b\in B: b\gamma\varepsilon(b)\gamma^{-1}\in Z\}$

Proof: Suppose first that ε is trivial. By the Jordan decomposition, an element $\gamma'\in\mathcal{O}'$ can be written as $\delta^{-1}n\gamma\delta$, where $\delta\in G$ and $n\in N_\gamma$, $n\neq 1$. Suppose that $\gamma'=\delta'^{-1}n'\gamma\delta'$, with $\delta'\in G$, $n'\in N_\gamma$. Set $h=\delta\delta'^{-1}$. Then h centralizes γ and $hn'h^{-1}=n$, by the uniqueness of the Jordan decomposition. Furthermore, h lies in B. For if $h\notin B$, then $h=bwb'$ where $b,\ b'\in B$, by the Bruhat decomposition, and $h^{-1}Nh\cap N=w^{-1}Nw\cap N=\{1\}$. Hence δ is unique modulo B_γ and the first sum in (7.2.2) is equal to

$$\sum_{\delta\in B_\gamma\backslash G}\sum_{\substack{n\in N_\gamma\\ n\neq 1}} f(g^{-1}\delta^{-1}n^{-1}\gamma\delta g)$$

Since γ is singular, $\mathcal{O}_{\mathrm{st}}\cap M=\{\gamma\}$. We have $B_\gamma=MN_\gamma$ and $B_\gamma\backslash B=N_\gamma\backslash N$. Hence the sums over δ and η can be combined into single sum over $B_\gamma\backslash G$ and the second term in (7.2.2) can be written as

$$\sum_{\delta\in B_\gamma\backslash G}\tau(H(g)-T)\int_{\mathbf{N}_\gamma} f(g^{-1}\delta^{-1}n^{-1}\gamma\delta g)dn\ .$$

The sum over $B_\gamma\backslash G$ can be eliminated if we integrate over $\mathbf{Z}B_\gamma\backslash\mathbf{G}$ rather than $\mathbf{Z}G\backslash\mathbf{G}$. Proposition follows in this case since $B'_\gamma=B_\gamma$.

Now suppose that ε is non-trivial. Modulo Z, each element γ'' in $\mathcal{O}_{\mathrm{st}}\cap M$ is of the form $m^{-1}\gamma\varepsilon(m)$ for some $m\in M$, and m is unique modulo $M'=$

$\{m \in M : m^{-1}\varepsilon(m) \in Z\}$. By the Jordan decomposition for $G \rtimes \langle\varepsilon\rangle$, modulo Z an element $\gamma' \in \mathcal{O}'$ can be written in the form $\delta^{-1}n^{-1}\gamma\varepsilon(\delta)$ where $\delta \in G$ and $n \in N_{\gamma\varepsilon}$. Suppose that $\gamma' = \delta'^{-1}n'^{-1}\gamma\varepsilon(\delta')$, where $\delta' \in G$ and $n' \in N_{\gamma\varepsilon}$, and set $h = \delta\delta'^{-1}$. Then $h \in G_{\gamma\varepsilon}$ and $hn'^{-1}h = n^{-1}$. It follows, as before, that $h \in B$ and the first sum in (7.2.2) is equal to

$$\sum_{\delta \in B'_{\gamma\varepsilon}\backslash G} \sum_{\substack{n \in N_{\gamma\varepsilon} \\ n \neq 1}} f(g^{-1}\delta^{-1}n^{-1}\gamma\varepsilon(\delta g)) .$$

We have $B'_{\gamma\varepsilon} = M'N_{\gamma\varepsilon}$ and the sums over γ'' and η in the second term can be eliminated as before if we let δ range over $B'_{\gamma\varepsilon}\backslash G$. This proves the proposition.

Lemma 7.1.1 can now be applied to (7.2.3) to evaluate $J_G^T(\mathcal{O}_{\rm st}, f)$. Recall that a finite set S_0 of places was chosen in §6. Let $S = \{m \in M : \alpha_3(m) = 1\}$ and define

$$C = \frac{1}{2}\lambda_0 m(\mathbf{Z}S\backslash\mathbf{S}) - m(\mathbf{Z}M\backslash\mathbf{M}^1) \sum_{v \in S_0} \zeta'_v(1)\zeta_v(1)^{-1}$$

Below, δ_0 will denote a fixed element of E^0. We use the map $t \to t\delta_0$ to transfer α to an additive measure on $\mathbf{E}^0$.

PROPOSITION 7.2.2: *Let $G = U(3)$ and let $\gamma \in M$ be a singular non-central element. Let $\mathcal{O}_{\rm st} = \mathcal{O}_{\rm st}(\gamma)$. Then $J_G^T(\mathcal{O}_{\rm st}, f)$ is equal to the sum of the following terms:*

(a) $\frac{1}{2}m(\mathbf{Z}G_\gamma\backslash\mathbf{G}_\gamma)\sum_{\kappa \in \mathcal{R}(G_\gamma/F)} \Phi^\kappa(\gamma, f)$

(b) $C\Phi^M(\gamma, f)$

(c) $\frac{1}{2}m(\mathbf{Z}\backslash\mathbf{S}) \int_{\mathbf{N}_\gamma\backslash\mathbf{N}} \int_{I_F} f^K(n^{-1}\gamma n(t\delta_0)n)\omega_{E/F}(t)|t|\, d^*t\, dn$

(d) $\frac{1}{2}m(\mathbf{Z}M\backslash\mathbf{M}^1)J_M^T(\gamma, f)$

Proof: By Proposition 3.8.1(a), we can identify a set $\mathcal{C}$ of representatives for the conjugacy classes within $\mathcal{O}_{\rm st}$ with F^*/NE^*. Let $\mathcal{C}_{\mathbf{A}}$ be a set of representatives for the $\mathbf{G}$-conjugacy classes within $\mathcal{O}_{\rm st}(\gamma/\mathbf{A})$. Then $\mathcal{C}_{\mathbf{A}}$ is naturally identified with $\oplus_v(F_v^*/NE_v^*)$ and $\mathcal{R}(G_\gamma/F)$ with the group of order two of characters of $\oplus_v(F_v^*/NE_v^*)$ which are trivial on the image of F^*/NE^*, which we view as a group of characters of $\mathcal{C}_{\mathbf{A}}$. The sum corresponding to (7.2.1) is equal to

$$\sum_{\gamma' \in \mathcal{C}} \Phi(\gamma', f) = \sum_{\mathcal{X} \in \mathfrak{X}(G_\gamma/F)} \sum_{\gamma' \in \mathcal{C}_{\mathbf{A}}} \kappa(\gamma') \Phi(\gamma', f)$$

and this yields (a).

Let $U = \{u(x) : x \in E\}$. Then $\mathbf{B} = \mathbf{N}_\gamma \mathbf{M} \mathbf{U}$ and $|\alpha_3(m)|^{-1} dndmdu$ is a left Haar measure on $\mathbf{B}$ with respect to this decomposition. Since $B_\gamma = MN_\gamma$ and $m(N_\gamma \backslash \mathbf{N}_\gamma) = 1$, (7.2.3) is equal to the integral over $\mathbf{Z}M \backslash \mathbf{M}$ and $U \backslash \mathbf{U}$ of:

$$\Bigg[\sum_{\substack{n \in N_\gamma \\ n \neq 1}} f^K(u(x)^{-1} m^{-1} \gamma n m u(x)) \\ - \tau(H(m) - T) \int_{\mathbf{N}_\gamma} f^K(u(x)^{-1} \gamma n m u(x)) dn \Bigg] |\alpha_3(m)|^{-1}.$$

If we integrate over $U \backslash \mathbf{U}$, we are left with the integral over $\mathbf{Z}M \backslash \mathbf{M}$ of

$$\left[\sum_{t \in F^*} \psi(\alpha_3(m)^{-1} t \delta_0) - \tau(H(m) - T) |\alpha_3(m)| \hat{\psi}(0) \right] |\alpha_3(m)|^{-1}$$

where

$$\psi(w) = \int_{\mathbf{A}_E} f^K(u(x)^{-1} \gamma n(w) u(x)) \; dx \; .$$

for $w \in E^0$. The measure dn on $\mathbf{N}$ is assumed to be $dxdw$ with respect to the coordinates $u(x)n(w)$. Recall that $A_j = 1 - \alpha_j^{-1}(\gamma)$. We have

$$u(x)^{-1} \gamma n(w) u(x) = \gamma u(A_1 x) n \left(w + \frac{1}{2} x \overline{x} (\alpha_2(\gamma)^{-1} - \alpha_1(\gamma)^{-1}) \right),$$

and hence

$$\begin{aligned} \hat{\psi}(0) &= \int_{\mathbf{A}_E} \int_{\mathbf{E}^0} f^K \left(\gamma u(A_1 x) n \left(w + \frac{1}{2} x \overline{x} (\alpha_2(\gamma)^{-1} - \alpha_1(\gamma)^{-1}) \right) \right) \; dxdw \\ &= \int_{\mathbf{N}} f^K(\gamma n) \; dn = \Phi^M(\gamma, f) \; , \end{aligned}$$

by change of variables. Now α_3 defines an isomorphism of $M\mathbf{S} \backslash \mathbf{M}$ with $NE^* \backslash NI_E$ where $N = N_{E/F}$. We embed $NE^* \backslash NI_E$ in $F^* \backslash I_F$ and write

(7.2.3) as:

$$m(\mathbf{Z}S\backslash\mathbf{S}) \int\limits_{F^*\backslash F^* N I_E} \left[\sum_{t\in F^*} \psi(at\delta_0) - |a|^{-1}\tau(ln|a|^{-1} - T)\hat{\psi}(0)\right] |a|\ d^*a\ .$$

This is equal to the sum over the characters χ of $F^*NI_E\backslash I_F$ of

$$\frac{1}{2}m(\mathbf{Z}S\backslash\mathbf{S}) \int\limits_{F^*\backslash I_F} \left[\sum_{t\in F^*} \psi(at\delta_0) - |a|^{-1}\tau(ln|a|^{-1} - T)\hat{\psi}(0)\right] \chi(a)|a|d^*a.$$

We now evaluate using Lemma 7.1.1. If χ is non-trivial, we obtain 7.2.2(c). Suppose χ is trivial. We can assume S_0 chosen so that ψ_v is a multiple of the characteristic function of $\mathcal{O}_v\delta_0$ if $v \notin S_0$. We have

$$m(M\mathbf{S}\backslash\mathbf{M}^1) = \frac{1}{2}m(F^*\backslash I_F^1) = \frac{1}{2}\lambda_{-1}$$

and hence

$$\frac{1}{2}\lambda_{-1}m(\mathbf{Z}S\backslash\mathbf{S}) = m(\mathbf{Z}M\backslash\mathbf{M}^1)\ . \tag{7.2.4}$$

By Lemma 7.1.1 and the remark following it, we obtain the sum of the terms

(i) $m(\mathbf{Z}M\backslash\mathbf{M}^1)T\Phi^M(\gamma, f)$

(ii) $\frac{1}{2}\lambda_0 m(\mathbf{Z}S\backslash\mathbf{S})\Phi^M(\gamma, f)$

(iii) $m(\mathbf{Z}M\backslash\mathbf{M}^1)\sum_{v\in S_0}\left(\Pi_{w\neq v}\hat{\psi}_w(0)\right)\left(\int\limits_{E_v^0}\psi_v(w)ln|w|_v\ dw\right)$

(iv) $-m(\mathbf{Z}M\backslash\mathbf{M}^1)\left(\sum_{v\in S_0}\zeta_v'(1)\ \zeta_v(1)^{-1}\right)\ \Phi^M(\gamma, f)$

Here, for $w \in E_v^0$, we set $|w|_v = \|w\|_v^{1/2}$. The sum of (ii) and (iv) is equal to 7.2.2(b). It remains to show that the sum of (i) and (iii) is 7.2.2(d). After a change of variables $x \to A_1^{-1}x$, (iii) becomes the sum over $v \in S_0$ of:

$$m(\mathbf{Z}M\backslash\mathbf{M}^1) \int\limits_{\mathbf{A}_E}\int\limits_{\mathbf{E}^0} f^K(\gamma u(x)\ n(w'))\ ln|w_v|_v\ dxdw$$

where $w' = w + \frac{1}{2}A_1^{-1}A_2^{-1}x\overline{x}(\alpha_2(\gamma)^{-1} - \alpha_1(\gamma)^{-1})$. One sees directly that $u(x)n(w') = n(x, z)$, where $z = w + A_2^{-1}x\overline{x}$, and hence (iii) is equal to (7.2.5)

$$m(\mathbf{Z}M\backslash\mathbf{M}^1)\sum_{v\in S_0}\left(\prod_{w\neq v}\Phi^M(\gamma, f_w)\right)\left(\int\limits_{E_v}\int\limits_{E_v^0} f^K(\gamma u(x, z))\ ln|w|_v\ dw\ dx\right).$$

Since $A_1A_2^{-1} = -\alpha_1(\gamma)^{-1}$, we have

$$A_1 w = A_1 z + \alpha_1(\gamma)^{-1} x\bar{x}$$

$$ln|w|_v = \frac{1}{2} ln\|A_1 z + \alpha_1(\gamma)^{-1} x\bar{x}\|_v - \frac{1}{2} ln\|A_1\|_v \, .$$

Note that S_0 has been chosen so that $\sum_{v\in S_0} ln\|A_1\|_v = 0$ when (7.2.5) is non-zero. Therefore (7.2.5) is equal to

$$\frac{1}{2} m(\mathbf{Z}M\backslash\mathbf{M}^1) \sum_{v\in S_0} \left(\prod_{w\neq v} \Phi^M(\gamma, f_w)\right)\left(\int_{\mathbf{N}_v} f_v^K(\gamma n) ln\|\beta(\gamma,n)\|_v dn\right)$$

by (6.3.1). By Proposition 6.3.1(b), the sum of (i) and (iii) is equal to 7.2.2(d).

7.3 Contribution of central elements. Let $G = U(3)$, $U(2)$, or $U(2)\times U(1)$. Let $\mathcal{O} = \mathcal{O}(\gamma)$ where $\gamma \in G$ is central. By Proposition 7.2.1, $J_G^T(\mathcal{O}, f)$ is the sum of $m(\mathbf{Z}G\backslash\mathbf{G})f(\gamma)$ and

$$\int_{\mathbf{Z}B\backslash\mathbf{G}} \left[\sum_{\substack{u\in N\\ u\neq 1}} f(g^{-1}\gamma u g) - \tau(H(g)-T)\int_{\mathbf{N}} f(g^{-1}\gamma\, ng)\, dn\right] dg$$

Using the decomposition $\mathbf{G} = \mathbf{NMK}$ and $dg = |\delta_B(m)|^{-1} dndmdk$, and changing variables $m \to m^{-1}$, we write this as the integral over $\mathbf{Z}M\backslash\mathbf{M}$ and $N\backslash\mathbf{N}$ of:

$$\left[\sum_{\substack{u\in N\\ u\neq 1}} f^K(mn^{-1}\gamma u n m^{-1})\right.$$

(7.3.1)

$$\left. - |\delta_B(m)|^{-1}\tau(H(m^{-1})-T)\int_{\mathbf{N}} f^K(\gamma n')dn'\right] |\delta_B(m)|.$$

Suppose that $G = U(2)$ or $U(2)\times U(1)$. Then $n^{-1}\gamma u n = \gamma u$. Let $S = \{m \in M : \alpha(m) = 1\}$ (if $G = U(2)$, then $S = Z$). As in the previous section, the integral is equal to the sum over the characters χ of $F^* N I_E \backslash I_F$

of:

$$m(\mathbf{Z}S\backslash\mathbf{S}) \int\limits_{F^*\backslash I_F} \Big[\sum_{t\in F^*} f^K(\gamma n(at)) - |a|^{-1}\tau(H(a^{-1})-T) \int\limits_{\mathbf{N}} f^K(\gamma n')dn' \Big] \chi(a)|a|d^*a.$$

If $\chi = \omega_{E/F}$, we obtain 7.3.1(b) below by Lemma 7.1.1.

Suppose that χ is trivial. Then (7.2.4) still holds and by Lemma 7.1.1, we obtain the sum of $C\Phi^M(\gamma, f)$, where

$$C = \frac{1}{2}\lambda_0 m(\mathbf{Z}S\backslash\mathbf{S}) - m(\mathbf{Z}M\backslash\mathbf{M}^1) \sum_{v\in S_0} \zeta_v'(1)\zeta_v(1)^{-1}$$

and the two terms

$$T\, m(\mathbf{Z}M\backslash\mathbf{M}^1)\Phi^M(\gamma, f)$$

$$m(\mathbf{Z}M\backslash\mathbf{M}^1) \sum_{v\in S_0} \int\limits_{\mathbf{A}_F} f^K(\gamma n(x)) ln|x|_v\, dx = \frac{1}{2} m(\mathbf{Z}M\backslash\mathbf{M}^1) \sum_{v\in S_0} \mathcal{I}_M(\gamma, f, v)\,.$$

since $\beta(\gamma, n(x)) = x$ by (6.3.2). The sum of these two terms is equal to 7.3.1(d).

PROPOSITION 7.3.1: *Let $G = U(2)$ or $U(2)\times U(1)$. Let $\gamma \in Z$ and let $\mathcal{O} = \mathcal{O}(\gamma)$. Then $J_G^T(\mathcal{O}, f)$ is equal to the sum of the terms:*

(a) $m(\mathbf{Z}G\backslash\mathbf{G})f(\gamma)$

(b) $\frac{1}{2} m(\mathbf{Z}S\backslash\mathbf{S}) \int\limits_{I_F} f^K(\gamma n(a))\omega_{E/F}(a)|a|\, d^*a$

(c) $C\Phi^M(\gamma, f)$

(d) $\frac{1}{2} m(\mathbf{Z}M\backslash\mathbf{M}^1) J_M^T(\gamma, f)$

Now let $G = U(3)$ and set $S' = \{m \in M : \alpha_1(m) = 1\}$. Let $\zeta_E(s)$ be the zeta-function of E, written as a product $\Pi\zeta_v^E(s)$ over the places of E, and let

$$\zeta^E(s) = \lambda_{-1}^E(s-1)^{-1} + \lambda_0^E + \cdots$$

be the Laurent expansion of $\zeta_E(s)$ about $s = 1$. Set:

$$D = \lambda_0^E m(\mathbf{Z}S'\backslash\mathbf{S}') - m(\mathbf{Z}M\backslash\mathbf{M}^1) \sum (\zeta_v^E)'(1)\zeta_v^E(1)^{-1}\,.$$

PROPOSITION 7.3.2: *Let $G = U(3)$ and let $\mathcal{O} = \mathcal{O}(\gamma)$ where γ is central in G. Then $J_G^T(\mathcal{O}_{st}, f)$ is equal to the sum of the following terms:*

(a) $m(\mathbf{Z}G\backslash\mathbf{G})f(\gamma)$
(b) $D\Phi^M(\gamma,f)$
(c) $\frac{1}{2}m(\mathbf{Z}S\backslash\mathbf{S})\int_{I_F} f^K(\gamma n(t\delta_0))\omega_{E/F}(t)\ |t|^2\ d^*t$
(d) $\frac{1}{2}m(\mathbf{Z}S\backslash\mathbf{S})\int_{I_F} f^K(\gamma n(t\delta_0))\ |t|^2\ d^*t$
(e) $\frac{1}{2}m(\mathbf{Z}M\backslash\mathbf{M}^1)\ J_M^T(\gamma,f)$

Proof: Let $N^r = \{u(x)n(w) :\ x \in E^*, w \in E^0\}$. The integrand (7.3.1) is the sum of two terms:

$$|\delta_B(m)| \sum_{t\in F^*} f^K(\gamma n(\alpha_3(m)t\delta_0)) \tag{7.3.2}$$

$$\Big[\sum_{u\in N^r} f^K(mn^{-1}\gamma unm^{-1}) - |\delta_B(m)|^{-1}\tau(H(m^{-1})-T)\int_{\mathbf{N}} f^K(\gamma n')dn'\Big]|\delta_B(m)|. \tag{7.3.3}$$

These two terms are separately integrable. The integral of (7.3.2) over $\mathbf{Z}M\backslash\mathbf{M}$ and $N\backslash\mathbf{N}$ can, since $m(N\backslash\mathbf{N}) = 1$ be rearranged to give:

$$m(\mathbf{Z}S\backslash\mathbf{S}) \sum_{a\in F^*/NE^*} \int_{NI_E} f^K(\gamma n(at\delta_0))|t|^2 d^*t\ .$$

This is equal to the sum over characters χ of $F^*NI_E\backslash I_F$ of

$$\frac{1}{2}m(\mathbf{Z}S\backslash\mathbf{S})\int_{I_F} f^K(\gamma n(t\delta_0))\chi(t)|t|^2 d^*t$$

and yields the terms 7.3.2(c) and 7.3.2(d).

Using the relation

$$mu(-y)[u(x)n(w)]u(y)m^{-1} = u(\alpha_1(m)x)n(\alpha_3(m)[w + x\overline{y} - \overline{x}y]),$$

we write the integral of (7.3.3) as the integral over $m \in ZM\backslash\mathbf{M}$ and $u(y) \in U\backslash\mathbf{U}$ of

$$\left[\sum_{\substack{x\in E^*\\ w\in E^0}} f^K(\gamma u(\alpha_1(m)x)\; n(\alpha_3(m)[w + x\overline{y} - \overline{x}y])) \right.$$
$$\left. - |\delta_B(m)|^{-1}\tau(ln|\alpha_3(m)|^{-1} - T)\int_{\mathbf{N}} f^K(\gamma n')dn'\right] |\delta_B(m)| \;.$$

Set:

$$\psi(x) = \int_{\mathbf{E}^0} f^K(\gamma u(x)n(w))\; dw \;.$$

Choose $\varepsilon \in E$ so that $\lambda = x\overline{\varepsilon} - \overline{x}\varepsilon$ is non-zero, and let $y = (r - \frac{w}{\lambda})\varepsilon + sx$, where $r, s \in \mathbf{A}_F$. Then $w + x\overline{y} - \overline{x}y = r\lambda$. We can assume that $dy = drds$. Then

$$\int_{U\backslash\mathbf{U}} \sum_{w\in E^0} f^K(\gamma u(\alpha_1(m)x)\; n(\alpha_3(m)[w + x\overline{y} - \overline{x}y]))$$
$$= \int_{F\backslash\mathbf{A}_F} \left[\sum_{w\in E^0} f^K(\gamma u(\alpha_1(m)x)n(\alpha_3(m)r\lambda))\right] dr = |\alpha_3(m)|^{-1}\psi(\alpha_1(m)x).$$

and the integral of (7.3.3) is equal to:

$$\int_{\mathbf{Z}M\backslash\mathbf{M}} \left[\sum_{x\in E^*} \psi(\alpha_1(m)x) - |\alpha_3(m)|^{-1}\tau(ln|\alpha_3(m)|^{-1} - T)\hat{\psi}(0)\right] |\alpha_3(m)|\; dm \;.$$

Since $|\alpha_3(m)| = |\alpha_1(m)\alpha_2(m)| = \|\alpha_1(m)\|$, this is equal to

$$m(\mathbf{Z}S'\backslash\mathbf{S}') \int_{E^*\backslash I_E} \left[\sum_{x\in E^*} \psi(ax) - \|a\|^{-1}\tau(ln\|a\|^{-1} - T)\hat{\psi}(0)\right] \|a\|\; d^*a \;.$$

and Lemma 7.1.1 can be applied. Using that $\lambda_{-1}^E m(\mathbf{Z}S'\backslash\mathbf{S}') = m(\mathbf{Z}M\backslash\mathbf{M}^1)$ and $\hat{\psi}(0) = \Phi^M(\gamma, f)$, we obtain 7.3.2(b) and the sum of the terms

(i) $T\; m(\mathbf{Z}M\backslash\mathbf{M}^1)\; \Phi^M(\gamma, f)$

(ii) $m(\mathbf{Z}M\backslash\mathbf{M}^1) \sum_{v\in S_0} \left(\Pi_{w\neq v}\hat{\psi}_v(0)\right) \left(\int_{E_v} \psi_v(x)\; ln\|x\|_v\; dx\right) \;.$

If $n = u(x)n(w)$, then $\beta(\gamma, n) = x\bar{x}$ by (6.3.1) since γ is central. Hence

$$\int_{E_v} \psi_v(x)\ ln\|x\|_v\ dx = \frac{1}{2}\int_{N_v} f_v^K(\gamma n)\ ln\|\beta(\gamma,n)\|_v\ dn\ .$$

and (ii) is equal to $\frac{1}{2}m(\mathbf{Z}M\backslash\mathbf{M}^1)\mathcal{J}_M(\gamma, f)$. By Proposition 6.3.1, the sum of (i) and (ii) is $\frac{1}{2}m(\mathbf{Z}M\backslash\mathbf{M}^1)J_M^T(\gamma, f)$. This completes the proof of Proposition 7.3.2.

7.4. Singular non-central elements, twisted case. Let $G = U(3)$ and let $\widetilde{G} = \mathrm{Res}_{E/F}(G)$.

PROPOSITION 7.4.1: *Let $\gamma \in M$ be singular but non-central and suppose that $N(\delta) = \gamma$, where $\delta \in \widetilde{M}$. Let $\mathcal{O}_{st} = \mathcal{O}_{st}(\delta)$. Then $J_{\widetilde{G}}^T(\mathcal{O}_{st}, f)$ is equal to the sum of the terms:*

(a) $\frac{1}{2}m(\mathbf{Z}G_\gamma\backslash\mathbf{G}_\gamma)\sum_{\kappa\in\mathfrak{K}(G_\gamma/F)}\Phi_\varepsilon^\kappa(\delta, f)$

(b) $C\Phi_\varepsilon^M(\delta,\phi)$

(c) $\frac{1}{2}m(\mathbf{Z}S\backslash\mathbf{S})\ \int_{\widetilde{\mathbf{Z}\mathbf{M}}\backslash\widetilde{\mathbf{M}}}$

$$\left(\int_{I_F}\int_{\mathbf{N}_\gamma\backslash\widetilde{\mathbf{N}}}\phi^{\widetilde{K}}(m^{-1}n^{-1}\delta n(t\delta_0)\varepsilon(nm))\omega_{E/F}(t)|t|d^*t du\right)\|\alpha_3(m)\|^{-2}dm$$

(d) $\frac{1}{2}m(\mathbf{Z}M\backslash\mathbf{M}^1)J_{\widetilde{M}}^T(\delta,\phi)$

Proof: We may assume, modulo $\widetilde{Z}$, that $\delta = d(\sigma, 1, \sigma)$ where $\sigma \in E^*$. Then $\widetilde{G}_{\delta\varepsilon} = G_\gamma$ and $N_\gamma = N_{\delta\varepsilon}$. The inclusion of $\mathcal{D}(\gamma/F)$ in $\mathcal{D}_\varepsilon(\delta/F)$ in Proposition 3.11.2(b) is an isomorphism. It follows as in the proof of Proposition 7.2.2 that (7.2.1) is equal to (a).

We perform the integration of (7.2.3) by taking $g = n_1 m_1 n m k$ with $n_1 \in \mathbf{N}_\gamma$, $m_1 \in \mathbf{M}$, $n \in \mathbf{N}_\gamma\backslash\widetilde{\mathbf{N}}$, $m \in \widetilde{\mathbf{Z}\mathbf{M}}\backslash\widetilde{\mathbf{M}}$, and $k \in \widetilde{\mathbf{K}}$. Let

$$\psi(m,w) = \|\alpha_3(m)\|^{-2}\int_{\mathbf{N}_\gamma\backslash\widetilde{\mathbf{N}}}\phi^{\widetilde{K}}(m^{-1}n^{-1}\delta n(w)\varepsilon(nm))dn\ .$$

for $m \in \widetilde{\mathbf{M}}$ and $w \in \mathbf{E}^0$, and let $\hat{\psi}(m,0)$ be the Fourier transform of the function $t \to \psi(m, t\delta_0)$ at 0. Then (7.2.3) is equal to $m(\mathbf{Z}S\backslash\mathbf{S})$ times the integral of

$$\int_{MZ\backslash \mathbf{M}} \left[\sum_{t\in F^*} \psi(m,\alpha_3(m_1)t\delta_0) - |\alpha_3(m_1)|^{-1}\tau(H(mm_1)-T)\hat{\psi}(m,0)\right] |\alpha_3(m_1)| dm_1$$

over $m \in \widetilde{\mathbf{Z}}\mathbf{M}\backslash\widetilde{\mathbf{M}}$. If $m \in \widetilde{\mathbf{M}}^1$, then $H(mm_1) = H(m_1)$. The inclusion $\widetilde{\mathbf{Z}}\mathbf{M}^1\backslash\widetilde{\mathbf{M}}^1 \to \widetilde{\mathbf{Z}}\mathbf{M}\backslash\widetilde{\mathbf{M}}$ is an isomorphism and we can write (7.2.3) as

$$m(\mathbf{Z}S\backslash\mathbf{S}) \int_{\widetilde{\mathbf{Z}}\mathbf{M}^1\backslash\widetilde{\mathbf{M}}^1} \Psi(m)\, dm$$

where

$$\Psi(m) = \int_{MZ\backslash \mathbf{M}} \left[\sum_{t\in F^*} \psi(m,\alpha_3(m_1)t\delta_0) - |\alpha_3(m_1)|^{-1}\tau(H(m_1)-T)^{-1}\hat{\psi}(m,0)\right] |\alpha_3(m_1)| dm_1$$

As in the proof of Proposition 7.2.2, we write $\Psi(m)$ as the sum over characters χ of $F^*NI_E\backslash I_F$ of:

$$\text{(7.4.1)}\quad \frac{1}{2}\int_{F^*\backslash I_F} \left[\sum_{t\in F^*} \psi(m,at\delta_0) - |a|^{-1}\tau(ln|a|^{-1}-T)\hat{\psi}(m,0)\right] \chi(a)|a|da^*$$

and apply Lemma 7.1.1. If $\chi = \omega_{E/F}$, we obtain 7.4.1(c).

Suppose that χ is trivial. A change of variables shows that the integral of $\hat{\psi}(m,0)$ over $\widetilde{\mathbf{Z}}\mathbf{M}\backslash\widetilde{\mathbf{M}}$ is equal to $\Phi^M_\varepsilon(\delta,\phi)$. By Lemma 7.1.1, (7.4.1) is equal to the sum of $\frac{1}{2}(\lambda_{-1}T+\lambda_0)\hat{\psi}(m,0)$ and

$$\text{(7.4.2)}\qquad \frac{1}{2}\lambda_{-1}\sum_v \Phi_v(m)$$

where

$$\Phi_v(m) = \int_{\mathbf{E}^0}\int_{\mathbf{N}_\gamma\backslash\widetilde{\mathbf{N}}} \phi^{\widetilde{K}}(m^{-1}n^{-1}\delta n(w)\varepsilon(nm))(ln|w|_v - \zeta_v'(1)^{-1}\zeta_v(1))\, dn\, dw\,.$$

Although the sum (7.4.2) defines a function on $\widetilde{\mathbf{Z}\mathbf{M}}^1\backslash\widetilde{\mathbf{M}}^1$, the individual terms $\Psi_v(m)$ are not invariant under $\mathbf{M}^1$. We therefore replace the integral of (7.4.2) by an integral of a sum over $v \in S_0$.

Let $\delta_m = m^{-1}\delta\varepsilon(m)$ where $m \in \widetilde{\mathbf{M}}^1$, and write

$$m^{-1}n^{-1}\delta n(w)\varepsilon(nm) = \delta_m n'$$

where $n' = n'(n,w) = \varepsilon(m)^{-1}\delta^{-1}n^{-1}\delta n(w)\varepsilon(n)\varepsilon(m)$. If

$$n = \begin{pmatrix} 1 & x & z \\ 0 & 1 & y \\ 0 & 0 & 1 \end{pmatrix}, \quad n' = \begin{pmatrix} 1 & x' & z' \\ 0 & 1 & y' \\ 0 & 0 & 1 \end{pmatrix}$$

then,

$$x' = \overline{\alpha}_2(m)^{-1}(\overline{y} - \alpha_1(\delta)^{-1}x)$$
$$y' = \overline{\alpha}_1(m)^{-1}(\overline{x} - \alpha_2(\delta)^{-1}y)$$

$$z' = \overline{\alpha}_3(m)^{-1}\left[\overline{(xy-z)} + (xy-z) - \alpha_1(\delta)^{-1}x\overline{x} + w\right] .$$

and

$$\begin{aligned}\overline{z}' - \alpha_3(\delta_m)z' &= \alpha_3(m)^{-1}[x\overline{x}(\alpha_1(\delta)^{-1} - \overline{\alpha}_1(\delta)^{-1}) - 2w] \\ &= -2\alpha_3(m)^{-1}w + \alpha_3(m)^{-1}\alpha_1(\delta)^{-1}\overline{A}_1(\gamma)x\overline{x} \\ &= -2\alpha_3(m)^{-1}w \\ &\quad + \alpha_2(\delta_m)A_1(\gamma)^{-1}(\overline{\alpha}_2(\delta_m)^{-1}x' + \overline{y}')(\alpha_2(\delta_m)^{-1}\overline{x}' + y')\end{aligned}$$

since $A_1(\gamma)x = \alpha_1(m)(\overline{\alpha}_2(\delta_m)^{-1}x' + \overline{y}')$, as follows from (6.4.1). It follows that

$$\begin{aligned}2A_1(\gamma)\alpha_3(m)^{-1}w &= -A_1(\gamma)(\overline{z}' - \alpha_3(\delta_m)z') \\ &\quad + \alpha_2(\delta_m)(\overline{\alpha}_2(\delta_m)^{-1}x' + \overline{y}')(\alpha_2(\delta_m)^{-1}\overline{x}' + y')\end{aligned}$$

and hence $\beta(\delta_m, n') = 2A_1(\gamma)\alpha_3(m)^{-1}w$ by (6.4.3). By the product formula,

$$\sum_v ln\|2A_1(\gamma)\alpha_3(m)^{-1}\|_v = 0 .$$

since $\|\alpha_3(m)\| = 1$. We can write (7.4.2) as $\lambda_{-1}\sum\Phi'_v(m)$ where

$$\Phi'_v(m) = \int_{\mathbf{E}_0}\int_{\mathbf{N}_\gamma\backslash\widetilde{\mathbf{N}}} \phi^{\widetilde{K}}(\delta_m n')\left[\frac{1}{2}ln\|\beta(\delta_m,n')\|_v - \frac{\zeta'_v(1)}{\zeta_v(1)}\right] dndw.$$

We now show that $\Phi'_v(m) = 0$ unless $v \in S_0$. Suppose that $v \notin S_0$. If $\Phi'_v(m) \neq 0$, then $\gamma \in \Omega$, ϕ_v is the unit in $\widetilde{\mathcal{H}}_v$, and $A_j(\gamma) \in \mathcal{O}_v^*$ for $j = 1, 2$. We may suppose that S_0 contains all places of F of residual characteristic 2. If $\phi^{\widetilde{K}}(\delta_m n') \neq 0$, then $\alpha_j(\delta_m)$ is a unit at v for $j = 1, 2, 3$ and $n'_v \in \widetilde{K}_v$. Observe that $n'_v = n'(n,0)_v n(\overline{\alpha}_3^{-1}(m)w_v)$ and if $n'_v \in \widetilde{K}_v$, then $\overline{\alpha}_3(m)^{-1}w_v$ is integral by (7.4.3). Hence, if $\delta_{mv} \in \widetilde{Z}_v\widetilde{K}_v$, then $n'_v \in \widetilde{K}_v$ if and only if $n'(n,0)_v \in \widetilde{K}_v$ and $\overline{\alpha}_3(m)^{-1}w_v$ is integral. If

$$\int_{E_v^0} \phi_v^{\widetilde{K}}(\delta_{mv}n'_v)\left[\frac{1}{2}ln\|\beta(\delta_{mv}, n'_v)\|_v - \frac{\zeta'_v(1)}{\zeta_v(1)}\right] dw$$

is non-zero, it is equal to

$$\int_{\{w \in E_v^0 : \overline{\alpha}_3^{-1}(m)w \text{ is integral}\}} \left[\frac{1}{2}ln\|\beta(\delta_m, n')\|_v - \frac{\zeta'_v(1)}{\zeta_v(1)}\right] dw.$$

If v remains prime in E, we obtain

$$\|\alpha_3(m)\|_v^{1/2} \int_{\mathcal{O}_v^0} \left(ln|w|_v - \frac{\zeta'_v(1)}{\zeta_v(1)}\right) dw$$

where $\mathcal{O}_v^0 = \{w \in E_v^0 : w \text{ is integral}\}$ and this is zero for all but finitely many v, the set of exceptional v depending only on the choice of dw. We can assume it vanishes if $v \notin S_0$. If v splits in E, identify E_v with $F_v \times F_v$. Let $m_v = d(a^{-1}, 1, b^{-1})$, where $a = (a_1, a_2), b = (b_1, b_2)$. Then $\overline{\alpha}_3(m)^{-1} = \left(\frac{a_2}{b_2}, \frac{a_1}{b_1}\right)$, and the integral is equal to

$$\int_{\{w \in F_v : |\frac{a_1}{b_1}w|_v \leq 1, |\frac{a_2}{b_2}w|_v \leq 1\}} \left[\frac{1}{2}ln\left|\frac{a_1a_2}{b_1b_2}w^2\right|_v - \frac{\zeta'_v(1)}{\zeta_v(1)}\right] dw.$$

However, $\delta_m = d(a\bar{b}\sigma, 1, \bar{a}b\sigma)$ and $\alpha_3(\delta_m) = \left(\frac{a_1b_2}{a_2b_1}, \frac{a_2b_1}{a_1b_2}\right)$. Since $\alpha_3(\delta_m)$ is a unit, $\left|\frac{a_1}{b_1}\right| = \left|\frac{a_2}{b_2}\right|$ and we obtain

$$\int_{\{w \in F_v : |\frac{a_1}{b_1}w|_v \leq 1\}} \left(ln\left|\frac{a_1}{b_1}w\right|_v - \frac{\zeta'_v(1))}{\zeta_v(1)}\right) dw$$

and this is again zero if dw assigns measure 1 to $\mathcal{O}_v$, which we may assume for $v \notin S_0$. This shows that $\Phi'_v(m) = 0$ for $v \notin S_0$.

The integral defining $\Phi'_v(m)$ can be re-written as

$$\int_{\widetilde{\mathbf{N}}} \phi^{\widetilde{K}}(\delta_m n')\left[\frac{1}{2}ln\|\beta(\delta_m, n')\| - \frac{\zeta'_v(1)}{\zeta_v(1)}\right] dn' .$$

The Jacobian of the change of variables $n' = n'(n, w)$ is equal to 1 since $\|\alpha_3(m)\| = 1$. Now (7.4.2) is equal to the sum of the terms

$$\frac{1}{2}\lambda_{-1}\sum_{v\in S_0}\int_{\widetilde{\mathbf{N}}} \phi^{\widetilde{K}}(\delta_m n)\left(\frac{1}{2}ln\|\beta(\delta_m, n)\|\right) dn$$

$$-\frac{1}{2}\lambda_{-1}\left(\sum_{v\in S_0}\frac{\zeta'_v(1)}{\zeta_v(1)}\right)\int_{\widetilde{\mathbf{N}}} \phi^{K}(\delta_m n)dn .$$

Each term is a function on $\widetilde{\mathbf{Z}}\mathbf{M}\backslash\widetilde{\mathbf{M}}$, so we can integrate over this quotient rather than $\mathbf{ZM}'\backslash\widetilde{\mathbf{M}}^1$. Since $\frac{1}{2}\lambda_{-1}m(\mathbf{Z}S\backslash\mathbf{S}) = m(\mathbf{Z}M\backslash\mathbf{M}^1)$, the contribution for χ trivial is equal to $m(\mathbf{Z}M\backslash\mathbf{M}^1)$ times the sum of $T\Phi_\varepsilon^M(\delta,\phi)$, $C\Phi_\varepsilon^M(\delta,\phi)$, and $\frac{1}{2}\mathcal{I}_{\widetilde{M}}(\delta,\phi)$. The sum of these last two terms is 7.4.1(d) by Proposition 6.4.1.

7.5. central elements, twisted case. In this section, we calculate $J^T_{\widetilde{G}}(\mathcal{O}_{\rm st},\phi)$ for $\mathcal{O}_{\rm st}$ containing a central element.

PROPOSITION 7.5.1: *Let $G = U(3)$ and let $\mathcal{O}_{\rm st} = \mathcal{O}_{\rm st}(\delta)$ where $\delta \in \widetilde{Z}$. Then $J^T_{\widetilde{G}}(\mathcal{O}_{\rm st},\phi)$ is equal to the sum of the following terms:*

(a) $m(\mathbf{Z}G\backslash\mathbf{G})\Phi_\varepsilon^{\rm st}(\delta,\phi)$

(b) $D\Phi_\varepsilon^M(\delta,\phi)$

(c) $\displaystyle\frac{1}{2}m(\mathbf{Z}S\backslash\mathbf{S})\int_{\widetilde{\mathbf{Z}}\mathbf{M}\backslash\widetilde{\mathbf{M}}}\int_{\mathbf{N}\backslash\widetilde{\mathbf{N}}}\int_{I_F} \phi^{\widetilde{K}}(m^{-1}n^{-1}\delta n(t\delta_0)\varepsilon(nm))\omega_{E/F}(t)|t|^2 d^*t dn\|\alpha_3(m)\|^{-2}dm$

(d) $\displaystyle\frac{1}{2}m(\mathbf{Z}S\backslash\mathbf{S})\int_{\widetilde{\mathbf{Z}}\mathbf{M}\backslash\widetilde{\mathbf{M}}}\int_{\mathbf{N}\backslash\widetilde{\mathbf{N}}}\int_{I_F} \phi^{\widetilde{K}}(m^{-1}n^{-1}\delta n(t\delta_0)\varepsilon(nm)))|t|^2 d^*t dn\|\alpha_3(m)\|^{-2}dm$

(e) $\displaystyle\frac{1}{2}m(\mathbf{Z}M\backslash\mathbf{M}^1)J^T_{\widetilde{M}}(\delta,\phi)$.

Proof: We have $\widetilde{G}_{\delta\varepsilon} = G$. By Proposition 3.13.2, $\mathcal{D}_\varepsilon(\delta/F) = H^1(F, G_{\mathrm{ad}})$ and by Hasse principle for Hermitian forms, $H^1(F, G_{\mathrm{ad}}) = \Pi H^1(F_v, G_{\mathrm{ad}})$, where the product is over the real places of F. The set of ε-conjugacy classes within $\mathcal{O}_{\varepsilon-\mathrm{st}}(\delta)$ is therefore finite modulo F^* and if $\{\delta'_v\}$ is a collection of ε-classes such that δ'_v is ε-stably conjugate to δ for all v, there exists $\delta' \in \mathcal{O}_{\varepsilon-\mathrm{st}}(\delta)$ such that δ'_v is ε-conjugate to δ' for all v. Since $a(\delta) = 1$, (7.2.1) is equal to

$$m(\mathbf{Z}G\backslash\mathbf{G}) \sum \Phi_\varepsilon(\delta', \phi) = m(\mathbf{Z}G\backslash\mathbf{G}) \Phi_\varepsilon^{\mathrm{st}}(\delta, \phi)$$

where δ' ranges over a set of representatives for the ε-conjugacy classes in $\mathcal{O}_{\varepsilon-\mathrm{st}}(\delta)$ modulo F^*. This is (a).

Observe that $\widetilde{N}_{\delta\varepsilon} = \mathbf{N}$ and $\widetilde{B}_{\delta\varepsilon} = B$. As in the non-twisted case, the integrand in (7.2.3) is a sum of two separately integrable parts. The first part is

$$\int\limits_{\widetilde{\mathbf{Z}}B\backslash\widetilde{\mathbf{G}}} \left[\sum_{n \in F^*} \phi(g^{-1}\delta n(t\delta_0)\varepsilon(g)) \right] dg.$$

We integrate with respect to the decomposition $g = n_1 m_1 n m k$, where $n_1 \in N\backslash\mathbf{N}, m_1 \in M\backslash\mathbf{M}$, $n \in \mathbf{N}\backslash\widetilde{\mathbf{N}}$, $m \in \widetilde{\mathbf{Z}\mathbf{M}}\backslash\widetilde{\mathbf{M}}$, $k \in \widetilde{\mathbf{K}}$, and obtain the integral over $m \in \widetilde{\mathbf{Z}\mathbf{M}}\backslash\widetilde{\mathbf{M}}$ of $\|\alpha_3(m)\|^{-2}$ times:

$$\int\limits_{N\backslash\widetilde{\mathbf{N}}} \int\limits_{\mathbf{Z}M\backslash\mathbf{M}} \left(\sum_{t \in F^*} \phi^{\widetilde{K}}(m^{-1}n^{-1}\delta n(\alpha_3(m_1)t\delta_0)\varepsilon(nmk)) \right) |\alpha_3(m_1)|^{-2} dm_1 dn$$

We obtain terms (c) and (d) by replacing the integral over $\mathbf{Z}M\backslash\mathbf{M}$ by an integral over $F^*\backslash I_F$ and a sum over the characters of $F^*NI_E\backslash I_F$.

The remaining part of $J^T_{\widetilde{G}}(\mathcal{O}_{\mathrm{st}}, \phi)$ is equal to:

$$\int\limits_{\mathbf{Z}B\backslash\widetilde{\mathbf{G}}} \Big(\sum \phi(g^{-1}\delta u(x)n(w)\varepsilon(g))$$

(7.5.1)

$$- \tau(H(g) - T) \int\limits_{\mathbf{N}} \phi(g^{-1}\delta n'\varepsilon(g)) dn' \Big) dg .$$

The sum is over $x \in E^*$ and $w \in E^0$. This time, we integrate with respect to the decomposition $g = u m_1 n m k$, where $u \in N\backslash\mathbf{N}$, $m_1 \in$

$\mathbf{Z}M\backslash\mathbf{M}$, $n \in \mathbf{N}\backslash\widetilde{\mathbf{N}}$, $m \in \mathbf{M}^1\backslash\widetilde{\mathbf{M}}^1$, and $k \in \widetilde{\mathbf{K}}$ and the measure $|\delta_B(m_1)|^{-1}$ $du dm_1 dn dm dk$. Let

$$\psi(x,n,m) = \int_{\mathbf{E}^0} \phi^{\widetilde{K}}(m^{-1}n^{-1}\delta u(x)n(w)\varepsilon(nm))\, dw$$

and let $\hat{\psi}(0,n,m)$ be the Fourier transform at 0 of $x \to \psi(x,n,m)$. As in the proof of Proposition 7.3.1, (7.5.1) is equal to:

$$\int_{\widetilde{\mathbf{Z}\mathbf{M}}^1\backslash\widetilde{\mathbf{M}}^1} \int_{\mathbf{N}\backslash\widetilde{\mathbf{N}}} \int_{\mathbf{Z}M\backslash\mathbf{M}} \Big[\sum_{x\in E^*} \psi(\alpha_1(m_1)x,n,m)$$

$$- |\alpha_3(m_1)|^{-1}\tau(ln(|\alpha_3(m_1)|^{-1}) - T)\hat{\psi}(0,n,m)\Big] |\alpha_3(m_1)| dm_1 dn dm,$$

which we rewrite as

$$m(\mathbf{Z}S'\backslash\mathbf{S}') \int_{\widetilde{\mathbf{Z}\mathbf{M}}^1\backslash\widetilde{\mathbf{M}}^1} \int_{\mathbf{Z}N\backslash\mathbf{N}} \int_{E^*\backslash I_E}$$

$$\Big[\sum_{x\in E^*} \psi(ax,n,m) - \|a\|^{-1}\tau(ln(\|a\|^{-1} - T)\hat{\psi}(0,n,m)\Big] \|a\|\, dn\, dn. \tag{7.5.3}$$

By Lemma 7.1.1, this is equal to the sum of

$$\lambda_{-1}^E m(\mathbf{Z}S'\backslash\mathbf{S}') \int_{\mathbf{Z}\mathbf{M}^1\backslash\widetilde{\mathbf{M}}^1} \int_{\mathbf{N}\backslash\widetilde{\mathbf{N}}} \Big[\sum_v \int_{\mathbf{A}_E} \int_{\mathbf{E}^0} \phi^K(m^{-1}n^{-1}\delta u(x)n(w)\varepsilon(nm))$$

$$[ln\|x\|_v - (\zeta_v^E)'(1)\zeta_v^E(1)^{-1}] dw dx\Big] dndm \tag{7.5.3}$$

and

$$(\lambda_{-1}^E T + \lambda_0^E) m(\mathbf{Z}S'\backslash\mathbf{S}') \int_{\widetilde{\mathbf{Z}\mathbf{M}}^1\backslash\widetilde{\mathbf{M}}^1} \int_{\mathbf{N}\backslash\widetilde{\mathbf{N}}} \hat{\psi}(0,n,m)\, dn$$

The second term is equal to the sum of

$$m(\mathbf{Z}M\backslash\mathbf{M}^1)T\Phi_\varepsilon^M(\delta,\phi) \tag{7.5.4}$$

$$\lambda_0^E m(\mathbf{Z}S'\backslash\mathbf{S}')\Phi_\varepsilon^M(\delta,\phi). \tag{7.5.5}$$

To evaluate (7.5.3), let

$$n = \begin{pmatrix} 1 & 0 & n_3 \\ 0 & 1 & n_2 \\ 0 & 0 & 1 \end{pmatrix}$$

where $n_2 \in \mathbf{A}_E$, $n_3 \in \mathbf{A}_F$. Then $\{n\}$ is a set of representatives for $\mathrm{N}\backslash\widetilde{\mathrm{N}}$. Let

$$m^{-1}n^{-1}\delta u(x)n(w)\varepsilon(nm) = \delta_m n'$$

where

$$n' = \begin{pmatrix} 1 & x' & z' \\ 0 & 1 & y' \\ 0 & 0 & 1 \end{pmatrix}$$

Then

$$x = \frac{1}{2}\alpha_1(m)(\overline{\alpha}_2(\delta_m)^{-1}x' + \overline{y}')$$

$$x\overline{x} = \frac{1}{4}\alpha_3(m)\alpha_2(\delta_m)(\overline{\alpha}_2(\delta_m)^{-1}x' + \overline{y}')(\alpha_2(\delta_m)^{-1}\overline{x}' + y')$$

since $\alpha_2(\delta) = 1$, and $\beta(\delta_m, n') = 4x\overline{x}\alpha_3(m)^{-1}$, where $\beta(\delta_m, n)$ is as in §6.4. By the product formula, we can replace the integrand in (7.5.3) by

$$\phi^{\widetilde{K}}(\delta_m n')[\frac{1}{2}ln\|\beta(\delta_m, n')\| - (\zeta_v^E)'(1)\zeta_v^E(1)^{-1}]$$

since $\|\alpha_3(m)\| = 1$.

Suppose that $v \notin S_0$. Then ϕ_v is the unit in the Hecke algebra. Suppose that $\phi^K(\delta_m n') \neq 0$. Then $n'_v \in \widetilde{K}_v$, and since

$$n^{-1}\delta u(x)n(w)\varepsilon(n) = \begin{pmatrix} 1 & x + \overline{n}_2 & \frac{1}{2}x\overline{x} + w - 2n_3 \\ 0 & 1 & \overline{x} - n_2 \\ 0 & 0 & 1 \end{pmatrix},$$

the elements

$$\overline{\alpha}_2(m_v)^{-1}(x_v + \overline{n}_{2v}), \overline{\alpha}_1(m)^{-1}(\overline{x}_v - n_2), \overline{\alpha}_3(m)^{-1}(\frac{1}{2}x_v\overline{x}_v + w_v - 2n_{3v})$$

are integral. Now $\overline{\alpha}_2(m_v)^{-1}\alpha_1(m_v)$ is a unit since $m_v^{-1}\varepsilon(m_v) \in \widetilde{Z}_v\widetilde{K}_v$, and hence $2\overline{\alpha}_2(m_v)^{-1}x_v$ is integral. Since we assume that S_0 contains the places of F of residual characteristic 2, $\overline{\alpha}_2^{-1}(m_v)^{-1}x_v$ is integral. Furthermore, it follows that $\phi_v^K(\delta_{mv}n'_v) \neq 0$ if and only if $m_v^{-1}\varepsilon(m_v) \in \widetilde{Z}_v\widetilde{K}_v$ and both $\overline{\alpha}_2(m_v)^{-1}x_v$ and $\overline{\alpha}_3(m_v)^{-1}(w_v - 2n_{3v})$ are integral.

We have $\|\overline{\alpha}_2(m_v)^{-1}x_v\| = \|\overline{\alpha}_1(m_v)^{-1}\overline{x}_v\|$, and

$$\frac{1}{2}ln\|\beta(\delta_m,n')\|_v = \frac{1}{2}ln\|\alpha_3^{-1}(m_v)x_v\overline{x}_v\|,$$

and thus

$$\int_{E_v} \phi_v^K(\delta_m n')\left[\frac{1}{2}ln\|\beta(\delta_m,n')\|_v - (\zeta_v^E)1(1)\zeta_v^E(1)^{-1}\right] dx$$

is a multiple of $\int_{\mathcal{O}_{E_v}} [ln\|x\|_v - (\zeta_v^E)'(1)\ \zeta_v^E(1)^{-1}]\,dx$, which we may assume is zero for $v \notin S_0$. The sum in (7.5.3) may therefore be taken over $v \in S_0$. We may also change variables, integrating with respect to n' instead of x, w, n. The Jacobian is equal 1 since $\|\alpha_3(m)\| = 1$. Therefore (7.5.3) is the sum of the terms

$$(7.5.6) \qquad m(\mathbf{Z}M\backslash\mathbf{M}^1)\sum_{v\in S_0}\ \int_{\widetilde{\mathbf{Z}\mathbf{M}^1}\backslash\widetilde{\mathbf{M}^1}}\ \int_{\widetilde{\mathbf{N}}} \phi^{\widetilde{K}}(\delta_m n')\frac{1}{2}ln\|\beta(\delta_m,n')\|dn'dm$$

$$(7.5.7) \qquad -m(\mathbf{Z}M\backslash\mathbf{M}^1)\sum_{v\in S_0}(\zeta_v^E)'(1)\zeta_v^E(1)^{-1}\int_{\widetilde{\mathbf{Z}\mathbf{M}^1}\backslash\widetilde{\mathbf{M}^1}}\ \int_{\widetilde{\mathbf{N}}} \phi^{\widetilde{K}}(\delta_m n')\ dn'dm$$

The integrals over $\widetilde{\mathbf{Z}\mathbf{M}^1}\backslash\widetilde{\mathbf{M}^1}$ can now be taken over $\widetilde{\mathbf{Z}\mathbf{M}}\backslash\widetilde{\mathbf{M}}$. The sum of (7.5.5) and (7.5.7) is equal to (b) and the sum of (7.5.4) and (7.5.6) is equal to (e). This concludes the proof of Proposition 7.5.1.

PROPOSITION 7.5.2: *Let $G = U(2)$ and let $\mathcal{O}_{st} = \mathcal{O}_{st}(\delta)$ where $\delta \in \widetilde{Z}$. Then $J_{\widetilde{G}}^T(\mathcal{O}_{st},\phi)$ is equal to the sum of the terms:*

(a) $m(\mathbf{Z}G\backslash\mathbf{G})\sum\Phi^{\kappa}(\delta,\phi)$

(b) $2C\Phi_{\varepsilon}^M(\delta,\phi)$

(c) $\frac{1}{2}a(\delta)m(\mathbf{Z}M\backslash\mathbf{M}^1)J_{\widetilde{M}}^T(\delta,\phi)$

Proof: In this case, $a(\delta) = 2$ (Lemma 5.5.1) and (7.2.1) is equal to

$$2m(\mathbf{Z}G\backslash\mathbf{G})\sum_{\{\delta\}}\Phi_{\varepsilon}(\delta,\phi)\ .$$

The above sum is equal to $\frac{1}{2}\sum\Phi^{\kappa}(\delta,\phi)$, where κ ranges over the characters of I_F/F^*NI_E, and this gives (a).

To integrate (7.2.3), we use the decomposition $g = dumk$ where $d \in \widetilde{\mathbf{Z}}M''\backslash\mathbf{M}''$, $u \in \mathbf{N}\backslash\widetilde{\mathbf{N}}$, $m \in \widetilde{\mathbf{Z}}\mathbf{M}''\backslash\widetilde{\mathbf{M}}$, and $k \in \widetilde{\mathbf{K}}$. Let $d = d(a,b)$. Then $d \in M'$ if and only if $a/b \in F^*$, and if $a/b \in F^*$, then $d^{-1}\delta n(t)\varepsilon(d) = (\overline{a}b)^{-1}\delta n\left(\frac{b}{a}t\right)$. Since ϕ is invariant under F^*,

$$\phi^K(m^{-1}u^{-1}d^{-1}un(t)\varepsilon(dum)) = \phi^K(m^{-1}u^{-1}\delta n\left(\frac{b}{a}t\right)\varepsilon(um)) .$$

For $m \in \widetilde{\mathbf{M}}$ and $t \in \mathbf{A}_E$, set

$$\psi(m,t) = \int\limits_{\mathbf{N}\backslash\widetilde{\mathbf{N}}} \phi^{\widetilde{K}}(m^{-1}u^{-1}\delta n(t)\varepsilon(um))du$$

and let $\hat{\psi}(m,0)$ be the Fourier transform of $t \to \psi(m,t)$ at 0. The map $\widetilde{\mathbf{Z}}M''^1\backslash\mathbf{M}'' \to F^*\backslash I_F$ sending d to (a/b) is an isomorphism and (7.2.3) is equal to

$$\int\limits_{\widetilde{\mathbf{Z}}\mathbf{M}''^1\backslash\widetilde{\mathbf{M}}^1} \left(\int\limits_{F^*\backslash I_F} \left[\sum_{t\in F^*} \psi(m,at) - \tau(|a|^{-1} - T)|a|^{-1}\hat{\psi}(m,0)\right] |a|\, da \right) dm .$$

We use again that the map $\widetilde{\mathbf{Z}}\mathbf{M}''^1\backslash\widetilde{\mathbf{M}}^1 \to \widetilde{\mathbf{Z}}\mathbf{M}''\backslash\widetilde{\mathbf{M}}$ is an isomorphism. By Lemma 7.1.1, this is equal to the sum of

$$(\lambda_{-1}T + \lambda_0)\Phi_\varepsilon^M(\delta,\phi) \tag{7.5.7}$$

and

$$\lambda_{-1}\int\int\left[\sum_v \int\limits_{\mathbf{A}_F} \phi^{\widetilde{K}}(\delta_m n(t'))\left[ln|t|_v - \frac{\zeta_v'(1)}{\zeta_v(1)}\right] dt\right] du\, dm, \tag{7.5.8}$$

where $\delta_m n(t') = m^{-1}n^{-1}\delta n(t)\varepsilon(um)$. If $u = n(x)$, then

$$t' = \overline{\alpha}(m)^{-1}(t + \overline{x} - x)$$

and, by (6.4.4),

$$\beta(\delta_m, n(t')) = \overline{\alpha}(\delta_m)^{-1}t' + \overline{t}' = 2\alpha(m)^{-1}t.$$

since $\overline{\alpha}(\delta_m)^{-1} = \overline{\alpha}(m)\alpha(m)^{-1}$.

We argue now as in the proof of Proposition 7.5.2. Since $\|\alpha(m)\| = 1$, the product formula allows us to replace the integrand in (7.5.8) by

$$\phi^{\widetilde{K}}(\delta_m n(t')) \left[\frac{1}{2} ln\|\beta(\delta_m, n)\|_v - \frac{\zeta_v'(1)}{\zeta_v(1)}\right] .$$

Again, $\phi_v^K(\delta_{mv} n(t_v'))$ is non-zero if and only if $\alpha(m)^{-1}\overline{\alpha}(m)$ is a unit at v and $\beta(\delta_m, n)$ is integral at v. The integrals over t in the terms in (7.5.8) corresponding to $v \notin S_0$ vanish if S_0 is chosen large enough and by a change of variables, we can write (7.5.7) and (7.5.8) as the sum of the terms

$$\lambda_{-1} T\Phi_\varepsilon^M(\delta, \phi) + \frac{1}{2}\lambda_{-1} \sum_{v \in S_0} \int_{\widetilde{\mathbf{M}}'' \backslash \widetilde{\mathbf{M}}} \int_{\widetilde{\mathbf{N}}} \phi^{\widetilde{K}}(\delta_m n) ln\|\beta(\delta_m, n)\|_v dn\, dm$$

$$\left(\lambda_0 - \lambda_{-1}\left(\sum_{v \in S_0} \zeta_v'(1)\, \zeta_v(1)^{-1}\right)\right) \Phi_\varepsilon^M(\delta, \phi)$$

Now $\lambda_{-1} = 2m(\mathbf{Z}M\backslash\mathbf{M}^1)$ because $\mathbf{Z}M\backslash\mathbf{M}^1$ is isomorphic to the image of $NE^*\backslash NI_E^1$ in $F^*\backslash I_F^1$. The Proposition follows.

7.6 The fine $\mathcal{O}$-expansion. Let $G = U(2)$, $U(2) \times U(1)$, or $U(3)$. By Proposition 6.3.1(a), the $\mathcal{O}$-expansion of $T_G(f)$ can be written as

$$T_G(f) = \sum_{\mathcal{O}_{st}} J_G^T(\mathcal{O}_{st}, f) + m(\mathbf{Z}M\backslash\mathbf{M}^1) \sum_{\{\gamma'\}} \varepsilon(\gamma')^{-1} J_M^T(\gamma', f) \tag{7.6.1}$$

where $\mathcal{O}_{st}$ ranges over a set of representatives modulo Z for the stable elliptic conjugacy classes and $\{\gamma'\}$ is a set of representatives in M for the regular conjugacy classes modulo Z which intersect M. Observe that $\varepsilon(\gamma) = 1$ or 2 and that $\varepsilon(\gamma) = 2$ cannot occur unless $G = U(2)$. Set

$$J_M(f) = \frac{1}{2} m(\mathbf{Z}M\backslash\mathbf{M}^1) \sum_{\gamma \in Z\backslash M} J_M^T(\gamma, f) .$$

The second sum in (7.6.1) is equal to $\frac{1}{2}\sum J_M^T(\gamma, f)$ where γ ranges over the regular elements in $Z\backslash M$.

Let $\mathcal{O}_{st}$ be a singular stable elliptic class containing $\gamma \in M$. By Propositions 7.2.2, 7.3.1, and 7.3.2, $\frac{1}{2} J_M^T(\gamma, f)$ appears in the expression for $J_G^T(\mathcal{O}_{st}, f)$. We now remove it from $J_G^T(\mathcal{O}_{st}, f)$ by defining

$$J_G(\mathcal{O}_{st}, f) = J_G^T(\mathcal{O}_{st}, f) - \frac{1}{2} m(\mathbf{Z}M\backslash\mathbf{M}^1) J_M^T(\gamma, f).$$

and use its appearance in $J_M(f)$ to obtain the fine $\mathcal{O}$-expansion decomposition

$$T_G(f) = J_G(f) + J_M(f)$$

described in §5.2. Here $J_G(f) = \sum J_G(\mathcal{O}_{\mathrm{st}}, f)$. The distribution $J_G(f)$ is supported on the elliptic set and it is not hard to check that it is invariant.

We now do the same for $\widetilde{G} = \mathrm{Res}_{E/F}(G)$, where $G = U(2)$ or $U(3)$. We have

$$T_G(\phi) = \sum_{\mathcal{O}_{\mathrm{st}}} J_G^T(\mathcal{O}_{\mathrm{st}}, \phi) + m(\mathbf{Z}M\backslash\mathbf{M}^1) \sum_{\{\delta'\}} a(\delta')\varepsilon(\delta')^{-1} J_M^T(\delta', \phi)$$

where $\mathcal{O}_{\mathrm{st}}$ ranges over a set of representatives modulo $\widetilde{Z}$ for the stable ε-elliptic conjugacy classes and $\{\delta'\}$ is a set of representatives in $\widetilde{\mathbf{M}}$ for the regular conjugacy classes modulo $\widetilde{Z}$ which intersect $\widetilde{\mathbf{M}}$. The norm map gives a bijection between the ε-conjugacy classes modulo $\widetilde{Z}$ and the conjugacy classes modulo Z which intersect M. Furthermore, if $N(\delta') = \gamma'$, then $\varepsilon(\delta') = \varepsilon(\gamma')$ by Lemma 5.5.1(b). As above, define

$$J_{\widetilde{M}}(\phi) = \frac{1}{2} m(\mathbf{Z}M\backslash\mathbf{M}^1) \sum_{\gamma \in Z\backslash M} a(\delta) J_{\widetilde{M}}^T(\delta, \phi) .$$

where δ denotes an element of $\widetilde{M}$ such that $N(\gamma) = \delta$. If $\mathcal{O}_{\mathrm{st}}$ is a singular ε-stable elliptic class containing δ, set

$$J_{\widetilde{G}}(\mathcal{O}_{\mathrm{st}}, \phi) = J_{\widetilde{G}}^T(\mathcal{O}_{\mathrm{st}}, \phi) - \frac{1}{2} m(\mathbf{Z}M\backslash\mathbf{M}^1) J_{\widetilde{M}}^T(\delta, \phi) .$$

Then the decomposition

$$T_{\widetilde{G}}(\phi) = J_{\widetilde{G}}(\phi) + J_{\widetilde{M}}(\phi)$$

holds.

CHAPTER 8

Germ expansions and limit formulas

The transfer of functions from one group to another is defined by matching orbital integrals over ε-regular semisimple classes. In this chapter, we use Shalika germ expansions in the p-adic case and limit formulas in the real case to obtain relations between orbital integrals over classes which are not ε-regular semisimple. For stable singular elliptic classes $\mathcal{O}_{\mathrm{st}}$, the results of §7 describe $J_G^T(\mathcal{O}_{\mathrm{st}}, f)$ as the sum of invariant part $J_G(\mathcal{O}_{\mathrm{st}}, f)$ and a non-invariant part involving $J_M^T(\gamma, f)$. Our goal is to match up some of the terms making up $J_G(\mathcal{O}_{\mathrm{st}}, f)$ with terms coming from the endoscopic groups of G. If H is an endoscopic group and γ_H is a (G, H)-regular but possibly singular element in H which transfers to a stable class $\mathcal{O}_{\mathrm{st}}$ containing γ in G, then $\Phi^\kappa(\gamma, f)$ appears in a term of $J_G(\mathcal{O}_{\mathrm{st}}, f)$ and this matches with a term involving $\Phi^{\mathrm{st}}(\gamma_H, f^H)$ (cf. Proposition 8.1.3). However, if γ_H is not (G, H)-regular, the matching is more subtle. For example, if $G = U(3)$ and $H = U(2) \times U(1)$, then there exist elements γ_H which are regular in H and such that γ is singular in G. In this case, $\Phi^{\mathrm{st}}(\gamma_H, f^H)$ matches with an invariant term in $J_G(\mathcal{O}_{\mathrm{st}}, f)$ which is supported on a non-semisimple conjugacy class (cf. Proposition 8.2.1(c)).

8.1 Germ expansion. In this section, let F be a p-adic field. The automorphism ε may be trivial or non-trivial. Let γ be an ε-semisimple element in G and let $\{u_j\}$ be a set of representatives for the (ordinary) conjugacy classes of unipotent elements in $G_{\gamma\varepsilon}$. For $f \in C(G, \omega)$, define $\Phi_\varepsilon(u_j\gamma, f)$ relative to the quotient of dg by a fixed Haar measure on $G(u_j\gamma\varepsilon)'$. If $\mathcal{U}$ is an open set in G, let U^r denote the subset of ε-regular elements in $\mathcal{U}$. The next proposition gives Shalika germ expansion in the twisted case.

PROPOSITION 8.1.1: *There exists functions $\Gamma_u^G(\delta, \mu_\delta, \gamma)$ depending on an element $\delta \in G^r$ and a choice μ_δ of Haar measure on $G_{\delta\varepsilon}$ with the following property. For all $f \in C(G, \omega)$, there is an open neighborhood $\mathcal{N}$ of γ such* *that for all* $\delta \in \mathcal{N}^r$,

$$\Phi_\varepsilon(\delta, f) = \sum_j \Phi_\varepsilon(u_j\gamma, f)\Gamma^G_{u_j}(\delta, \mu_\delta, \gamma) .$$

where $\Phi_\varepsilon(\delta, f)$ *is defined relative to the quotient of* dg *by* μ_δ.

Proof: We use an argument due to Howe ([H]) and Gelfand-Kazhdan ([GK]). Assume that the u_j are numbered so that the dimension d_j of $\mathcal{O}(u_j)$ is non-decreasing in j. If γ is contained in the closure of $\mathcal{O}_\varepsilon(\gamma')$, then γ is the ε-semisimple part of γ' and, by the Jordan decomposition in $G \rtimes \langle\varepsilon\rangle$, γ' belongs to $\mathcal{O}_\varepsilon(u_k\gamma)$ for some k. The closure of $\mathcal{O}_\varepsilon(u_k\gamma)$ is contained in the union of the $\mathcal{O}_\varepsilon(u_j\gamma)$ for $j \leq k$. It follows easily that the distributions $f \to \Phi_\varepsilon(u_k\gamma, f)$ are linearly independent and we can choose $f_j \in C(G,\omega)$ such that $\Phi_\varepsilon(u_k\gamma, f_j) = \delta_{jk}$. Set $\Gamma^G_{u_j}(\delta, \mu_\delta, \gamma) = \Phi_\varepsilon(\delta, f_j)$, where $\Phi_\varepsilon(\delta, f)$ is defined relative to μ_δ.

Let $F = f - \sum \Phi_\varepsilon(u_j\gamma, f)f_j$. To prove the proposition, we must show that $\Phi_\varepsilon(\delta, F) = 0$ for all ε-regular δ in an open neighborhood of γ. For any function φ, let $\varphi^g(x) = \varphi(g^{-1}x\varepsilon(g))$ and let C_0 be the span in $C(G,\omega)$ of functions of the form $(\varphi^g - \varphi)$. Then C_0 is in the kernel of all ε-invariant distributions on $C(G,\omega)$.

Let $x \in G$ and let $C(x)$ be the space of smooth functions φ on $Z\mathcal{O}_\varepsilon(x)$ such that $\varphi(zy) = \omega(z)^{-1}\varphi(y)$ for $z \in Z_G$ and $\text{supp}(\varphi)$ is compact modulo Z. Let $C_0(x)$ be the subspace of $C(x)$ spanned by functions of the form $(\varphi^g - \varphi)$ with $\varphi \in C(x)$. Observe that $C_0(x) = C(x)$ if $g^{-1}x\varepsilon(g) = zx$ for some $g \in G$ and $z \in Z'$ such that $\omega(z) \neq 1$. If no such g and z exist, then, by the uniqueness of the invariant measure on $G'_{x\varepsilon}\backslash G$, there is a unique non-zero linear functional T_x on $C(x)$ such that $T_x(\varphi^g) = \varphi$. In this case, the codimension of $C_0(x)$ in $C(x)$ is one (cf. [H], Prop. 2).

If the restriction ψ_x of some $\psi \in C(G,\omega)$ to $Z\mathcal{O}_\varepsilon(x)$ has compact support modulo Z and satisfies $T_x(\psi_x) = 0$, then there exist $\varphi_j \in C(x)$ such that $\psi_x = \sum(\varphi_j^g - \varphi_j)$. Each φ_j extends to a function in $C(G,\omega)$ and therefore ψ, being locally constant, coincides with an element of C_0 in an open neighborhood of $\mathcal{O}_\varepsilon(x)$. Suppose that $F = F_0 + F_1$ where $F_0 \in C_0$ and F_1 vanishes in a neighborhood of $\mathcal{O}_\varepsilon(u_k\gamma)$ for $k < N$. The restriction of F_1 to $\mathcal{O}_\varepsilon(u_N\gamma)$ has compact support and $\Phi_\varepsilon(u_N\gamma, F_1) = 0$. Hence $F_1 = F_{00} + F_{01}$ where $F_{00} \in C_0$ and F_{01} vanishes in an open neighborhood of $\overline{\mathcal{O}_\varepsilon(u_N\gamma)}$. This shows, by induction on N, that F coincides with an element of C_0 in an open invariant neighborhood of γ and the proposition follows.

The functions $\Gamma_u^G(\delta, \mu_\delta, \gamma)$, which are called (Shalika) germs, depend on the choice of measures on $G_{u_j\gamma\varepsilon}$ but they are independent of dg. They are unique as germs of functions defined by δ near γ. They also satisfy the descent and homogeneity properties described in the next proposition. These properties are due to Harish-Chandra ([H$_1$]) in the non-twisted case. Since the proofs were omitted in [H$_1$], we give them here.

Set $H = (G'_{\gamma\varepsilon})^0$. If δ is regular in H and is close to 1, then $\delta\gamma$ is ε-regular and $H_\gamma = G'_{\delta\gamma\varepsilon}$. Let T be a Cartan subgroup of H. The exponential map is defined in a small neighborhood $\mathfrak{T}$ of 0 in Lie(T). Let $\mathfrak{T}'$ be the (open) set of H in $\mathfrak{T}$ such that exp(Y) is regular in H. We can assume that $\mathcal{O}_F\mathfrak{T} \subset \mathfrak{T}$ and that $t\mathfrak{T}' \subset \mathfrak{T}'$ if $t \in \mathcal{O}_F$, $t \neq 0$.

PROPOSITION 8.1.2: *Let $\delta \in H$ be a regular element. Let u be a unipotent element in H and let $\Gamma_u^H(\delta, \mu_\delta, 1)$ be the germ on H associated to ε trivial.*

(a) $\Gamma_u^G(\delta\gamma, \mu_{\gamma\delta}, \gamma) = \Gamma_u^H(\delta, \mu_\delta, 1)$ *if δ is sufficiently close to 1 (in particular, $H_\gamma = G_{\delta\gamma\varepsilon}$) and $\mu_{\gamma\delta} = \mu_\delta$.*

(b) *Let $Y \in \mathfrak{T}'$. For all $t \in \mathcal{O}_F$, $t \neq 0$,*

$$\Gamma_u^G(\exp(t^2Y)\gamma, \mu, \gamma) = |t|^{-d(u)}\Gamma_u^G(\exp(Y)\gamma, \mu, \gamma)$$

where $d(u) = \dim(G_{\gamma\varepsilon}/G_{u\gamma\varepsilon})$ and μ is a Haar measure on T.

Proof: It follows from Lemma 4.12.1 that there exists a neighborhood V of 1 in H such that the set

$$S = \{g \in G : f(g^{-1}\delta\gamma\varepsilon(g)) \neq 0 \quad \text{for some} \quad \delta \in V\}$$

is compact modulo H. Let α be a smooth compactly supported function on G such that

$$\int_H \alpha(hg)dh$$

is equal to 1 for all $g \in S$ and define

$$f_1(\delta) = \int_G \alpha(g)f(g^{-1}\delta\gamma\varepsilon(g))dg.$$

for $\delta \in V$. Then $f_1 \in C(H, \omega)$ and $\Phi_\varepsilon(\delta\gamma, f) = \Phi_H(\delta, f_1)$ for $\delta \in V$ sufficiently close to 1 so that $H_\gamma = G_{\delta\gamma\varepsilon}$, provided that the measure μ_δ on H_δ used to define the two orbital integrals coincide. In particular,

$$\Phi_\varepsilon(\delta\gamma, f) = \sum \Phi_\varepsilon(u_j\gamma, f)\Gamma_{u_j}^H(\delta, \mu_\delta, 1)$$

for δ close to 1 and (a) follows.

We now prove (b). By (a), we can assume that ε is trivial and that $\gamma = 1$. Let U be an ad(G)-invariant open neighborhood of 0 in Lie(G) on which the exponential map is defined. Assume that $\mathcal{O}_F U \subset U$. Set $f_t(\exp(Y)) = f(\exp(t^2 Y))$ for $Y \in U$. Let U^r be the subset of $Y \in U$ such that $\exp(Y)$ is regular. If $|t|$ is small but non-zero, then

$$\Phi(\exp(t^2 Y), f) = \Phi(\exp(Y), f_t)$$

for all $Y \in U'$. Choose f such that $\Phi(u, f) \neq 0$ and set

$$\lambda(u,t) = \Phi(u, f_t)/\Phi(u, f).$$

Replacing u by a conjugate if necessary, we can assume that $u = \exp(X)$ for some $X \in U$. By the Jacobson-Morosow theorem, X forms part of an $s\ell_2$-triple $\{X, \tau, X'\}$ where τ is semisimple and

$$\exp(t\tau)\exp(X)\exp(-t\tau) = \exp(t^2 X).$$

Let $h = \exp(t\tau)$. Then

$$\begin{aligned} \Phi(u, f_t) &= \int_{G_u\backslash G} f(g^{-1}\exp(t^2 X) g)dg \\ &= \int_{G_u\backslash G} f((h^{-1}g^{-1}h)\exp(X)(h^{-1}gh))dg \end{aligned}$$

since dg is right G-invariant. By a change of variables $\lambda(u,t)$ is equal to the determinant of $ad(h)$ acting on Lie$(G)/$Lie(G_u). In particular, $\lambda(u,t)$ is independent of f and by the uniqueness of germs as germs of functions, we have

$$\Gamma_u^G(\exp(t^2 Y), \mu, \gamma) = \lambda(u,t)\Gamma_u^G(\exp(Y), \mu, \gamma)$$

where μ is a measure on the centralizer of $\exp(Y)$. Since G is unimodular,

$$\lambda(u,t)^{-1} = |\det(\mathrm{ad}(\exp(t\tau))|\,\mathrm{Lie}(G_u))|.$$

Let M be the connected subgroup of G such that Lie(M) is spanned by $\{X, \tau, X'\}$ and let Lie$(G) = \oplus W_i$ be a decomposition into irreducible components of the adjoint representation of M on Lie(G). Let X_i be a non-zero vector in W_i such that $\mathrm{ad}(X)X_i = 0$. Then X_i is unique up to scalars and is a highest weight vector for the centralizer T_1 of τ in M. Let ξ be the character of T_1 such that $\mathrm{ad}(t_1)X = \xi(t_1)X$. If $\mathrm{ad}(t_1)X_i = \xi_i(t_1)X_i$, then

$\xi_i(t_1)^2 = \xi(t_1)^{d(i)}$, where $\dim(W_i) = d(i) + 1$. The $\{X_i\}$ form a basis of $\mathrm{Lie}(G_u)$, hence $|\det(\mathrm{ad}(\exp(t\tau))|\,\mathrm{Lie}(G_u))|$ is equal to $|t|^N$ where

$$N = \sum d(i) = \dim(G/G_u)\ .$$

and $\lambda(u,t) = |t|^{-N}$.

Assume that ε is trivial and that γ is elliptic semisimple. Let H be an endoscopic group for G and let $\kappa \in \mathfrak{K}(I/F)$ be the element corresponding to (H, s, η). For simplicity, in the next lemma we assume that G_{der} is simply connected.

PROPOSITION 8.1.3: *Let $\gamma_H \in H$ be an elliptic (G,H)-regular element and suppose that $\gamma_H \to \gamma$. Then*

$$\Delta(\gamma_H, \gamma)\Phi^\kappa(\gamma, f) = \Phi^{\mathrm{st}}(\gamma_H, f^H)$$

if $f \to f^H$.

Proof: The argument is similar to that used in [Kt6], §3. Since γ_H is (G,H)-regular, its connected centralizer is isomorphic to the centralizer I of γ. Let T be an elliptic Cartan subgroup of H containing γ_H. The endoscopic datum (H, s, η) determines an embedding of T in G up to stable conjugacy. We fix one embedding, regard T as a subgroup of G, and identify γ with the image of γ_H. We identify $\mathcal{D}_G(T/F)$ with a set $\{j\}$ of representatives for the G-conjugacy classes of embeddings $j : T \to G$ such that $j(T)$ is $\overline{F}$-conjugate to T. Choose $Y \in \mathrm{Lie}(T)$ close to 0, so that $\exp(tY)$ is defined and is regular for all non-zero $t \in \mathcal{O}_F$, and set $\delta_t = \exp(tY)$. Now consider the germ expansion of

$$\Phi^\kappa(\gamma\delta_t, f) = \sum_{\{j\}} \kappa(\mathrm{inv}(\gamma\delta_t, j(\gamma\delta_t))\Phi(j(\gamma\delta_t), f).$$

where we also denote by κ the element of $\mathfrak{K}(T/F)$ corresponding to (H, s, η). Let $I(j) = G_{j(\gamma)}$. By Proposition 8.1.2(a),

$$\Gamma_1^G(j(\gamma\delta_t), \mu, j(\gamma)) = \Gamma_1^{I(j)}(j(\delta_t), \mu, 1)$$

where μ is a measure on T.

Let $d(j)$ be the formal degree of the Steinberg representation, defined with respect to a measure dx_j on $I(j)$ and dz on $Z_{I(j)}$. The germ $\Gamma_1^{I(j)}(\delta, \mu, 1)$ depends on μ and a choice of measure on the centralizer of 1 in $I(j)$, which is $I(j)$. Assume the measure chosen on $I(j)$ is dx_j and that the measure of $Z_{I(j)}\backslash I(j)_\delta$ with respect to the quotient of μ by dz is 1. Then $\Gamma_1^{I(j)}(\delta,\mu,1) = (-1)^{q(I(j))}d(j)^{-1}$, where $q(I(j))$ is the split rank of $I(j)_{\mathrm{der}}$, by [R₁]. The groups $I(j)$ are inner forms of each other and we can assume that the dx_j are compatible measures. The centers of the $I(j)$ are canonically isomorphic and we take the same measure dz on all of them. Then $d(j)$ is known to be independent of j ([Kt₆], §1). Writing d for the common value of $d(j)^{-1}$, we obtain

$$\text{(8.1.1)} \quad \Phi^\kappa(\gamma\delta_t, f) = d\sum_{\{j\}}(-1)^{q(I(j))}\kappa(\mathrm{inv}(\gamma\delta_t, j(\gamma\delta_t))\Phi(j(\gamma), f) + A(t)$$

where $A(t)$ is a linear combination of functions $|t|^{-a}$ with $a > 0$, by Propositions 8.1.1 and 8.1.2(b). In particular, $A(t)$ is either identically 0 or tends to infinity as t goes to 0. We have

$$\kappa(\mathrm{inv}(\gamma\delta_t, j(\gamma\delta_t))) = \kappa(z_t)$$

where z_t is the image of $\mathrm{inv}(\gamma\delta_t, j(\gamma\delta_t))$ in $A(I)$ under the map $A(T) \to A(I)$, since κ arises from s via the inclusions $Z(\widehat{H}) \subset Z(\hat{I}) \subset \widehat{T}$. Suppose that $j(t) = g^{-1}tg$ for all $t \in T$, for some $g \in G(\overline{F})$. Then z_t is the image of the cocycle $\{\tau(g)g^{-1}\}$ in $A(I)$ and this is equal to $\mathrm{inv}(\gamma, j(\gamma))$, since $j(\gamma) = g^{-1}\gamma g$. The constant term in (8.1.1) is thus equal to

$$\text{(8.1.2)} \quad d\sum_{\{j\}}(-1)^{q(I(j))}\kappa(\mathrm{inv}(\gamma, j(\gamma)))\Phi(j(\gamma), f)$$

Since T is elliptic, $H^1(F, T)$ maps surjectively onto $H^1(F, I)$ by [Kt₄], §10. If $j_1, j_2 \in \{j\}$, then $j_1(\gamma)$ is conjugate to $j_2(\gamma)$ if and only if j, and j_2 have the same image in $H^1(F, I)$. We have canonical identifications $H^1(F, T) = A(T)$ and, since F is p-adic, $H^1(F, I) = A(I)$, and the map $A(T) \to A(I)$ is a homomorphism of groups. Hence, the number of $j_2 \in \{j\}$ such that $j_1(\gamma)$ is conjugate to $j_2(\gamma)$ is equal to $|\ker(H^1(F,T) \to H^1(F, I))|$. It follows from the definition of $\Phi^\kappa(\gamma, f)$ that (8.1.2) is equal to:

$$d(-1)^{q(I')}|\ker(H^1(F,T) \to H^1(F,I))|\Phi^\kappa(\gamma, f)$$

where I' is a quasi-split form of I. Similarly, the constant term in the germ expansion of $\Phi^{\mathrm{st}}(\gamma\delta_t, f^H)$ is equal to

$$d(-1)^{q(I')}|\ker(H^1(F,T) \to H^1(F,I))|\Phi^{\mathrm{st}}(\gamma, f^H).$$

The proposition follows.

PROPOSITION 8.1.4: *For all semisimple elements $\gamma \in G$, $\Phi^{\mathrm{st}}(\gamma, f)$ is a stable distribution.*

Proof: If γ is regular, then this is a definition. In the general case, this is Proposition 1 in [Kt6].

8.2 Singular non-central elements

For the rest of this chapter, unless otherwise stated, we let $G = U(3)$ and $H = U(2) \times U(1)$.

PROPOSITION 8.2.1: *Let $\gamma_0 \in M$ and suppose that γ_0 is singular but not central in G. If E/F is a quadratic extension, let κ be the non-trivial element of $\mathfrak{D}(G_{\gamma_0}/F)$ and let κ be trivial otherwise.*

(a) *Then $\Delta_{G/H}(\gamma_0)\Phi^\kappa(\gamma_0, f) = f^H(\gamma_0)$.*

(b) *If F is global, then $\Phi^\kappa(\gamma_0, f) = f^H(\gamma_0)$.*

(c) *Suppose that F is global. Let γ_1 be a regular element in H which is stably conjugate to γ_0 in G. Then 7.2.2(c) is equal to $\frac{1}{4}m(\mathbf{Z}T\backslash\mathbf{T})\Phi_H^{\mathrm{st}}(\gamma_1, f^H)$.*

(d) *If F is local, then $\Phi^{\mathrm{st}}(\gamma_0, f)$ is a stable distribution.*

Proof: Part (b) follows from (a) and the product formula $\Pi\Delta_{G_v/H_v}(\gamma_{0v}) = 1$ for $\gamma_0 \in M$. If E/F is not a quadratic extension, then (a) follows from Lemma 4.13.1 since γ_0 is (G, H)-regular and (d) is automatic. If E/F is a p-adic quadratic extension, then (a) and (d) follow from Propositions 8.1.3 and 8.1.4.

Suppose that $E/F = \mathbf{C}/\mathbf{R}$ and let T be the compact Cartan subgroup $\{\gamma(\theta, \varphi, \psi)\}$ of G contained in H, where

$$\gamma(\theta, \varphi, \psi) = \begin{pmatrix} e^{i\theta}\cos\psi & 0 & e^{i\theta}\sin\psi \\ 0 & e^{i\varphi} & 0 \\ -e^{i\theta}\sin\psi & 0 & e^{i\theta}\cos\psi \end{pmatrix}.$$

The eigenvalues of $\gamma(\theta, \varphi, \psi)$ are $\{e^{i(\theta+\psi)}, e^{i\varphi}, e^{i(\theta-\psi)}\}$. Set

$$\gamma = \gamma(\theta, \varphi, \psi) \quad , \quad \gamma_1 = \gamma(\theta, \varphi, -\psi)$$
$$\gamma_2 = \gamma((\theta + \varphi + \psi)/2,\ \theta - \psi,\ (\theta - \varphi + \psi)/2)\ .$$

The elements $\gamma, \gamma_1, \gamma_2$ form a set of representatives for the conjugacy classes within $\mathcal{O}_{\mathrm{st}}(\gamma)$, and γ is stably conjugate to γ_1 in H_v. The element $\kappa \in \mathfrak{K}(T/F)$ defined by the endoscopic datum (H, s, ξ_H) is trivial on $\mathcal{E}_H(T/F)$ and

$$\Phi^\kappa(\gamma, f) = \Phi(\gamma, f) + \Phi(\gamma_1, f) - \Phi(\gamma_2, f)$$

$$\Phi_H^{\mathrm{st}}(\gamma, f^H) = \Phi_H(\gamma, f^H) + \Phi_H(\gamma_1, f^H) .$$

We can assume that $\gamma_0 = \gamma(\theta, \varphi, 0)$, where $e^{i\theta} \neq e^{i\varphi}$, since γ_0 is non-central. The centralizer of γ_0 is H. As $\psi \to 0, \gamma$ and γ_1 approach γ_0 while γ_2 approaches a singular element γ_0' whose centralizer H' is the compact inner form of H. We have $e(\gamma_0) = 1$, $e(\gamma_0') = -1$ and $\{\gamma_0, \gamma_0'\}$ is a set of representatives for the conjugacy classes within $\mathcal{O}_{\mathrm{st}}(\gamma_0)$. Hence

$$\Phi^\kappa(\gamma_0, f) = \Phi(\gamma_0, f) + \Phi(\gamma_0', f)$$

Let $g(\psi) = 2\sin(\psi)\Phi_H(\gamma, f^H)$. By Harish-Chandra's limit formula ([A$_1$], Lemma 7.1), there is a non-zero constant c such that

$$\lim_{\psi \to 0} \frac{\partial}{\partial \psi} g(\psi) = c f^H(\gamma_0) .$$

We have $D_H(\gamma) = |2\sin(\psi)|$, hence

$$D_H(\gamma)\Phi_H^{\mathrm{st}}(\gamma, f^H) = g(\psi) - g(-\psi)$$

if $0 \leq \psi < \pi$, and

$$\lim_{\psi \to 0+} \frac{\partial}{\partial \psi}(D_H(\gamma)\Phi_H^{\mathrm{st}}(\gamma, f^H)) = 2c f^H(\gamma_0) .$$

Consider the equality.

$$\frac{\partial}{\partial \psi}(\tau(\gamma)D_G(\gamma)\Phi^\kappa(\gamma, f)) = \frac{\partial}{\partial \psi}(D_H(\gamma)\Phi_H^{\mathrm{st}}(\gamma, f^H)) . \tag{8.2.1}$$

There exist functions F and F' are functions on H and H', respectively, such that for γ near γ_0

$$\Phi_H^{\mathrm{st}}(\gamma, F) = \Phi(\gamma, f) + \Phi(\gamma_1, f), \;\; \Phi_{H'}(\gamma_2, F') = \Phi(\gamma_2, f),$$

and hence

$$\Phi^\kappa(\gamma, f) = \Phi_H^{\mathrm{st}}(\gamma, F) - \Phi_{H'}(\gamma_2, F')$$

(cf. the argument of §4.12). Assume that compatible measure on H and H' are used. Then $\Phi^\kappa(\gamma_0, f) = F(\gamma_0) - F'(\gamma_0')$. Furthermore

$$\lim_{\psi \to 0+} \frac{\partial}{\partial \psi}(|2\sin(\psi)| \cdot \Phi_{H'}(\gamma_2, F')) = -2cF'(\gamma_0') .$$

In other words, the constant in the limit formula for H' differs by a sign from that in the limit formula for H. This can be deduced from the comparison of orbital integrals on GL_2 and quaternion algebras (cf. [Ca$_2$], §5). In fact, $\Phi_H^{\mathrm{st}}(\gamma, F)$ and $\Phi_{H'}(\gamma_2, F')$ are stable orbital integrals on H and H', respectively, and can be written as integrals over GL_2 or the multiplicative group of a quaternion algebra.

To compute the limit of the left-hand side of (8.2.1) as $\psi \to 0+$, we apply the limit formulas for H and H'. By definition, $\tau(\gamma)|A_1(\gamma)A_2(\gamma)|$ is equal to:

$$\mu(e^{i\varphi})\mu^{-1}((e^{i(\varphi-\theta-\psi)}-1)(1-e^{i(\varphi-\theta+\psi)}))|(e^{i(\varphi-\theta-\psi)}-1)(1-e^{i(\varphi-\theta+\psi)})|$$

For some integer t, $\mu^{-1}(z) = z|z|^{-1}(z/\overline{z})^t$, and $\tau(\gamma)|A_1(\gamma)A_2(\gamma)|$ is equal to

$$\mu(e^{i\varphi})e^{2i(\varphi-\theta)t}(e^{i(\varphi-\theta-\psi)}-1)(1-e^{i(\varphi-\theta+\psi)})$$

This is smooth near $\psi = 0$ and its derivative with respect to ψ vanishes at zero. We have $D_G(\gamma) = |2\sin(\psi)| \cdot |A_1(\gamma)A_2(\gamma)|$ and it follows that the limit of the left-hand side of (8.2.1) as $\psi \to 0+$ is equal to $\tau(\gamma_0)|A_1(\gamma_0)A_2(\gamma_0)|$ times

$$\lim_{\psi\to 0}\frac{\partial}{\partial\psi}(|2\sin(\psi)| \cdot [\Phi_H^{\mathrm{st}}(\gamma, F) - \Phi_{H'}(\gamma_2, F')])$$
$$= 2c[F(\gamma_0) + F'(\gamma_0')] = 2c\Phi^\kappa(\gamma_0, f)$$

and (a) follows if $E/F = \mathbf{C}/\mathbf{R}$.

This calculation also shows that the derivative of the stable orbital integral

$$D_G(\gamma)\Phi^{\mathrm{st}}(\gamma, f_v) = D_G(\gamma)[\Phi(\gamma, f_v) + \Phi(\gamma_1, f_v) + \Phi(\gamma_2, f_v)]$$

at $\psi = 0$ is equal to a non-zero constant times $\Phi^{\mathrm{st}}(\gamma_0, f)$. This proves (d) for the case $E/F = \mathbf{C}/\mathbf{R}$.

To prove (c), we reduce to a problem on SL_2 which has been dealt with in [LL]. Suppose first that E/F is a quadratic extension. Let T be a Cartan subgroup of type (1) contained in H. By Proposition 3.8.2 there exist elements $\gamma_1, \gamma_2 \in T$ which are regular in H such that $\{\gamma_1, \gamma_2\}$ is a set of representatives for the conjugacy classes within the stable class of γ_0 in G. Relabelling if necessary, we may assume that $G(\gamma_1)$ is isomorphic to H and that $G(\gamma_2)$ is isomorphic to the compact inner form of H. Then γ_1 is conjugate to γ_0 in G, but not in H.

Let $\gamma \in T^r$ and let $X = \{\gamma, \gamma', \gamma'', \gamma'''\}$ be a set of representatives for the conjugacy classes within $\mathcal{O}_{\mathrm{st}}(\gamma)$ (if $E/F = \mathbf{C}/\mathbf{R}$, then γ''' does not occur). It follows from the proof of Proposition 8.1.3 that as $\gamma \to \gamma_1$, two of the elements in X approach a G-conjugate of γ_1 and two approach a

G-conjugate of γ_2 (except in the case $E/F = \mathbf{C}/\mathbf{R}$). The elements of X may therefore be labelled so that $\gamma'' \to \gamma_1$ and $\gamma', \gamma''' \to \gamma_2$ as $\gamma \to \gamma_1$ and γ' is stably conjugate to γ in H. We may assume that $\gamma'' \in G(\gamma_1)$. Then γ and γ'' are stably conjugate in $G(\gamma_1)$. It now follows that if $\kappa \in \mathfrak{K}(T/F)$ is associated to the endoscopic group H, then

$$\Phi^\kappa(\gamma, f) = \Phi(\gamma, f) + \Phi(\gamma', f) - \Phi(\gamma'', f) - \Phi(\gamma''', f)$$

Again, the term $\Phi(\gamma''', f)$ does not occur if $E/F = \mathbf{C}/\mathbf{R}$. Now $D_{G/H}(\gamma_1) = 0$ since γ_1 is singular in G, and since $G(\gamma_2)$ is compact, we have

$$\lim_{\gamma \to \gamma_1} \tau(\gamma) D_{G/H}(\gamma)[\Phi(\gamma', f) - \Phi(\gamma''', f)] = 0.$$

Taking the limit of the relation $\Delta_{G/H}(\gamma)\Phi^\kappa(\gamma, f) = \Phi^{\mathrm{st}}(\gamma, f^H)$, we obtain

$$\lim_{\gamma \to \gamma_1} \Delta_{G/H}(\gamma)(\Phi(\gamma, f) - \Phi(\gamma'', f)) = \Phi^{\mathrm{st}}(\gamma_1, f^H)\ .$$

If E/F is not a quadratic extension, this formula remains valid, but the term involving γ'' does not occur.

Now suppose that F is global. Let $\{\alpha, \alpha, \beta\}$ be the set of eigenvalues of γ_0 and let $\{x, y, z\}$ be the set of eigenvalues of γ, labelled so that $x, y \to \alpha$ and $z \to \beta$ as $\gamma \to \gamma_1$. Then

$$\Delta_{G/H}(\gamma)_v = c_v(\gamma)\mu_v^{-1}(x^{-1} - y^{-1})\|x - y\|_v^{1/2}$$

where $c_v(\gamma)$ is smooth in a neighborhood of γ_1 and $\Pi c_v(\gamma) = 1$. Let $H' = G(\gamma_1)$, and let F_v be a function on H' such that $\Phi(\gamma, f_v) = \Phi_{H'}(\gamma, F_v)$ for γ near γ_1. Then

$$\lim_{\gamma \to \gamma_1} c_v(\gamma)\mu_v^{-1}(x^{-1} - y^{-1})\|x - y\|_v^{1/2}(\Phi_{H'}(\gamma, F_v) - \Phi_{H'}(\gamma'', F_v)) = \Phi^{\mathrm{st}}(\gamma_1, f_v^H).$$

The derived group of H' is isomorphic to SL_2 and $H^* = \mathrm{GL}_2(F_v)$ acts on H' by conjugation. The elements γ, γ'' are conjugate under the action of $\mathrm{GL}_2(F_v)$ since they are stably conjugate in H and

$$\{\mathrm{Ad}(g) : g \in H'\} = \{\mathrm{Ad}(g) : g \in H^*(F_v), \det(g) \in NE_v^*\}\ .$$

It follows that

$$\begin{aligned} &\mu_v^{-1}(x^{-1} - y^{-1})\|x - y\|_v^{1/2}[\Phi_{H'}(\gamma, F_v) - \Phi_{H'}(\gamma'', F_v)] \\ (8.2.2)\qquad & \\ &= \mu_v^{-1}(x^{-1} - y^{-1})\|x - y\|_v^{1/2} \int_{T'\backslash H_v^*} F(g^{-1}\gamma g)\omega_v(\det(g))\, dg \end{aligned}$$

where $\omega = \omega_{E/F}$ and T' is the centralizer of T in H_v^*, for a suitable choice of measure dg. Let η and η' be non-semisimple elements in H_v^* with semisimple part γ_1 such that η is conjugate to $n(\delta_0)$ in G_v and $\{\eta, \eta'\}$ is a set of representatives for the conjugacy classes in $\mathcal{O}_{st}(\eta)$. The behavior of the right-hand side of (8.2.2) as γ approaches a singular element is dealt with in [LL], §5. It is shown that if the integrals are defined with respect to unnormalized Tamagawa measures, there are constants a_v such that $\Pi a_v = 1$ and the limit of (8.2.2) as $\gamma \to \gamma_1$ is equal to

$$a_v L(1, \omega_v)^{-1} \int_{H^*(\eta)_v \backslash H_v^*} F_v(g^{-1}\eta g)\omega_v(\det(g))\, dg$$
$$= a_v L(1, \omega_v)^{-1}[\Phi_{H'}(\eta, F_v) - \Phi_{H'}(\eta', F_v)].$$

This is equal to

$$\text{(8.2.3)} \qquad a_v L(1, \omega_v)^{-1} \int_{F_v^*} \int_{N(\gamma_0)_v \backslash N_v} f_v^K(u^{-1}\gamma_0 n(t\delta_0)u)\omega_v(t)|t| d^*t\, dn.$$

and $\Phi^{st}(\gamma_1, f^H)$ is equal to the product over all v of the term (8.2.3). Observe that $\mathbf{Z}S\backslash\mathbf{S}$ and $\mathbf{Z}T\backslash\mathbf{T}$ are isomorphic to $E^1\backslash\mathbf{E}^1$ and $(E^1\backslash\mathbf{E}^1)^2$, respectively, since T is of type (1). By Ono's formula for the unnormalized Tamagawa number of a torus (cf. [Kt$_6$]), $m(E^1\backslash\mathbf{E}^1) = 2L(1, \omega)$, hence

$$\frac{1}{2}L(1, \omega)m(\mathbf{Z}S\backslash\mathbf{S}) = \frac{1}{4}m(\mathbf{Z}T\backslash\mathbf{T})$$

and (c) follows.

8.3. Central elements. To state the next proposition, recall that locally and globally we have decompositions $dg = dm dn dk$ and $dh = dm dn_H dk_H$. Assume that the same measure dm is used in both dg and dh. We can write $dn = dx dw$ with respect to the parametrization $n = u(x, w)$, where dx and dw are additive measures on E and E^0, respectively, and $dn_H = dt$ where dt is an additive measure on F. For $w \in E^0$, set $|w|_E = \|w\|^{1/2}$.

PROPOSITION 8.3.1: *Assume that F is p-adic and let γ_0 be a central element in G. Then*

$$\text{(8.3.1)} \qquad c(E/F) \int_{E^0} f^K(\gamma_0 u(0, w))|w|_E dw = \mu(\gamma_0)^{-1} f^H(\gamma_0)\,.$$

where $c(E/F) = N(\delta_{E/F})^{1/2}\dfrac{(1+q^{-1})m(\mathcal{O}_E)}{m(\mathcal{O}_F)m(K_H)}$ *and* $m(\mathcal{O}_E)$, $m(\mathcal{O}_F)$, *and* $m(K_H)$ *are defined with respect to* dx, dt, *and* dk_H, *respectively. If* E *is not a field, then* $N(\delta_{E/F}) = 1$, *and if* E *is a field, then* $N(\delta_{E/F}) = \mathrm{Card}(\mathcal{O}_E/\delta_{E/F})$, *where* $\delta_{E/F}$ *is the relative different of* E/F.

Proof: Since $f^K(\gamma_0) = \omega^{-1}(\gamma_0)f^K(1)$, $\mu(\gamma_0)^{-1}f^H(\gamma_0) = \omega^{-1}(\gamma_0)f^H(1)$, it suffices to consider the case $\gamma_0 = 1$. If E/F is a quadratic extension of p-adic fields, (8.3.1) is essentially contained in [LS$_2$]. Observe that if (8.3.1) holds with one choice of measures dg, dh, then it holds for all choices and it will therefore suffice to check it for the choice of measures used in [LS$_2$]. Let N' (resp., N'_H) be the unipotent radical of the Borel subgroup containing M which is opposite to B (resp., B_H). In [LS$_2$], dg, and dh are chosen by fixing a measure dm on M and setting $dg = dmdndn'$ and $dh = dmdn_Hdn'_H$. Suppose that dn_H and dn'_H are the measures associated to a one-form du on F. Choose $\alpha \in E^*$ so that $\mathcal{O}_E = \mathcal{O}_F \oplus \mathcal{O}_F\alpha$ and write $y = u + u'\alpha$ with $u, u' \in F$. Let dx be the measure on E associated to the form $dyd\bar{y} = (\overline{\alpha} - \alpha)dudu'$ and assume that $dn = dxdw$. With these choices,

$$\int\limits_{N'}\int\limits_{E^0} f(n'^{-1}u(0,w)n')|w|_E dwdn' = f^H(1)\ .$$

by the results of [LS$_2$], Appendix II. Now dk is such that $dmdndk = dmdndn'$ and hence

$$\int\limits_{N'}\int\limits_{E^0} f(n'^{-1}u(0,w)n')|w|_E dwdn' = \int\limits_{E^0} f^K(u(0,w))|w|_E dw\ .$$

It remains to check that

$$N(\delta_{E/F})^{1/2}(1+q^{-1})m(\mathcal{O}_E) = m(K_H)m(\mathcal{O}_F)$$

with these choices. Now $|\overline{\alpha} - \alpha| = N(\delta_{E/F})^{-1/2}$ and the measure dx is such that $m(\mathcal{O}_E) = |\overline{\alpha} - \alpha|m(\mathcal{O}_F)^2$. It will therefore suffice to verify that $m(K_H) = (1+q^{-1})m(\mathcal{O}_F)$. Let I be the Iwahori subgroup of matrices in K_H whose reduction modulo π is upper-triangular (where π is a prime element in $\mathcal{O}_F$). If $mnk \in I$, where $m \in M$, $n \in N_H$, and $k \in K_H$ then $mn \in I$. Calculating $m(I)$ with respect to $dmdn_Hdk_H$, we obtain

$$m(I) = m(M \cap K_H)m(N \cap K_H)m(K_H)[K_H : I]^{-1}$$

On the other hand, if $n'mn \in I$, where $n' \in N'_H$, $m \in M$, $n \in N_H$, then $mn \in I$ and $n' \in N' \cap I$, and the calculation of $m(I)$ with respect to $dmdn'_H dn_H$ yields

$$m(I) = m(M \cap K_H)m(N \cap K_H)m(N'_H \cap I) .$$

It follows that $m(K_H) = [K_H : I]m(N' \cap I) = (q+1)q^{-1}m(\mathcal{O}_F)$ as required.

If $E = F \oplus F$, then G is isomorphic to $\mathrm{GL}_3(F)$ and the left hand side of (8.3.1) defines an invariant distribution with support on the subregular unipotent class, so it is clear from the definition of f^H that (8.3.1) holds for some constant $c(E/F)$. Let f be the unit in the Hecke algebra for G, i.e., f is the characteristic function of K divided by the measure $m_G(K)$ of K with respect to dg. Then f^H is the unit in the Hecke algebra for H and, since $f^K = m(K)f$, (8.3.1) gives

$$c(E/F)\frac{m(K)}{m_G(K)}\frac{1}{1-q^{-2}}(1-q^{-1})m(\mathcal{O}_F) = \frac{1}{m_H(K_H)}$$

where $m(K)$ (resp., $m_H(K_H)$) is the measure of K (resp., K_H) with respect to dk (resp. dh). Now

$$\frac{m_G(K)}{m(K)} = \frac{m_H(K_H)}{m(K_H)}\frac{m(N \cap K)}{m(N_H \cap K_H)}, \frac{m(N \cap K)}{m(\mathcal{O}_F)m(N_H \cap K_H)} = \frac{m(\mathcal{O}_E)}{m(\mathcal{O}_F)}$$

and the proposition follows in this case.

PROPOSITION 8.3.2: *Assume that F is archimedean and let γ_0 be a central element in G. Let $dg = dmdndn'$ and $dh = dmdn_H dn'_H$. Suppose that dn_H and dn'_H are defined by the standard additive measure on F. Let $dn = dxdw$, where dx is the standard additive measure on E and define dn' similarly. Then*

$$\int_{E^0} f^K(\gamma_0 u(0,w))|w|_E dw = \mu(\gamma_0)^{-1} f^H(\gamma_0) . \tag{8.3.2}$$

Proof: We may assume that $\gamma_0 = 1$. The case $E/F = \mathbf{C}/\mathbf{R}$ is treated in [L_5]. If $E = F \oplus F$, identify G with $\mathrm{GL}_3(F)$ and H with the Levi factor of the parabolic subgroup P of G of type (2,1). Let N_1 be the unipotent radical of P. Let dk and dk' be the measures on K such that $dg = \delta_M(m)^{-1}dndmdk$ and $dg = \delta_P(h)^{-1}dn_1 dhdk'$, where dn_1 is the standard measure on N_1. Let

$$p(t_1, t_2) = \begin{pmatrix} 1 & 0 & t_1 \\ 0 & 1 & t_2 \\ 0 & 0 & 1 \end{pmatrix} .$$

The left-hand side of (8.3.2) is equal to the orbital integral of f over $u = p(1,0)$, which, using $dg = \delta_P(h)^{-1} dn_1 dh dk'$, can also be written as

$$\int_{H_u \backslash H} f^{K'}(h^{-1}p(1,0)h)\delta_P(h)^{-1} dh$$

where

$$f^{K'}(g) = \int_K f(kgk^{-1})dk' .$$

Let $S = M_u$. Integrating with respect to $dh = dn_H dn'_H dm$ gives

$$\int_{S\backslash M}\int_{N'_H} f^{K'}(m^{-1}n'^{-1}p(1,0)n'm)|\alpha_3(m)\alpha_2(m)|^{-1} dn'dm$$

$$= \int_{S\backslash M}\int_F f^{K'}(p(\alpha_3^{-1}(m), \alpha_2^{-1}(m)t))|\alpha_3(m)\alpha_2(m)|^{-1} dtdm$$

$$= \int_{F^*}\int_F f^{K'}(p(a,t))|a|dtd^*a = \int_{N_1} f^{K'}(n_1)dn_1 .$$

This is equal to $f^H(1)$ by the definition of f^H.

PROPOSITION 8.3.3: *Let $G = U(3)$ and let γ_0 be a central element in G. Then*

(a) 7.3.2(d) *is equal to* $\frac{1}{2}m(\mathbf{Z}H\backslash\mathbf{H})f^H(\gamma_0)$.

(b) *The integral*

$$\int_{I_F} f^K(\gamma_0 n(t\delta_0))\omega_{E/F}(t)\ |t|^2 d^*t$$

defines a stable distribution.

Proof: We use Tamagawa measures, and hence $m(\mathbf{Z}S\backslash S) = m(\mathbf{Z}H\backslash\mathbf{H})$ by the formula of §5.3. Let $L_v(s) = L(s, \omega_v)$ where $\omega = \omega_{E/F}$. Suppose that dn_H and dn'_H are defined by the standard additive measure on $\mathbf{A}_F$. Let $dn = dxdw$, where dx is the standard additive measure on

$\mathbf{A}_E$ and define dn' similarly. The measure $dg = \otimes dg_v$, where $dg_v = L_v(1)dm_v^0 dn_v dn'_v$ and dm_v^0 is a measure on M_v defined by an invariant form on M defined over F, is a Tamagawa measure. Similarly, $dh = \otimes dh_v$, where $dh_v = L_v(1)^2 dm_v^0 dn_{H_v} dn'_{H_v}$ is a Tamagawa measure on H. Let g_v^H denote the transfer of f_v with respect to the measures $L_v(1)^{-1}dg_v$ and $L_v(1)^{-2}dh_v$ on G and H, respectively. Let dk'_v be the measure on K such that $dm_v^0 dn_v dn'_v = dm_v^0 dn_v dk'_v$. It follows from Propositions 8.3.1 and 8.3.2 that

$$c(E_v/F_v) \int_{K_v} \int_{E_v^0} f(k^{-1}\gamma_0 u(0,w)k)|w|_{E_v} dw dk' = g_v^H(\gamma_0)$$

(where $c(E_v/F_v) = 1$ for v archimedean). Now $g_v^H = L_v(1) f_v^H$. Furthermore, $dg = dm_v dn_v dk_v$ where $dm_v = \zeta_v(1)L_v(1)^2 dm_v^0$ is a Tamagawa measure on M_v, and hence $dk' = \zeta_v(1)L_v(1)dk$. We obtain

$$c(E_v/F_v)\zeta_v(1) \int_{E_v^0} f_v^K(\gamma_0 n(0,w))|w|_{E_v} dw = f_v^H(1).$$

For finite v, $c(E_v/F_v) = N(\delta_{E_v/F_v})^{1/2} \frac{(1+q^{-1})m(\mathcal{O}_{E_v})}{m(\mathcal{O}_{F_v})m(K_{H_v})}$, where $m(K_{H_v})$ is computed relative to the measure dk'' on K_H such that $dm_v^0 dn_{H_v} dn'_{H_v} = dm_v^0 dn_{H_v} dk''$. As seen in the proof of Proposition 8.3.1, $m(K_{H_v}) = (1 + q^{-1})m(\mathcal{O}_{F_v})$, and hence $c(E_v/F_v) = N(\delta_{E_v/F_v})^{1/2} m(\mathcal{O}_{E_v}) m(\mathcal{O}_{F_v})^{-2}$. Since $\Pi_{v<\infty} m(\mathcal{O}_{F_v}) = |D_E|^{-1/2}$, where D_F is the discriminant of F, we see that

$$\prod_{v<\infty} c(E_v/F_v) = N_{E/\mathbf{Q}}(\delta_{E/F})^{1/2}|D_E|^{-1/2}|D_F|$$
$$= N_{E/\mathbf{Q}}(\delta_{E/F}\delta_{E/\mathbf{Q}}^{-1}\delta_{F/\mathbf{Q}})^{1/2} = 1.$$

This gives

$$\prod_v \left[\zeta_v(1) \int_{E_v^0} f_v^K(\gamma_0 u(0,w))|w|_{E_v} dw \right] = f^H(1)$$

and the left-hand side is equal to $\int_{I_F} f^K(\gamma_0 n(t\delta_0))|t|^2 d^*t$. This proves (a).

Part (b) amounts to the assertion that the local integral

$$\int_{F_v^*} f(\gamma_0 n(t))|t|^2 \omega_{E/F}(t) d^*t$$

defines a stable distribution when v remains prime in E. This is shown in [L_5].

Remark: In [L_5], the real and p-adic cases are both treated by the same method, based on Igusa theory. In the case $E/F = \mathbf{C}/\mathbf{R}$, the more traditional method of limit formulas can also be used to prove that (8.3.4) is stable. In the notation of the previous section, let $\gamma(t) = \gamma(0,0,t)$. Then $\gamma(t)$ is a regular element in the Cartan subgroup $T = \{\gamma(\theta,\varphi,\psi)\}$. By a result of Barbasch ([Ba]), (8.3.4) is equal to a constant times

$$\lim_{t\to 0+} \frac{d}{dt}(t^3[\Phi(\gamma(t),f) - \Phi(\gamma(-t),f)] \,. \tag{8.3.5}$$

The Fourier transform of $\Phi(\gamma, f)$ for regular γ has been calculated by Sally-Warner ([SW]). The result is a formula for $\Phi(\gamma, f)$ as a sum of two terms. One term is an integral of $\mathrm{Tr}(\pi(f))$ over the space principal series representations and the other term is a sum of $\mathrm{Tr}(\pi(f))$ where π ranges over discrete series and limits of discrete series representations. The Fourier transform of (8.3.5) can be calculated using term by term differentiation of the Sally-Warner formula, and one sees that only the principal series contribution remains. Hence (8.3.5) is equal to a continuous integral of principal series characters, which is stable and furthermore supported on the hyperbolic set. This additional information will be used in the proof of Lemma 14.5.2.

Similarly, by [Ba], the integral appearing in (8.3.2) is a non-zero constant times

$$\lim_{t\to 0+} \frac{d}{dt}(t^3[\Phi(\gamma(t),f) + \Phi(\gamma(-t),f))] \,. \tag{8.3.6}$$

Let $\gamma'(t) = \gamma\left(\frac{t}{2}, -t, \frac{t}{2}\right)$. The formula of Sally-Warner shows that

$$-\lim_{t\to 0} \frac{d}{dt}(t^3\Phi(\gamma'(t),f))$$

is equal to (8.3.6). Now

$$\Phi^\kappa(\gamma(t),f) = \Phi(\gamma(t),f) + \Phi(\gamma(-t),f) - \Phi(\gamma'(t),f)$$

and so $\lim_{t\to 0+} \frac{d}{dt}(t^3\Phi^\kappa(\gamma(t),f))$ is a non-zero multiple of the left-hand side of (8.3.2). It is also equal to $c\lim_{t\to 0+} \frac{d}{dt}(t\Phi^{st}(\gamma(t),f^H))$ which, by the limit formula, is $c' f^H(1)$, where $c, c' \neq 0$. Thus (8.3.2) follows up to the problem of calculating the constant.

8.4. The twisted case. In the first part of this section, let $G = U(3)$ and $\widetilde{G} = \mathrm{Res}_{E/F}(G)$. Let $\gamma_0 \in M$ and let $\delta_0 \in \widetilde{M}$ be an element such that $N(\delta_0) = \gamma_0$ and $\widetilde{G}_{\delta_0\varepsilon} = G_{\gamma_0}$.

PROPOSITION 8.4.1: *Let F be a local field.*

(a) *If γ_0 is central, Then $\Phi_\varepsilon^{\mathrm{st}}(\delta_0, \phi) = f(\gamma_0)$.*

(b) *Suppose that γ_0 is singular but not central in G. Then $\Phi_\varepsilon^{\mathrm{st}}(\delta_0, \phi) = \Phi^{\mathrm{st}}(\gamma_0, f)$. Let κ be the non-trivial element of $\mathfrak{R}(G_{\gamma_0}/F)$ if E/F is a quadratic extension and the trivial element otherwise. Then $\widetilde{\Delta}(\delta_0)\Phi_\varepsilon^\kappa(\delta_0, \phi) = \phi^H(\gamma_0)$.*

Proof: If E/F is not a quadratic extension, then (a) and (b) follow from Proposition 4.13.2.

Now suppose that E/F is a quadratic extension. For (a), it suffices to consider the case $\delta_0 = \gamma_0 = 1$. If v is finite, there is only one ε-conjugacy class in the stable ε-conjugacy class of δ_0 and (a) follows from the construction of the correspondence $\phi \to f$. To treat the case $E/F = \mathbf{C}/\mathbf{R}$, it is convenient to take the Hermitian form Φ to be:

$$\Phi = \begin{pmatrix} 1 & & \\ & 1 & \\ & & -1 \end{pmatrix}.$$

Let

$$\delta_1 = \begin{pmatrix} -1 & & \\ & 1 & \\ & & 1 \end{pmatrix}, \delta_2 = \begin{pmatrix} 1 & & \\ & -1 & \\ & & 1 \end{pmatrix}, \delta_3 = \begin{pmatrix} 1 & & \\ & 1 & \\ & & -1 \end{pmatrix}.$$

Then $\mathrm{N}(\gamma_j) = 1$ for $0 \leq j \leq 3$, where now $N(g) = g\Phi^t\overline{g}^{-1}\Phi$ with Φ as above. The ε-centralizers of δ_0, δ_1, and δ_2 are all isomorphic to $G = U_3(\mathbf{R})$ and the ε-centralizer G' of δ_3 is isomorphic to $U_3(\mathbf{R})$. By §3.13, $\mathcal{O}_{\varepsilon-\mathrm{st}}(\delta_0)$ contains two ε-conjugacy classes. Hence $\{\delta_0, \delta_3\}$ is a set of representatives for conjugacy classes and the elements $\delta_0, \delta_1, \delta_2$ are ε-conjugate. Let T be the diagonal subgroup of G and let $\delta = \delta(\theta_1, \theta_2, \theta_3)$ be the element of T with entries $e^{i\theta_1}, e^{i\theta_2}, e^{i\theta_3}$. The group T is contained in G and G'. Assume that δ is ε-regular. Then $\widetilde{G}_{\delta\varepsilon} = T$. Observe that $H^1(F, T) = (F^*/NE^*)^3$. By the discussion of §3.11, a set of representatives for the ε-conjugacy classes within $\mathcal{O}_{\varepsilon-\mathrm{st}}(\delta)$ is given by $\{\delta_\eta\}$, where $\eta = d(\eta_1, \eta_2, \eta_3)$ and the η_j range through a set of representatives for F^*/NE^*. Hence $\{\delta\delta_0, \delta\delta_1, \delta\delta_2, \delta\delta_3\}$ is a set of representatives for the ε-conjugacy classes within $\mathcal{O}_{\varepsilon-st}(\delta)$ modulo F^*. Let $\gamma = N(\delta) = \delta^2$ and let $\{\gamma, \gamma', \gamma''\}$ be a set of representatives for the conjugacy classes within $\mathcal{O}(\gamma)$. We have

$$\Phi_\varepsilon^{st}(\delta, \phi) = \Phi_\varepsilon(\delta\delta_0, \phi) + \Phi_\varepsilon(\delta\delta_1, \phi) + \Phi_\varepsilon(\delta\delta_2, \phi) + \Phi_3(\delta\delta_3, \phi)$$
$$\Phi^{st}(\gamma, f) = \Phi(\gamma, f) + \Phi(\gamma', f) + \Phi(\gamma'', f)$$

To obtain a relation between $f(\gamma_0)$ and $\Phi_\varepsilon^{st}(\delta_0, \phi)$ we use the Harish-Chandra limit formula. Let $\rho(\gamma) = e^{i(\theta_1 - \theta_3)}$ and set

$$'\Delta(\gamma) = (1 - e^{i(\theta_2 - \theta_1)})(1 - e^{i(\theta_3 - \theta_2)}(1 - e^{i(\theta_3 - \theta_1)}) .$$

According to $[H_2]$, Lemma 17.5, there is a differential operator ω on T such that the function $\omega[\rho(\gamma)'\Delta(\gamma)\Phi_G(\gamma, f)]$ is continuous at γ_0 and its value there is equal to $c_G f(\gamma_0)$, where c_G is a constant given in $[H_2]$, Lemma 27.5. A similar formula holds for G'. The operator ω is skew-symmetric, as is $\rho(\gamma)'\Delta(\gamma)$ and hence the value of $\omega[\rho(\gamma)\rho\gamma)'\Delta(\gamma)\Phi^{st}(\gamma, f)]$ at γ_0 is $3c_G f(\gamma_0)$.

For $0 \leq j \leq 2$, we can choose functions φ_j on G such that $\Phi_\varepsilon(\delta\delta_j, \phi) = \Phi_G(\delta^2, \varphi_j)$ for δ near 1 and $\Phi_\varepsilon(\delta_j, \phi) = \varphi_j(\gamma_0)$. Then $\varphi_j(\gamma_0) = \Phi_\varepsilon(\delta_0, \phi)$ for $j = 0, 1, 2$ since $\delta_0, \delta_1, \delta_2$ are ε-conjugate. Similarly, let φ_3 be a function on G' such that $\Phi_\varepsilon(\delta\delta_3, \phi) = \Phi_{G'}(\delta^2, \varphi_3)$ and $\Phi_\varepsilon(\delta_3, \phi) = \varphi_3(\gamma_0)$. We have

$$\sum_{j=0}^{2} \Phi_G(\gamma, \varphi_j) + \Phi_G(\gamma, \varphi_3) = \Phi^{st}(\gamma, f)\Phi_\varepsilon^{st}(\delta_0, \phi) = \Phi_\varepsilon(\delta_0, \phi) + \Phi_\varepsilon(\delta_3, \phi)$$

and the limit formula gives

$$3c_G\Phi_\varepsilon(\delta_0, \phi) + 3c_{G'}\Phi_\varepsilon(\delta_3, \phi) = 3c_G f(\gamma_0)$$

The proposition will follow if we verify that $c_{G'} = 3c_G$ when the measure on the inner form G' is obtained from the measure on G via inner twisting in the standard way.

In $[H_2]$, §7 a Haar measured dL is defined for any closed subgroup of a connected real reductive group L. With respect to the measures dG and dG', $c_G = (2\pi)^r v(T)\mathrm{Card}(\Omega_G(T))$ and $c_{G'} = (2\pi)^r v(T)\mathrm{Card}(\Omega_{G'}(T))$ where $v(T)$ is the measure of T with respect to dT,

$$r = \frac{1}{2}\dim(G/T) = \frac{1}{2}\dim(G'/T),$$

and $\Omega_G(T)$ (resp. $\Omega_{G'}(T)$) is the Weyl group of T in G (resp. G'), by Lemmas 37.2 and 37.3 of [H$_2$]. Since $3\mathrm{Card}(\Omega_G(T)) = \mathrm{Card}(\Omega_{G'}(T))$, it will suffice to show that dG and dG' correspond via inner twisting. Define a Cartan involution $\theta(X) = -{}^t\overline{X}$. Let $\mathcal{G} = \mathrm{Lie}(G)$ and let $\mathcal{G} = \mathcal{K} + \mathcal{P}$ be the associated Cartan decomposition, where $\mathcal{K}$ and $\mathcal{P}$ are the $+1$ and -1 eigenspaces of $\mathcal{G}$, respectively. Let B denote the Killing form on $\mathcal{G}$. Then dG is the invariant measure associated to a form $dX_1 \wedge dX_2 \ldots \wedge dX_n$ on $\mathcal{G}$ where $X_1, \ldots, X_n$ is a basis of $\mathcal{G}$ which is orthonormal with respect to the positive-definite symmetric form $-B(X, \theta(Y))$.

Let $X_1, \ldots, X_\ell$ and $Y_1, \ldots, Y_m$ be orthonormal bases of $\mathcal{K}$ and $\wp$, respectively. Then dG is associated to $\omega = dX_1 \wedge \cdots \wedge dX_\ell \wedge dY_1, \wedge \cdots \wedge dY_m$. Let $\mathcal{G}' = \mathrm{Lie}(G')$ and identify $\mathcal{G} \otimes \mathbf{C}$ and $\mathcal{G}' \otimes \mathbf{C}$ with $M_3(\mathbf{C})$. We have $\mathcal{G}' = \mathcal{K} + i\wp$ and $X_1, \ldots, X_\ell, iY_1, \ldots, iY_m$ is an orthonormal base of $\mathcal{G}'$ with respect to $-B(X, \theta'(Y))$, where θ' is the Cartan involution on $\mathcal{G}'$ (in this case, θ' is the identity). Now dG' is associated to $\omega' = i^m\omega$ and hence dG and dG' correspond. This proves (a).

Let $\{\delta\delta_j\}$ be as above and suppose that $\delta' = \delta(\theta_1, \theta_2, \theta_1) \in T$ is an ε-singular such that $N(\delta') = \delta^{12}$ is not central. Then $\delta'\delta_0$ and $\delta'\delta_2$ (resp., $\delta'\delta_3$) are ε-conjugate modulo F^* and we identify their ε-centralizers with H (resp., with a compact form H' of H). There exist functions F and F' on H and H', respectively, such that for δ near δ',

$$\Phi_H^{\mathrm{st}}(\gamma, F) = \Phi_\varepsilon(\delta\delta_0, \phi) + \Phi_\varepsilon(\delta\delta_2, \phi)$$
$$\Phi_{H'}^{\mathrm{st}}(\gamma, F) = \Phi_\varepsilon(\delta\delta_1, \phi) + \Phi_\varepsilon(\delta\delta_3, \phi) \ .$$

where $\gamma = N(\delta)$, and $F(\gamma') = \Phi_\varepsilon(\delta'\delta_0, \phi)$, $F'(\gamma') = \Phi_\varepsilon(\delta'\delta_1, \phi)$, where $\gamma' = N(\delta')$. We then have

$$\Phi_\varepsilon^{\mathrm{st}}(\delta, \phi) = \Phi_H^{\mathrm{st}}(\gamma, F) + \Phi_{H'}^{\mathrm{st}}(\gamma, F') = \Phi^{\mathrm{st}}(\gamma, f)$$
$$\Phi_\varepsilon^{\mathrm{st}}(\delta, \phi) = F(\gamma') - F'(\gamma'), \Phi_\varepsilon^{\kappa}(\delta', \phi) = F(\gamma') + F'(\gamma')$$
$$\widetilde{\Delta}(\delta)\Phi_\varepsilon^{\kappa}(\delta, \phi) = \widetilde{\Delta}(\delta)[\Phi_H^{\mathrm{st}}(\gamma, F) - \Phi_{H'}^{\mathrm{st}}(\gamma, F')] = \Phi^{\mathrm{st}}(\gamma, \phi^H) \ .$$

The limit formula, as applied in the proof of Proposition 8.2.1, gives

$$F(\gamma') = F'(\gamma) = \Phi^{\mathrm{st}}(\gamma', f)$$

and

$$\widetilde{\Delta}(\delta')F(\gamma') + F'(\gamma') = \phi^H(\gamma') \ .$$

This proves (b) in the case $E/F = \mathbf{C}/\mathbf{R}$. The p-adic cases are proved similarly using the germ expansions.

PROPOSITION 8.4.2: *Let F be a global field.*

(a) *Suppose that γ_0 is singular but not central. Let γ_1 be a regular element in H which is stably conjugate to γ_0 in G. Then* 7.4.1(c) *is equal to* $\frac{1}{4}m(\mathbf{Z}T\backslash\mathbf{T})\Phi^{\mathrm{st}}_H(\gamma_1,\phi^H)$.

(b) *Suppose that γ_0 is central. Then* 7.5.1(c) *is equal to* 7.3.2(c) *and* 7.5.1(d) *is equal to* $\frac{1}{2}m(\mathbf{Z}H\backslash\mathbf{H})\phi^H(\gamma_0)$.

Proof: If γ_0 is singular but not central, then $\widetilde{G}_{\delta_0\varepsilon} = H$. Part (a) follows from the proof of Proposition 8.2.1(c) by a reduction to the non-twisted case as in the proof of the previous proposition.

To prove (b), we can assume that $\delta_0 = \gamma_0 = 1$. Assume that F is local and let $I(\omega)$ denote the local integral in 7.5.1(c) or (d), where ω is $\omega_{E/F}$ or the trivial character ω_0 of F^*/NE^* in the two cases, respectively. If E/F is a quadratic extension, let $n_j = n(w_j)$ for $j = 1,2$, where $w_j \in E^0$ and $w_1/w_2 \notin NE^*$. If $E = F\oplus F$, let $n_1 = n(w)$ for any $w \in E^0$. In both cases, $\widetilde{G}_{n_j\varepsilon} = G_{n_j}$ and

$$I(\omega) = \sum_j \omega(w_j/\delta_0) \int_{\widetilde{Z}G_{n_j}\backslash\widetilde{G}} \phi(g^{-1}n_j\varepsilon(g))dg$$

where, as in §7.5, $\delta_0 \in E^*$ is a fixed element of trace 0 and the quotient measure is defined with respect to a suitable choice of measure on G_{n_j}. The sum ranges over $j \in \{1,2\}$ or $j \in \{1\}$ according as E is or is not a field. We can re-write $I(\omega)$ as

$$\int_{\widetilde{Z}G\backslash\widetilde{G}} \int_{F^*} \int_K \phi(g^{-1}k^{-1}n(t\delta_0)k\varepsilon(g))\omega(t)|t|^2 d^*t\,dk\,dg. \tag{8.4.1}$$

LEMMA 8.4.3: (i) $I(\omega_{E/F}) = \omega_{E/F}(2)^{-1}|4|^{-1}\int_{F^*} f^K(n(t\delta_0))\omega_{E/F}(t)|t|^2 d^*t$.

(ii) $I(\omega_0) = c(E/F)^{-1}|4|^{-1}\phi^H(1)$, *where $c(E/F)$ is defined by* (8.3.1).

Proof: If E is not a field, we identify $\widetilde{G}$ with $\mathrm{GL}_3(F)\times\mathrm{GL}_3(F)$ and G with $\{(g,\varepsilon(g)) : g \in \mathrm{GL}_3(F)\}$. where $\varepsilon(g) = \Phi\,{}^tg^{-1}\Phi^{-1}$. In this case, $\omega_{E/F}$ is trivial and $I(\omega_{E/F}) = I(\omega_0)$. Assume that $\phi = \varphi\times\varphi'$ and

$$f(g) = \int_{Z\backslash \mathrm{GL}_3(F)} \varphi(gh^{-1})\varphi'(\varepsilon(h))\,dh$$

where Z is the center of $\mathrm{GL}_3(F)$. Taking the outside integral in (8.4.1) over $\{(1, \varepsilon(x)^{-1}) : x \in Z\backslash \mathrm{GL}_3(F)\}$, we obtain

$$\begin{aligned}
&\int_{Z\backslash G}\int_{F^*}\int_{K} \varphi(k^{-1}n(t)kx^{-1})\varphi'(\varepsilon(xk^{-1}n(t)k)\ |t|^2 d^*t\ dk\ dx \\
&= \int_{Z\backslash G}\int_{F^*}\int_{K} \varphi(k^{-1}n(2t)kx^{-1})\varphi'(\varepsilon(x))\ |t|^2 d^*t\ dk\ dx \\
&= |4|^{-1} \int_{Z\backslash G}\int_{F^*}\int_{K} \varphi(k^{-1}n(t)kx^{-1})\varphi'(\varepsilon(x))|t|^2 d^*t\ dk\ dx, \\
&= |4|^{-1} \int_{F^*}\int_{K} f(k^{-1}n(t)k)|t|^2 d^*t\ dk
\end{aligned}$$

where $n(t)$ is the projection of $n(t\delta_0)$ to $\mathrm{GL}_3(F)$ and this is equal to the right-hand side of (i). Part (ii) follows from Proposition 8.3.1 because $\phi^H(1) = f^H(1)$ in this case by §4.13.

Now suppose that E/F is a quadratic extension. As in §4.12, it suffices to treat the case of compactly-supported functions f and functions ϕ such that $\phi(zg) = \phi(g)$ for $z \in F^*$ and $\mathrm{supp}(\phi) \subset \{g^{-1}x\varepsilon(g) : x \in C_1, g \in C_2\}$, where C_1 is a small neighborhood of 1 in G and $C_2 \subset \widetilde{G}$ is compact modulo $\widetilde{Z}G$. Let $\alpha(g)$ be a compactly supported function on $\widetilde{Z}\backslash\widetilde{G}$ such that

$$\int_{Z\backslash G} \alpha(xg)dx = 1$$

for all $g \in C_2$, and for $x \in C_1$ and $h \in G$, set

$$f_1(h^{-1}x^2h) = \int_{\widetilde{Z}\backslash\widetilde{G}} \alpha(g)\phi(g^{-1}h^{-1}xh\varepsilon(g))\ dg\ .$$

Then

$$I(\omega) = \int_{\widetilde{Z}\backslash\widetilde{G}} \int_{F^*} \int_K \alpha(g)\phi(g^{-1}k^{-1}n(t)k\varepsilon(g))\omega(t)\ |t|^2 d^*t\ dk\ dg$$

$$= \int_{K^*} \int_K f_1(k^{-1}n(2t)k)\omega(t)\ |t|^2 d^*t\ dk$$

$$\omega(2)^{-1}|4|^{-1} \int_{F^*} \int_K f_1(k^{-1}n(t)k)\omega(t)\ |t|^2 d^*t\ dk\ dg$$

and similarly $\Phi_\varepsilon(\delta, \phi) = \Phi(\delta^2, f_1)$ for $\delta \in C_1$.

Suppose that $\delta \in C_1$ lies in a Cartan subgroup T of G. Let $\{g\}$ be a set of elements in $G(\overline{F})$ such that the cocycles $\{\tau(g)g^{-1}\}$ represent the elements of $\mathcal{D}(T/F)$. Then $\{g^{-1}\delta g\}$ is a set of representatives, modulo F^*, for the ε-conjugacy classes within the stable ε-class of δ which pass through the support of ϕ. For in the p-adic case, $\mathcal{D}(T/F) = \mathcal{E}(T/F)$ and if $E/F = \mathbf{C}/\mathbf{R}$, the ε-conjugacy classes in $\mathcal{O}_{\varepsilon-\mathrm{st}}(\delta)$ corresponding to elements of $\mathcal{E}(T/F) - \mathcal{D}(T/F)$ do not pass near 1 (cf. Lemma 4.12.2). It follows that:

$$\Phi_\varepsilon^{\mathrm{st}}(\delta, \phi) = \sum \Phi_\varepsilon(g^{-1}\delta g, \phi) = \Phi^{\mathrm{st}}(\delta^2, f_1)$$

and hence ϕ and f_1 correspond. If ω is non-trivial, the right-hand side of (i) is a stable distribution by Proposition 8.3.3(b), and (i) holds for all functions f associated to ϕ. We also have $\Phi_\varepsilon^\kappa(\delta, \phi) = \Phi_\varepsilon^\kappa(\delta^2, f_1)$ and $\Delta_{G/H}(\delta^2)\Phi^\kappa(\delta^2, f_1) = \Phi^{\mathrm{st}}(\delta^2, f_1^H)$ if $\delta \in H$ and δ^2 is (G,H)-regular. We can define ϕ^H by $\phi^H(\delta^2) = \mu(\det_0(\delta))^{-1} f_1^H(\delta)$ for $\delta \in H$ near 1, as in §4.12. By Propositions 8.3.1 and 8.3.2, there is a constant $c(E/F)$ such that

$$I(\omega_0) = c(E/F)^{-1}|4|^{-1} f_1^H(1) = c(E/F)^{-1}|4|^{-1}\phi^H(1)\ .$$

This proves (ii).

Part (b) of Proposition 8.4.2 now follows from the product formula and the equalilty $\Pi c(E_v/F_v) = 1$, which was checked in the proof of Proposition 8.3.3.

The next proposition follows from arguments similar to those used above. We omit the proof.

PROPOSITION 8.4.4: *Let $G = U(2)$ and let F be a local field. Let $\delta_0 \in \widetilde{G}$ and suppose that $\gamma_0 = N(\delta_0)$ is central in G. Then $\Phi_\varepsilon^{\mathrm{st}}(\delta_0, \phi) = f_1(\gamma_0)$ and $\Phi_\varepsilon^\kappa(\delta_0, \phi) = \mu(\det(\delta_0)) f_2(\gamma_0)$.*

CHAPTER 9

Singularities

The functions $\mathcal{J}_M(\gamma, f, v)$ defined in §6 are continuous but they are not, in general, smooth. The word singularity will be used to refer to the departure of a continuous function f from smoothness and φ will be called a germ of f if $f - \varphi$ is smooth. Following a lecture of Langlands [M], we explicitly compute the germs of $\mathcal{J}_M(\gamma, f, v)$.

9.1 Unitary group in two variables. Let $G = U(2)$ or $U(2) \times U(1)$ and suppose that F is p-adic. The integral

$$\mathcal{J}(\gamma, f) = 2 \int_N f^K(\gamma n(x)) \; ln(\max\{|A(\gamma)|, \; |x|\}) dn \; .$$

(cf. §6.3) is equal to:

$$2 \int_N f^K(\gamma n(x)) \; ln|x| \; dn + 2 \int_{|x| \leq |A(\gamma|} f^K(\gamma n(x)) \; ln \left| \frac{A(\gamma)}{x} \right| \; dn \; .$$

The first integral is smooth as a function of γ and the second integral is smooth unless γ is near a central element. If γ_0 is central and γ is close to γ_0, then $|A(\gamma)|$ is small and $f^K(\gamma n(x)) = m(K) f(\gamma_0)$ for $|x| \leq |A(\gamma)|$. Here, $m(K)$ denotes the measure of K with respect to the measure dk such that $dg = dmdndk$. The second integral is then equal to:

$$-2|A(\gamma)| \; m(K) \left(\int_{|x|<1} ln(|x|) \; dx \right) f(\gamma_0),$$

and this is the germ of $\mathcal{J}(\gamma, f)$ near γ_0. Set $Q = \dfrac{ln(q)}{q-1}$. Then

$$- \int_{|x|<1} ln|x|_v dx = Q m(\mathcal{O}_F),$$

where $m(\mathcal{O}_F)$ is the measure of $\mathcal{O}_F$ with respect to dx, and the next lemma follows. Let $c_H = 2Qm(\mathcal{O}_F)m(K)$.

LEMMA 9.1.1: *Let $G = U(2)$ or $U(2) \times U(1)$. The germ of $\mathcal{I}(\gamma, f)$ near a central element γ_0 is equal to $c_H|A(\gamma)|f(\gamma_0)$.*

9.2 The unitary group in three variables. Suppose that E/F is a p-adic quadratic extension and let $G = U(3)$. As in §6.3, let

$$A_j = A_j(\gamma) = (1 - \alpha_j(\gamma)^{-1}) .$$

The singularity of $\mathcal{I}(\gamma, f)$ is the same as that of the integral

(9.2.1)

$$\int_N f^K(\gamma u(x,z)) \; ln(\max\{\|A_1A_3\|, \|A_3x\|, \|\lambda^{-1}(A_1z + \alpha_1(\gamma)^{-1}x\overline{x})\|\})dn.$$

We now calculate the germ of $\mathcal{I}(\gamma, f)$ near a central element. Let $w = z - \frac{1}{2}x\overline{x}$ and set:

$$\xi = A_1z + \alpha_1(\gamma)^{-1}x\overline{x} = A_1w + \frac{1}{2}(1 + \alpha_1(\gamma)^{-1})x\overline{x} .$$

Recall that $\|\lambda\| \leq \|y\|$ for all $y \in E$ such that $\mathrm{Tr}_{E/F}(y) = 1$. We have

$$\mathrm{Tr}_{E/F}(A_1^{-1}\xi) = A_3A_1^{-1}A_2^{-1}x\overline{x} ,$$

and hence $\|\lambda^{-1}A_1^{-1}\xi\| \geq \|A_3A_1^{-1}A_2^{-1}x\overline{x}\|$. This shows that $\|\lambda^{-1}\xi\| \geq \|A_3x\| \cdot \|A_1^{-1}x\|$. If $\|x\| \geq \|A_1\|$, this gives

$$\|\lambda^{-1}\xi\| \geq \|A_3x\| \quad \text{and} \quad \|\lambda^{-1}\xi\| \geq \|A_1A_3\|,$$

and the maximum occurring in (9.2.1) is $\|\lambda^{-1}\xi\|$. If $\|\lambda^{-1}\xi\|$ is not the maximum in (9.2.1), then $\|x\| < \|A_1\|$, and the maximum is $\|A_1A_3\|$. In this case

$$\|x\| < \|A_1\| \; , \; \|z + A_1^{-1}\alpha_1(\gamma)^{-1}x\overline{x}\| \leq \|\lambda A_3\|,$$

and $\|z\| \leq \max(\|A_1\|, \|\lambda A_3\|)$, since $\alpha,(\gamma)$ is a unit if γ is near a central element.

Let $\gamma_0 \in Z$. If γ is sufficiently close to γ_0, then the A_j are small, and $f^K(\gamma u(x,z)) = f(\gamma_0)$ for all x, z such that $\|A_1A_3\|$ occurs as the maximum. Hence (9.2.1) is the sum of the integrals:

$$(9.2.2) \qquad \int_N f^K(\gamma u(x,z)) \; ln(\|\lambda^{-1}\xi\|) \; dn$$

$$(9.2.3)\qquad -\left(\iint\limits_{\substack{\|\lambda^{-1}\xi\|<\|A_1A_3\| \\ \|x\|<\|A_1\|}} ln\left(\|\frac{\lambda^{-1}\xi}{A_1A_3}\|\right)\,dn\right) f(\gamma_0)\,.$$

Let

$$t = A_1^{-1}\xi - \lambda A_3 A_1^{-1} A_2^{-1} x\overline{x} = w + \left[\frac{1}{2}(1+\alpha_1(\gamma)^{-1}) - \lambda A_3 A_2^{-1}\right] A_1^{-1} x\overline{x}\,.$$

Then $\mathrm{Tr}_{E/F}(t) = 0$ and $dn = dxdw = dxdt$. Furthermore, if $\|x\| < \|A_1\|$, then $\|\lambda^{-1}\xi\| < \|A_1A_3\|$ if and only if $\|\lambda^{-1}t\| < \|A_3\|$, and (9.2.3) is equal to:

$$-\left(\iint\limits_{\substack{\|\lambda^{-1}t\|<\|A_3\| \\ \|x\|<\|A_1\|}} ln\left(\|\frac{\lambda^{-1}t}{A_3}+\frac{x\overline{x}}{A_1A_2}\|\right) dxdt\right) f(\gamma_0) = |A_1A_2A_3| c_G f(\gamma_0),$$

where

$$c_G = -\iint\limits_{\substack{\|x\|<1 \\ \|\lambda^{-1}t\|<1}} ln(\|\lambda^{-1}t + x\overline{x}\|)\,dxdt.$$

To deal with (9.2.2), we use that the integral

$$\int\limits_N f^K(u(x,z))\;ln(\|x\overline{x}\|)\;dn$$

defines a smooth function of γ and hence can be added to (9.2.2) without affecting its germ. For the same reason, we drop the term $ln\|\lambda^{-1}\|$. Thus we are interested in the germ of:

$$\int\limits_N f^K(\gamma u(x,z))\;ln(\|\xi/x\overline{x}\|)\;dn.$$

Set $w_1 = 2A_1(1+\alpha_1(\gamma)^{-1})^{-1}w$. Then $w_1 + x\overline{x} = 2(1+\alpha_1(\gamma)^{-1})^{-1}\xi$ and $\|\xi\| = \|w_1 + x\overline{x}\|$, since $\|\frac{1}{2}(1+\alpha_1(\gamma)^{-1})\| = 1$ for γ near γ_0. The integral thus equals:

$$(9.2.4)\qquad \int\limits_N f^K(\gamma u(x,z))\;ln(\|(w_1+x\overline{x})/x\overline{x}\|)\;dn\,.$$

The function $f^K(\gamma u(x,z))$ is non-zero only for a bounded set of w and $ln(\|(w_1 + x\overline{x})/x\overline{x}\|) = 0$ if $\|x\overline{x}\| > \|w_1\|$. The conditions

$$\|x\overline{x}\| \leq \|w_1\| \quad \text{and} \quad f^K(\gamma u(x,z)) \neq 0$$

therefore imply that x is small and $f^K(\gamma u(x,z)) = f^K(\gamma u(0,w))$. We obtain:

$$(9.2.5) \qquad \int\limits_{E^0} f^K(\gamma u(0,w)) \left(\int\limits_{E} ln(\|w_1 + x\overline{x})/x\overline{x}\|) \, dx \right) dw,$$

where $dn = dxdw$ with respect to the decomposition $n = u(x)u(0,w)$.

For $x \in E^*$, define:

$$\alpha_+(x) = \int\limits_F ln \left(\|\frac{x+t}{t}\| \right) dt$$

$$\alpha_-(x) = \int\limits_F ln \left(\|\frac{x+t}{t}\| \right) \omega_{E/F}(t) \, dt \, .$$

Observe that there is a constant μ such that the formula $\int_E \varphi(x\overline{x})dx = \mu \int_{NE} \varphi(t)dt$ holds. Taking φ to be the characteristic function of O_F, we see that $\mu = m(\mathcal{O}_E)/m(N(\mathcal{O}_E))$, where $m(\mathcal{O}_E)$ and $m(N(\mathcal{O}_E))$ are calculated with respect to dx and dt, respectively. The inner integral in (9.2.5) is equal to

$$\frac{1}{2}\mu(\alpha_+(w_1) + \alpha_-(w_1)).$$

For $\lambda \in F^*$, a change of variables shows that

$$(9.2.6) \qquad \alpha_+(\lambda x) = |\lambda|\alpha_+(x), \quad \alpha_-(\lambda x) = \omega_{E/F}(\lambda)|\lambda|\alpha_-(x)$$

and $\alpha_+(\lambda + x) = \alpha_+(e)$. In particular, $\alpha_+(x) = \alpha_+ \left(\frac{1}{2}(x - \overline{x}) \right)$. We have

$$\frac{1}{2}(w_1 - \overline{w}_1) = 2wA_3(1 + \alpha_1(\gamma)^{-1}))^{-1}(1 + \alpha_2(\gamma)^{-1})^{-1}$$

and $|(1 + \alpha_1(\gamma)^{-1})(1 + \alpha_1(\gamma)^{-1})| = |4|$ for γ near γ_0, hence

$$\alpha_+(w_1) = |\frac{1}{2}A_3|\alpha_+(w).$$

Let $|\cdot|_E$ denote the absolute value on E defined by $|w|_E = \|w\|^{1/2}$.

LEMMA 9.2.1: *Let $w \in E^0$. Then*

$$\frac{1}{2}\mu|\frac{1}{2}|\alpha_+(w) = N(\delta_{E/F})^{1/2}\left(1+\frac{1}{q}\right) Qm(\mathcal{O}_E)|w|_E,$$

where $\delta_{E/F}$ is the relative different.

Proof: By (9.2.6), $\alpha_+(w) = c|w|_E$ for some constant c. To calculate c, we calculate $\alpha_+(e)$ for specific choices of $e \in E^0$. Since $ln\|\frac{e+t}{t}\| = 0$ if $\|t\| > \|e\|$,

$$\alpha_+(e) = \int\limits_{\|t\|<\|e\|} ln\|\frac{e}{t}\|\, dt + \int\limits_{\|t\|=\|e\|} \ln\|\frac{e+t}{t}\|\, dt$$

which we write as $\alpha_+(e) = A_1 + A_2$. We consider the unramified and ramified cases separately. Let p be the residual characteristic of F and let π be a prime in F.

Case 1: E/F unramified. In this case, there exists a unit e such that $\mathrm{Tr}_{E/F}(e) = 0$ and $c = \alpha_+(e)$. The integral A_1 is then equal to $2Qm(\mathcal{O}_F)$.

If $p \neq 2$, then $A_2 = 0$, for if $\|e+t\| < 1$, where $t \in \mathcal{O}_F^*$, then $\|-e+t\| < 1$ and $\|2e\| < 1$, contradicting that e is a unit. Suppose that $p = 2$. Then there is a unit $u \in \mathcal{O}_E^*$, such that $\mathrm{Tr}(u) = 1$ ([We], pg. 141), and we can take $e = \overline{u} - u = 1 - 2u$. If $\|t\| = 1$, then

$$\|t+e\| = \begin{cases} \|t+1\| & \text{if} \quad \|t+1\| > \|2\| \\ \|2\| & \text{if} \quad \|t+1\| \leq \|2\| \end{cases}$$

We only need to check that $\|t+e\| = \|2\|$ if $\|t+1\| = \|2\|$. But if $t+1 = 2\eta$, where $\eta \in \mathcal{O}_F^*$, then $t + e = 2(\eta - u)$ and $(\eta - u)$ is a unit, since $u \in \mathcal{O}_E^* - \mathcal{O}_F^*$ and E/F is unramified. Let $(2) = (\pi^\alpha)$. The integral of $ln\|e+t\|$ over $\{t \in \mathcal{O}_E^* : |t+1| = |\pi|^j\}$ for $1 \leq j < \alpha$ is $ln\|\pi^j\|m(\pi^j\mathcal{O}_F^*) = -(2j)ln(q)q^{-j}m(\mathcal{O}_F^*)$. Summing over j gives

$$2ln(q)m(\mathcal{O}_F)\left[(\alpha-1)q^{-\alpha} - \frac{1-q^{-\alpha+1}}{q-1}\right].$$

The integral over $\{\|t+1\| \leq \|2\|\}$ is $ln(\|2\|)m(2\mathcal{O}_F) = -2\alpha ln(q)q^{-\alpha}m(\mathcal{O}_F)$. Hence $A_2 = 2ln(q)m(\mathcal{O}F)\left[-q^{-\alpha} - \frac{1-q^{-\alpha+1}}{q-1}\right]$ and

$$c = 2ln(q)m(\mathcal{O}_F)\left[-q^{-\alpha} + \frac{q^{-\alpha+1}}{q-1}\right] = |2|\frac{2ln(q)}{q-1}m(\mathcal{O}_F).$$

In both cases, we obtain $c = 2Q|2|m(\mathcal{O}_F)$. It is easy to check that:

$$\frac{1}{2}\mu = \begin{cases} \frac{1}{2}(1+q^{-1})\frac{m(\mathcal{O}_E)}{m(\mathcal{O}_F)} & \text{if} \quad E/F \text{ is unramified} \\ \frac{m(\mathcal{O}_E)}{m(\mathcal{O}_F)} & \text{if} \quad E/F \text{ is ramified.} \end{cases}$$

The proposition follows in the unramified case.

Case 2: E/F is ramified. Let ω be a prime element of E. We consider two cases. Suppose first that ω can be chosen so that $\mathrm{Tr}_{E/F}(\omega) = 0$ and let $e = \omega$. Then $|e|_E = q^{-1/2}$, hence $A_2 = 0$, and

$$A_1 = \int_{\|t\|<1} ln\left\|\frac{e}{t}\right\| \, dt = \int_{\|t\|<1} ln\|e\| \, dt - \int_{\|t\|<1} ln\|t\| \, dt$$

$$= -ln(q)\, m(\pi\mathcal{O}_F) + \frac{2ln(q)}{q-1} m(\mathcal{O}_F) = \left(1 + \frac{1}{q}\right) Qm(\mathcal{O}_F)$$

and $c = q^{1/2}\left(1+\frac{1}{q}\right) Qm(\mathcal{O}_F)$. Now $N(\delta_{E/F})^{1/2} = |2|^{-1}q^{1/2}$ since $\delta_{E/F} = (\omega - \overline{\omega}) = (2\omega)$, and

$$\left|\frac{1}{2}\right|c = \mathbf{N}(\delta_{E/F})^{1/2}\left(1 + \frac{1}{q}\right) Qm(\mathcal{O}_F).$$

The proposition follows in this case since $\frac{1}{2}\mu = \dfrac{m(\mathcal{O}_E)}{m(\mathcal{O}_F)}$.

Finally, suppose there exists a unit e such that $\mathrm{Tr}_{E/F}(e) = 0$. Then $p = 2$. Write $e = \varepsilon_0 + \omega^r u$ where $\varepsilon_0 \in \mathcal{O}_F$, $u \in \mathcal{O}_E^*$, and r is maximal. Observe that r is odd. For if r is even, then $e = \varepsilon_0 + \pi^m u'$ for some $u' \in \mathcal{O}_E^*$ where $m = r/2$, and $e = \varepsilon_0 + p^m u_1 + p^m(u' - u_1)$ for $u_1 \in \mathcal{O}_F$. Since E/F is ramified, we can choose $u_1 \equiv u' (\mathrm{mod}\ \omega)$ and increase r. So let $r = 2m+1$, and choose ω so that $e = \varepsilon_0 + \pi^m\omega$. Then $2e = e - \overline{e} = \pi^m(\omega - \overline{\omega})$. Now $(\omega - \overline{\omega}) = \omega^n$, where $n = \mathrm{val}_\omega(\delta_{E/F})$, and hence $\alpha = 2m + n$, where $(2) = (\omega^\alpha)$.

We now calculate A_2. Note that for $t \in \mathcal{O}_F^*$,

$$\|t + e\| = \begin{cases} \|t + \varepsilon_0\| & \text{if} \quad \|t+\varepsilon_0\| \geq \|\omega^r\| = |\pi|^{2m+1} \\ \|\omega^r\| & \text{if} \quad \|t+\varepsilon_0\| < \|\omega^r\| = |\pi|^{2m+1} \end{cases}.$$

The integral of $ln\|e+t\|$ over $\{t \in \mathcal{O}_F : |t+\varepsilon_0| = |\pi|^j\}$ for $1 \leq j \leq m$ is $ln\|\pi^j\| m(p^j\mathcal{O}_F^*) = -(2j)ln(q)q^{-j}m(\mathcal{O}_F^*)$. Summing over j gives

$$-2ln(q)m(\mathcal{O}_F)\left(1 - \frac{1}{q}\right)\sum_{1\leq j\leq m} jq^{-j} = 2ln(q)m(\mathcal{O}_F)\left[mq^{-m-1} - \frac{1-q^{-m}}{q-1}\right].$$

The integral of $ln\|e+t\|$ over $\{t \in \mathcal{O}_F^* : |t+\varepsilon_0| < |\pi|^{m+(1/2)}\}$ is

$$ln\|\omega^r\| \; m(\pi^{m+1}\mathcal{O}_F) = -(2m+1)ln(q)q^{-m-1}m(\mathcal{O}_F),$$

hence $A_2 = -ln(q)m(\mathcal{O}_F)\left[q^{-m-1} + 2\frac{1-q^{-m}}{q-1}\right]$. Since $A_1 = \frac{2ln(q)}{q-1}m(\mathcal{O}_F)$, we have $c = q^{-m}\left(1+\frac{1}{q}\right) Qm(\mathcal{O}_F)$. Furthermore, $q^{-m} = N(\delta_{E/F})^{1/2}|2|$ and hence $\frac{1}{2}\mu\left|\frac{1}{2}\right|c = N(\delta_{E/F})^{1/2}(1+\frac{1}{q})Qm(\mathcal{O}_E)$. The completes the proof of the lemma.

The next proposition follows. The lemma shows that (9.2.4) is equal to the sum of the terms (b) and (c) below.

PROPOSITION 9.2.2: *For γ near a central element $\gamma_0 \in G$, the germ of $\mathcal{I}(\gamma, f)$ is equal to the sum of the following terms:*

(a) $c_G|A_1A_2A_3|f(\gamma_0)$

(b) $|A_3| \; N(\delta_{E/F})^{1/2}\left(1+\frac{1}{q}\right) Qm(\mathcal{O}_E) \int_{E_0} f^K(\gamma_0 u(0,w)) \; |w|_E \; dw$

(c) $\frac{1}{2}\mu \int_{E_0} f^K(\gamma_0 u(0,w)) \; \alpha_-(w_1) \; dw.$

Now suppose that $\gamma_0 \in M$ is singular but not central in G. Let $H = U(2) \times U(1)$ and identify H with $G(\gamma_0)$. Then $\|A_1(\gamma)\|$ and $\|A_2(\gamma)\|$ are constant near γ_0. Let $\mathcal{W}_G$ and $\mathcal{W}_H$ be the weight factors for G and H, respectively, as defined in §6.1.

LEMMA 9.2.3: *The function:*

$$|A_1A_2A_3| \int_{H\backslash G}\int_{M\backslash H} f(g^{-1}h^{-1}\gamma hg)(\mathcal{W}_G(hg) - \mathcal{W}_H(h)) \; dh \; dg$$

extends to a smooth function of γ near γ_0.

Proof: The integral can be written as:

(9.2.7)

$$|A_1A_2A_3| \int_{H\backslash G}\int_{N_H}\int_{K_H} f(g^{-1}k^{-1}n^{-1}\gamma nkg) \; (\mathcal{W}_G(nkg) - \mathcal{W}_H(n)) \; dn \; dk \; dg$$

$$= |A_1A_2| \int_{H\backslash G}\int_{N_H}\int_{K_H} f(g^{-1}k^{-1}\gamma nkg)(\mathcal{W}_G(n_1kg) - \mathcal{W}_H(n_1)) \; dn \; dk \; dg$$

where n_1 is defined by $n = \gamma^{-1} n_1^{-1} \gamma n_1$. We have

$$\mathcal{W}_G(n_1 kg) - \mathcal{W}_H(n_1) = -H(n_1 kg) - H(wn_1 kg) - H(wn_1) \tag{9.2.8}$$

and if $wn_1 = m_2 n_2 k(\gamma, n)$, where $m_2 \in M$, $n_2 \in N_H$, and $k(\gamma, n) \in K_H$, then (9.2.8) is equal to $-H(kg) - H(k(\gamma, n)kg)$. This function is bounded independently of γ, n, k for g in a compact set, and the set of g such that $f(g^{-1}k^{-1}\gamma nkg) \neq 0$ for some k and n and some γ near γ_0 is compact modulo H (cf. Lemma 4.12.1). It follows that (9.2.7) defines a smooth function near γ_0, and the lemma is proved.

Let $\alpha(g)$ be a smooth compactly-supported function on G such that $\int_{H\backslash G} \alpha(hg)dh = 1$ for all g such that $f(g^{-1}hg) \neq 0$ for some h near γ_0 (cf. §4.12). Define $\varphi(h) = \int_G \alpha(g) f(g^{-1}hg)dh$. Then $\Phi(\gamma, f) = \Phi_H(\gamma, \varphi)$ for $\gamma \in H$ near γ_0. In particular, $\varphi(\gamma_0) = \Phi(\gamma_0, f)$.

LEMMA 9.2.4: *Let $\gamma_0 \in M$ be a singular non-central element.*

(a) *$\mathfrak{I}(\gamma, f) - |A_1 A_2| \mathfrak{I}_H(\gamma, \varphi)$ is smooth for γ near γ_0.*

(b) *The germ of $\mathfrak{I}(\gamma, f)$ near γ_0 is equal to $c_H |A_1 A_2 A_3| \Phi(\gamma_0, f)$.*

Proof: We can write $\mathfrak{I}(\gamma, f)$ as the sum of the terms

$$ln|A_1 A_2| \Phi^M(\gamma, f),$$

$$|A_1 A_2 A_3| \int_{M\backslash G} f(g^{-1}\gamma g)(\mathcal{W}_G(g) - 2ln|A_3|)dg.$$

The first term is smooth near γ_0 and Lemma 9.2.3 implies that the second term differs from

$$|A_1 A_2 A_3| \int_{H\backslash G} \int_{M\backslash H} f(g^{-1}h^{-1}\gamma hg)(\mathcal{W}_H(h) - 2ln|A_3|)dh\, dg \tag{9.2.9}$$

by a function which is smooth near γ_0. Now (9.2.9) is equal to $|A_1 A_2| \mathfrak{I}_H(\gamma, \varphi)$. This proves (a) and (b) follows from (a) and Lemma 9.1.1.

Lemmas 9.2.3 and 9.2.4(a) are valid in the archimedean as well as the p-adic case. The proofs are the same.

Recall that, in the ordinary (resp., twisted) case, a function f is said to be stably equivalent to zero if $\Phi^{\mathrm{st}}(\gamma, f) = 0 (\Phi^{\mathrm{st}}_\varepsilon(\gamma, f) = 0)$ for all (ε)-regular semisimple γ.

Define:

$$S\mathcal{I}(\gamma, f) = \mathcal{I}(\gamma, f) - \frac{1}{2}\mu(\gamma)^{-1}\mathcal{I}_H(\gamma, f^H).$$

PROPOSITION 9.2.5: *Let $G = U(3)$ and let E/F be a p-adic quadratic extension. Let $\gamma_0 \in M$ be a singular element. Then the germ of $S\mathcal{I}(\gamma, f)$ is equal to*

(a) $\frac{1}{2}c_H|A_1A_2A_3|\Phi^{st}(\gamma_0, f)$ *if γ_0 is not central in G.*

(b) $c_G|A_1A_2A_3|\, f(\gamma_0)+\frac{1}{2}\mu \int_{E^0} f^K(\gamma_0 u(0,w))\alpha_-(w_1)\, dw$ *if γ_0 is central in G.*

If f is stably equivalent to 0, then $S\mathcal{I}(\gamma, f)$ extends to a smooth function on M.

Proof: If γ_0 is not central, the germ of $\mathcal{I}_H(\gamma, f^H)$ is equal to $c_H|A_3|f^H(\gamma_0)$ by Lemma 9.1.1. By Proposition 8.2.1(a), $f^H(\gamma_0) = \Delta_{G/H}(\gamma_0)\Phi^\kappa(\gamma_0, f)$ and $\Delta_{G/H}(\gamma_0) = \mu(\gamma_0)|A_1A_2|$ for $\gamma \in M$ near γ_0. By Lemma 9.2.4(b), we see that the germ of $S\mathcal{I}(\gamma, f)$ is equal to

$$c_H|A_1A_2A_3|[\Phi(\gamma_0, f) - \frac{1}{2}\Phi^\kappa(\gamma_0, f)] = \frac{1}{2}c_H|A_1A_2A_3|\Phi^{st}(\gamma_0, f).$$

Suppose that γ_0 is central. The germ of $\frac{1}{2}\mu(\gamma)^{-1}\mathcal{I}_H(\gamma, f^H)$ is equal to $\frac{1}{2}c_H\mu(\gamma_0)^{-1}|A_3|f^H(\gamma_0)$ and to prove (b), we must check that this is equal to

$$(9.2.10) \qquad |A_3|N(\delta_{E/F})^{1/2}(1+\frac{1}{q})Qm(\mathcal{O}_E)\int_{E^0} f^K(\gamma_0 u(0,w))|w|_E dw$$

by Proposition 9.2.2. However, (9.2.10) is equal to

$$|A_3|m(\mathcal{O}_F)m(K_H)Q\mu(\gamma_0)^{-1}f^H(\gamma_0)$$

by Proposition 8.3.1. Part (b) follows from the definition of c_H (§9.1).

To prove the last statement of Proposition 9.2.5, observe that in case (a), $\Phi^{st}(\gamma_0, f)$ is stable by Proposition 8.1.4. In case (b), $f(\gamma_0)$ is stable by Proposition 8.1.4 and the integral, which is a multiple of (8.3.4), is stable by Proposition 8.3.3(b).

9.3 The real case. In this section, assume E_v/F_v is $\mathbf{C}/\mathbf{R}$. We will write $\mathcal{Z}_G$ for the center of the enveloping algebra of G. Let $\Gamma : \mathcal{Z}_G \to \mathcal{M}$

be the Harish-Chandra homomorphism for G, where $\mathcal{M}$ is the enveloping algebra of the Lie algebra of M.

If $G = U(2)$ or $U(2) \times U(1)$, then

$$\mathcal{I}(\gamma, f) = \int_{\mathbf{R}} f^K(\gamma n(x)) \, ln(|A(\gamma)|^2 + x^2) \, dx$$

Let $\gamma = \gamma(t)$ denote $d(e^t e^{i\theta}, e^{-t} e^{i\theta})$ (resp., $d(e^t e^{i\theta}, e^{i\psi}, e^{-t} e^{i\theta})$) if $G = U(2)$ (resp., $G = U(2) \times U(1)$) and denote differentiation with respect to t by a prime. Then $A(\gamma) = |(1 - e^{-2t})|$ and following [A$_1$], §4, we have

$$\mathcal{I}(\gamma, f)' = \int_{-\infty}^{\infty} f^K(\gamma n(x)) 4e^{-2t}(|A(\gamma)|^2 + x^2)^{-1} dx$$
$$+ \int_{-\infty}^{\infty} \varphi(t, x) \ell n(|A(\gamma)|^2 + x^2) dx \ ,$$

where $\varphi(t, x) = \dfrac{d}{dt}(f^K(\gamma(t)n(x)))$. The second integral is differentiable at $t = 0$ and by a change of variables $x \to A(\gamma)x$ we can re-write the first integral as

$$4sgn(t)e^{-2t} \int_{-\infty}^{\infty} f^K(\gamma n((1 - e^{-2t})x))(1 + x^2)^{-1} dx.$$

As $t \to 0\pm$, this approaches

$$\pm 4 f^K(\gamma(0)) \int_{-\infty}^{\infty} (1 + x^2)^{-1} dx = \pm 4\pi f^K(\gamma(0))$$

by the dominated convergence theorem, and we obtain the jump relation:

$$\left[\lim_{t \to 0+} - \lim_{t \to 0-}\right] \mathcal{I}(\gamma, f)' = 8\pi f(\gamma(0)) \ . \tag{9.3.1}$$

PROPOSITION 9.3.1: *Let $G = U(2)$ or $U(2) \times U(1)$. If f is stably equivalent to 0, then $\mathcal{I}(\gamma, f)$ defines a smooth function on M.*

Proof: The function $\mathcal{I}(\gamma, f)$ is continuous. It will suffice to show that $\mathcal{I}(\gamma, f)$ has continuous derivatives to all order in t. If f is stably equivalent to zero, then $f(\gamma_0) = 0$ for all $\gamma_0 \in Z_G$ by the Harish-Chandra limit formula, and (9.3.1) shows that the $\mathcal{I}(\gamma, f)'$ is continuous.

We temporarily denote the weighted orbital $D_G(\gamma)\int_{M\backslash G} f(g^{-1}\gamma g)\mathcal{W}(g)dg$ by $I(\gamma,f)$. Let $z\in\mathcal{Z}_G$ by the results of Arthur ([A$_1$], §5), for γ regular,

$$I(\gamma,zf)-\Gamma(z)I(\gamma,f)$$

is equal to a derivative of $\Phi^M(\gamma,f)$ (in [A$_1$], only the case of the Casimir operator is considered, but the calculations extend to the general case). In particular, if f is stably equivalent to zero, then $I(\gamma,zf)=\Gamma(z)I(\gamma,f)$. In this case, we also have $\mathcal{I}(\gamma,f)=I(\gamma,f)$, i.e., the "correction term" added in §6.3 is equal to zero, hence $\mathcal{I}(\gamma,zf)=\Gamma(z)\mathcal{I}(\gamma,f)$.

Now if T is any Cartan subgroup and $F_f(\gamma)$ is defined for regular $\gamma\in T$ as in the usual way ([Kp], page 402), then $F_{\zeta f}(\gamma)=\Gamma_T(\zeta)F_f(\gamma)$, where Γ_T is the Harish-Chandra homomorphism relative to T, for all $\zeta\in\mathcal{Z}_G$ ([Kp], Proposition 11.9). It follows that zf is also stably equivalent to 0. Now suppose that $\Gamma(z)=\dfrac{d^2}{dt^2}$. Then $\mathcal{I}(\gamma,f)''=\mathcal{I}(\gamma,zf)$ is continuous and has a continuous first derivative. It follows by induction that $\mathcal{I}(\gamma,f)$ is smooth.

For the rest of this section, let $G=U(3)$ and $H=U(2)\times U(1)$.

PROPOSITION 9.3.2: *Let $G=U(3)$. If f is stably equivalent to 0, then $S\mathcal{I}(\gamma,f)$ defines a smooth function on M.*

Proof: As above, $\mathcal{I}(\gamma,zf)=\Gamma(z)\mathcal{I}(\gamma,f)$ for $z\in\mathcal{Z}_G$ and regular γ when $\Phi^M(\gamma,f)$ vanishes identically. Define $z_H\in\mathcal{Z}_H$ by the relation $\Gamma_H(z_H)=\mu(\gamma)\Gamma(z)\mu(\gamma)^{-1}$, where $\Gamma_H:\mathcal{Z}_H\to\mathcal{M}$ is the Harish-Chandra homomorphism for H. Note that $\mu(\gamma)$ is invariant under the Weyl group of H, hence $\mu(\gamma)\Gamma(z)\mu(\gamma)^{-1}$ is invariant under the Weyl group of H and lies in the image of Γ_H. Now $\Phi_H^M(\gamma,f^H)$ is also identically zero and hence $\Gamma(z)\mu(\gamma)^{-1}\mathcal{I}_H(\gamma,f^H)=\mu(\gamma)^{-1}\mathcal{I}_H(\gamma,z_Hf^H)$. By [S$_3$], Lemma 4.2.1, if $f\to f^H$, then $zf\to z_Hf^H$ and we obtain:

$$S\mathcal{I}(\gamma,zf)=\Gamma(z)S\mathcal{I}(\gamma,f) \tag{9.3.2}$$

where, in the left hand side, $(zf)^H$ is taken to be z_Hf^H.

We now show that for all $X\in\mathcal{M}$, $XS\mathcal{I}(\gamma,f)$ is bounded on $M^r\cap\Omega$, where Ω is any compact subset of M. From the formulas for $\mathcal{V}_G(\gamma,n,v)$ and $\mathcal{V}_H(\gamma,n,v)$, we derive an estimate

$$|XS\mathcal{I}(\gamma,f)|<C(X,f)|A_1A_2A_3|^{-r(X)}$$

for some non-negative integer $r(X)$ and constant $C(X,f)$ depending on X and f. Since $\mathcal{M}$ is finitely-generated as a module over $\Gamma(\mathcal{Z}_G)$, (9.3.2) implies that there exists a non-negative integer r such that the estimate holds with $r(X) = r$ for all $X \in \mathcal{M}$. The result follows from the next lemma.

LEMMA 9.3.3: *Let $\lambda_1 \dots, \lambda_n$ be linear functions on $\mathbf{R}^m$ and let φ be a smooth function on*

$$B'(t) = \{\xi \in \mathbf{R}^m : \|\xi\| \leq t, \lambda_j(\xi) \neq 0 \quad for \quad 1 \leq j \leq n\}.$$

Suppose that there is a non-negative integer r such that for any constant coefficient differential operator X on $\mathbf{R}^m$, there is a constant $C(X)$ such that

$$|X\varphi(\xi)| < C(X)| \prod_{j=1}^{n} \lambda_j(X)|^{-r}$$

for all $\xi \in B'(t)$. Then there exists $t' > 0$, $t' \leq t$ such that $X\varphi$ is bounded on $B'(t')$.

Proof: We may work within a connected component of $\{\lambda_i \neq 0\}$. Each component is a union of cones of the form $\{\alpha_j(x) > 0\}$ for some collection $\{\alpha_j\}$ of m independent linear functions. On such a cone, we have $\pm\lambda_i = \Sigma a_{ij}\alpha_j$ where $a_{ij} \geq 0$. Hence

$$|\Pi\alpha_j(x)| < C|\lambda_i(x)|$$

if $|\alpha_j(x)| \leq 1$ for all j. In other words, it is enough to consider an inequality of the form:

$$|X\varphi(x)| < C(X)\Pi|x_i|^{-m},$$

where the x_i are the coordinate functions, for some m. Let $\partial_i/\partial_i = \partial_i$. For ε small,

$$\begin{aligned}
&|X\varphi(x_1, \dots x_m)| \\
&\quad = \Big|X\varphi(\varepsilon, \dots, \varepsilon) + \int_{x_1}^{\varepsilon} \cdots \int_{x_m}^{\varepsilon} (\partial_1 \dots \partial_m)X\varphi(t_1, \dots, t_m)dt_1 \dots dt_m\Big| \\
&\quad < \begin{cases} C\Pi|x_i|^{-r+1} & \text{if} \quad r > 1 \\ C\Pi ln|x_i| & \text{if} \quad r = 1 \end{cases}
\end{aligned}$$

for $|x_i|$ sufficiently small, and the lemma follows by induction on r.

Let Ω be a small neighborhood of a singular element $\gamma_0 \in M$. The boundedness of the derivatives $XS\mathfrak{I}(\gamma, f)$ on $\Omega \cap M^r$ implies, by a calculus lemma ([V], Lemma 21, page 48) that for all $X, XS\mathfrak{I}(\gamma, f)$ extends to a continuous function in the closure of any connected component of $\Omega \cap M^r$. To prove that $\mathcal{I}(\gamma, f)$ is smooth, it suffices to show it is smooth at semi-regular points $\gamma_0 = d(e^{i\theta}, e^{i\psi}, e^{i\theta})$ with $\theta \neq \psi$. Let $\gamma = \gamma(t) = d(e^{t+i\theta}, e^{i\psi}, e^{-t+i\theta})$. The derivatives of $\mathcal{I}(\gamma, f)$ of all orders with respect to θ and ψ are continuous at γ_0. Hence we need to show that the derivatives with respect to t of all orders do not have a jump across $t = 0$.

In the notation of Lemma 9.2.4, it will suffice to consider $|A_1 A_2| \mathcal{I}_H(\gamma, \varphi) - \frac{1}{2}\mu(\gamma)^{-1}\mathcal{I}_H(\gamma, f^H)$. Note that $|A_1 A_2|$ and $\mu(\gamma)$ are smooth near γ_0. By Proposition 8.2.1,

$$f^H(\gamma_0) = \mu(\gamma_0)|A_1(\gamma_0)A_2(\gamma_0)|\Phi^\kappa(\gamma_0, f)$$

and since $\varphi(\gamma_0) = \Phi(\gamma_0, f)$, (9.3.1) implies that the jump in the first derivative with respect to t of $S\mathcal{I}(\gamma, f)$ is equal to

$$8\pi|A_1(\gamma_0)A_2(\gamma_0)|[\Phi(\gamma_0, f) - \frac{1}{2}\Phi^\kappa(\gamma_0, f)] = 4\pi|A_1(\gamma_0)A_2(\gamma_0)|\Phi^{st}(\gamma_0, f)\ .$$

This is zero since $\Phi^{st}(\gamma_0, f)$ is stable (Proposition 8.2.1(d)), and it follows that all derivatives of $S\mathcal{I}(\gamma, f)$ of order ≤ 1 in t are continuous across $t = 0$. Now consider (9.3.2) in the case that z is the Casimir operator. Up to a non-zero constant, $\Gamma(z)$ is a constant coefficient differential operator of the form $\frac{\partial^2}{\partial t^2} + P$ where P is a second order operator that has order ≤ 1 in t. As above, in the proof of Proposition 9.3.1, zf is stably equivalent to zero and so all derivatives of $S\mathcal{I}(\gamma, zf)$ which are at most first order in t are continuous across $t = 0$. Equation (9.3.2) implies the same for all derivatives of order ≤ 3 in t. By induction, all derivatives of $S\mathcal{I}(\gamma, f)$ are continuous across $t = 0$ and the proposition is proved.

Let G_1 be a compact unitary in three variables over $\mathbf{R}$. Let T be a Cartan subgroup of G_1, which we also view as a subgroup of G and H. In G_1, conjugacy and stable conjugacy coincide. In §14, two transfers $f_1 \to f$ and $f_1 \to f_1^H$ will be discussed, determined by the requirements:

$$\Phi(\gamma, f_1) = \Phi^{st}(\gamma, f), \tau(\gamma)D_{G/H}(\gamma)\Phi(\gamma, f_1) = \Phi^{st}_H(\gamma, f_1^H)$$

for regular $\gamma \in T$. Set $S\mathcal{I}(\gamma, f_1) = -\frac{1}{2}\mu_v(\gamma)^{-1}\mathcal{I}_H(\gamma, f_1^H)$.

COROLLARY 9.3.4: *If $f_1 \to f$ and $f_1 \to f_1^H$, then $S\mathcal{I}(\gamma, f) - S\mathcal{I}(\gamma, f_1)$ extends to a continuous function on M.*

Proof: With $\gamma = \gamma(t)$ as before, let $T(\gamma) = S\mathcal{I}(\gamma, f) - \frac{1}{2}\mu(\gamma)^{-1}\mathcal{I}_H(\gamma, f_1^H)$. By the proof of Proposition 9.3.2, the jump in $S\mathcal{I}(\gamma, f)'$ at $t = 0$ is equal to $4\pi|A_1(\gamma_0)A_2(\gamma_0)|\Phi^{st}(\gamma_0, f)$. The jump in $\mu(\gamma)^{-1}\mathcal{I}_H(\gamma, f_1^H)'$ is equal to $8\pi\mu(\gamma)^{-1}f_1^H(\gamma_0)$. On the other hand, for all regular γ, the relation

$$\tau(\gamma)D_{G/H}(\gamma)\Phi^{st}(\gamma, f) = \Phi_H^{st}(\gamma, f_1^H)$$

holds and the Harish-Chandra limit formula implies (cf. §8.2)

$$8\pi\mu(\gamma)^{-1}f_1^H(\gamma_0) = 8\pi|A_1(\gamma_0)A_2(\gamma_0)|\Phi^{st}(\gamma_0, f).$$

It follows that $T(\gamma)'$ is equal is continuous at $t = 0$ and hence all derivatives of $T(\gamma)$ involving t to the first order are continuous in a neighborhood of $\gamma_0 = \gamma(0)$ where $\theta \neq \psi$. The argument of Proposition 9.3.2 implies the corollary.

9.4. The case of split primes. In this section, assume that v splits. We may identify G with $\mathrm{GL}_n(F), n = 2$ or 3.

PROPOSITION 9.4.1: *Let v be a place that splits in E. If f is stably equivalent to zero, then $S\mathcal{I}(\gamma, f)$ defines a smooth function on M.*

Proof: In this case, there is no difference between conjugacy and stable conjugacy, and f is thus stably equivalent to zero if and only if all orbital integrals of f vanish.

Assume that $n = 3$. Let s_j be the simple reflection with respect to α_j and set

$$r_1 = s_2 s_3,\ r_2 = 1,\ r_3 = s_2\ .$$

Let H_j' be the projection onto $\mathcal{A}$ of the co-root associated to α_j. Under the identification of $\mathcal{A}$ with $\mathbf{R}$, $H_j' = 1$ for $j = 1, 2$ and $H_3' = 2$. Since $w = s_2 s_3 s_1$, we may write:

$$\mathcal{V}(\gamma, n, v) = -\sum_{1\leq j\leq 3}(s_j^{-1}r_j^{-1}H(r_j s_j n_1) - r_i^{-1}H(r_i n_1) - ln|A_j|H_j')$$

where n_1 is defined by the relation $n = \gamma^{-1}n_1^{-1}\gamma n_1$. We decompose $\mathcal{I}(\gamma, f)$ into a corresponding sum of 3 integrals $\mathcal{I}_i(\gamma)$ for $j = 1, 2, 3$, where:

$$\mathcal{I}_j(\gamma) = -\int_N f^K(\gamma n)(s_j^{-1}r_j^{-1}H(r_j s_j n_1) - r_j^{-1}H(r_j n_1) - ln|A_j|H_j')\,dn\ .$$

Let $B_j = r_j^{-1} B r_j$ and let N_j be the unipotent radical of B_j. Let $H_j(g)$ be the projection onto $\mathcal{A}_0$ of the analogue of the function H, defined with respect to the Iwasawa decomposition $G = B_j K$. Then $r_j^{-1} H(r_j g) = H_j(g)$. By a change of variables, $\mathcal{J}_j(\gamma)$ is equal to

$$-|\prod_{1 \le j \le 3} (1 - \alpha_j^{-1}(\gamma))| \int_N f^K(n_1^{-1}\gamma n_1)(s_j^{-1} H_j(s_j n_1) - H_j(n_1) - ln|A_j|H_j')dn_1$$

$$= -|\prod_{1 \le j \le 3} (1 - \alpha_j^{-1}(\gamma))| \int_{M\backslash G} f(g^{-1}\gamma g)(s_j^{-1} H_j(s_j g) - H_j(g) - ln|A_j|H_j')dg$$

$$= -|\rho(\gamma)^{-1}\rho_j(\gamma)| \int_{N_j} f^K(\gamma n)(s_j^{-1} H_j(s_j n) - ln|A_j|H_j')\, dn$$

where $|\rho_j(\gamma)|$ is the product of $|\alpha(\gamma)|^{1/2}$ over the roots of M which are positive with respect to B_j and $|\rho(\gamma)|$ is defined similarly with respect to B. The root α_j is simple and positive with respect to B_j. Let U_j be the one-dimensional root subgroup of N_j corresponding to α_j and let U^j be the product of the remaining root subgroups, so that $N_j = U_j U^j$. Then $H_j(s_j u_j u^j) = H_j(s_j u_j)$ for all $u_j \in U_j$, $u^j \in U^j$, and

$$\mathcal{J}_j(\gamma) = -|\rho(\gamma)^{-1}\rho_j(\gamma)| \int_{U_j} \left[\int_{U^j} f^K(\gamma u_j u^j) du^j \right] (s_j^{-1} H_j(s_j n) - ln|A_j|H_j')\, du_j\,.$$

Let M_j be the centralizer of $\{m \in M : \alpha_j(m) = 1\}$ and set

$$\phi_j(m) = \int_{U^j} f^K(mu^j)du^j$$

for $m \in M_j$. Note that M_j is isomorphic to $\mathrm{GL}_2(F) \times \mathrm{GL}_1(F)$. We have:

$$\mathcal{J}_j(\gamma) = -|\rho(\gamma)^{-1}\rho_j(\gamma)| \int_{U_j} \phi_j(\gamma u_j)(s_j^{-1} H_j(s_j n) - ln|A_j|H_j')du_j\,.$$

If $\alpha_j(\gamma_0) \neq 1$, then γ_0 is regular in M_j and $\mathcal{J}_j(\gamma)$ is clearly smooth in a neighborhood of γ_0. Assume that f is stably equivalent to zero. If $\gamma \in M_j$ is regular, then $\Phi(\gamma, f) = 0$ and hence $\Phi(\gamma, \phi_j) = 0$ and ϕ_j is stably equivalent to zero. In particular, $\phi_j(\gamma) = 0$ for central $\gamma \in M_j$ and the arguments of §9.1 and §9.3 applied to M_j, show that $\mathcal{J}_j(\gamma)$ defines a smooth function on M. This proves Proposition 9.4.1 in the case $n = 3$. If $n = 2$, then $G = \mathrm{GL}_2(F)$ and again the arguments of §9.1 and §9.3 apply.

PROPOSITION 9.4.2: *Let $G = \mathrm{GL}_3(F)$. If $\Phi(\gamma, f) = 0$ for all regular non-elliptic $\gamma \in G$, then $\mathcal{I}(\gamma, f)$ defines a smooth function on M.*

Proof: The germ of $\mathcal{I}_j(\gamma)$ near a singular element γ_0 such that $\alpha_j(\gamma_0) = 1$ is a multiple of $|\rho(\gamma)^{-1}\rho_j(\gamma)|\cdot|A_j(\gamma)|\phi_j(\gamma_0)$ by Lemma 9.1.1. If γ_0 is central, then $\phi_j(\gamma_0)$, up to normalization, is equal to $\Phi(\gamma_0 u, f)$, where $u \in U_j$ and $u \neq 1$. If $\Phi(\gamma, f) = 0$ for all regular non-elliptic $\gamma \in G$, then $\Phi(\gamma_0 u, f) = 0$ by the proof of Lemma 2.6 of [R3] and $\phi_j(\gamma_0) = 0$. Similarly, if γ_0 is not central, then α_j is the unique root such that $\alpha_j(\gamma_0) = 1$ and $\phi_j(\gamma_0) = \Phi(\gamma_0, f)$. Again, $\Phi(\gamma_0, f) = 0$ and the proposition follows.

9.5 Twisted case. In this section, we consider $\widetilde{G} = \mathrm{Res}_{E/F}(G)$ where $G = U(2)$ or $U(3)$. Let δ be an element of M and let $\gamma = N(\delta)$. In the global case, define $R(\gamma, \phi, v)$ by

$$\begin{array}{ll} \mathcal{I}_{\widetilde{M}}(\delta, \phi, v) - S\mathcal{I}_M(\gamma, f, v) - \frac{1}{2} S\mathcal{I}_M(\gamma, \phi^H, v) & \text{if} \quad G = U(3) \\ 2\mathcal{I}_{\widetilde{M}}(\delta, \phi, v) - [\mathcal{I}_M(\gamma, f_1, v) + \mu(\det(\delta))^{-1}\mathcal{I}_M(\gamma, f_2, v)] & \text{if} \quad G = U(2) \end{array}$$

for ε-regular δ (the above factor of 2 occurs because $a(\delta) = 2$ in the case of $U(2)$). Recall that $\mathcal{I}_{\widetilde{M}}(\delta, \phi, v)$ depends only on γ. The germ of $R(\gamma, \phi, v)$ depends only on ϕ, since the germs of $S\mathcal{I}_M(\gamma, f, v)$ and $\mathcal{I}_M(\gamma, f_j, v)$ are stably invariant.

PROPOSITION 9.5.1: (a) *For all v, $R(\gamma, \phi, v)$ extends to a smooth function on* $\mathbf{M}$.

(b) *Assume that ϕ_v is a unit in the Hecke algebra and that μ_v is unramified. Then $R(\gamma, \phi, v) = 0$ for $\gamma \in M$.*

Proof: By the definition and the relations between orbital integrals, $R(\gamma, \phi, v)$ is equal to the product of $\Pi_{w \neq v} \Phi_\varepsilon^M(\delta_w, \phi_w)$ and

(9.5.1)
$$\mathcal{I}_\varepsilon(\delta_v, \phi_v) - \mathcal{I}(\gamma_v, f_v) - \frac{1}{2}[\mathcal{I}(\gamma_v, \phi_v^H) - \mu_v(\gamma)^{-1}\mathcal{I}(\gamma_v, f_v^H)] \quad \text{if} \quad G = U(3)$$

(9.5.2) $\quad 2\mathcal{I}_\varepsilon(\delta_v, \phi_v) - \mathcal{I}(\gamma_v, f_{1v}) - \mu_v(\det(\delta_v))^{-1}\mathcal{I}(\gamma_v, f_{2v}) \quad \text{if} \quad G = U(2).$

Suppose that ϕ_v is a unit in $\mathcal{H}(\widetilde{G})$ and that μ_v is unramified. Then $f_v, f_v^H, \phi_v^H, f_{1v}$, and f_{2v} are the units in their respective Hecke algebras. The terms $\mathcal{I}(\gamma, f_v), \mathcal{I}(\gamma, \phi_v^H), \mathcal{I}(\gamma, f_v^H)$ and $\mathcal{I}(\gamma, f_{jv})$ vanish unless γ lies in a maximal compact subgroup. If γ does lie in a maximal compact, then $\mu_v(\gamma) = 1$ and $\mu_v(\det(\delta)) = 1$ in the two cases, respectively. By [Kt$_1$], §3, $\mathcal{I}_\varepsilon(\delta, \phi) = \mathcal{I}(\gamma, f)$ for the general base change situation, when ϕ and f are units in the Hecke algebra, and it follows that (9.5.1) and (9.5.2) vanish.

We first prove (a) in the case that v splits into two places w, w' in E. Identify $\widetilde{G}$ with $\mathrm{GL}_n(E_w) \times \mathrm{GL}_n(E_{w'})$ ($n = 2$) or 3. By the construction of §4.13, we can assume, that $\mathcal{I}(\gamma_v, \phi_v^H) = \mu_v(\gamma)^{-1}\mathcal{I}(\gamma_v, f_v^H)$ in (9.5.1) and $\mathcal{I}(\gamma_v, f_{1v}) = \mu_v(\det(\delta_v))^{-1}\mathcal{I}(\gamma_v, f_{2v})$ in (9.5.2). We need only show that $\mathcal{I}(\delta_v, \phi_v) - \mathcal{I}(\gamma_v, f_v)$ extends to a smooth function, where $\delta = (\delta_1, \delta_2)$ and $\gamma = \delta_1\varepsilon(\delta_2)$. Here $f_v = f_{1v}$ if $G = U(2)$ and $\varepsilon(\delta_2) = \Phi^t\overline{\delta_2}^{-1}\Phi^{-1}$. In this expression, the differences between the modified weight functions of §6 and the ordinary weight functions cancel. Assume for ease of notation, that $\mathcal{I}_\varepsilon(\delta_v, \phi_v)$ and $\mathcal{I}(\gamma_v, f_v)$ are defined by ordinary weight functions. Let $\lambda(g) = -(H(g) + H(wg))$ for $g \in \mathrm{GL}_3(E_w)$. If $g = (g_1, g_2)$, then $\mathcal{W}(g) = \frac{1}{2}(\lambda(g_1) + \lambda(\varepsilon(g_2)))$. Suppose that $\phi_v = \phi_1 \times \phi_2$ and that $f_v = \phi_1 * \phi_2$. Then $\mathcal{I}_\varepsilon(\delta, \phi)$ is a sum of the integrals

$$\frac{1}{2}\int_{\widetilde{Z}M\backslash\widetilde{G}} \phi_1(g_1^{-1}\delta_1\varepsilon(g_2))\phi_2(g_2^{-1}\delta_2\varepsilon(g_1))\lambda(g_1)dg\ .$$

$$\frac{1}{2}\int_{\widetilde{Z}M\backslash\widetilde{G}} \phi_1(g_1^{-1}\delta_1\varepsilon(g_2))\phi_2(g_2^{-1}\delta_2\varepsilon(g_1))\lambda(\varepsilon(g_2))dg\ .$$

By a change of variables, the first integral can be written as

$$\frac{1}{2}\int_{M\backslash G}\left[\int_{Z\backslash G} \phi_1(g_1^{-1}\gamma g_1 g_2)\phi_2(\varepsilon(g_2)^{-1})dg_2\right]\lambda(g_1)dg_1,$$

where we identify G with $\mathrm{GL}_3(E_w)$, and this is equal to $\frac{1}{2}\mathcal{I}(\gamma, f_v)$. Define $\tilde{f}_v(g) = \phi_2 * \phi_1(\varepsilon(g))$ and let $\gamma' = \varepsilon(\delta_2)\delta_1$. The second integral can be written as

$$\frac{1}{2}\int_{M\backslash G}\left[\int_{Z\backslash G} \phi_2(\varepsilon(g_2^{-1}\gamma' g_2 g_1))\phi_1(g_1^{-1})dg_1\right]\lambda(g_2)dg_2$$

and this is equal to $\frac{1}{2}\mathcal{I}(\gamma', \tilde{f}_v)$. Observe that γ and γ' are conjugate and the singularities of $\mathcal{I}(\gamma, f_v)$ and $\mathcal{I}(\gamma', \tilde{f}_v)$ coincide since the singularities are invariant. This proves (a) when v splits.

For the rest of this section, we assume that v does not split in E, and we drop v from the notation. Let $A_j = A_j(\gamma)$ (cf. §6.3). We first show that $R(\gamma, \phi, v)$ is smooth near a central element $\gamma_0 \in M$. We can assume that $\gamma_0 = 1$. If γ is close to γ_0, then there exists $\delta \in M$ such that $\delta^2 = N(\gamma) = \gamma$ and δ is near $\delta_0 = 1$.

PROPOSITION 9.5.2: *The integral*

$$|A_1 A_2 A_3| \int\limits_{\widetilde{Z}G\backslash\widetilde{G}} \int\limits_{M\backslash G} \phi(g^{-1}y^{-1}\delta y\varepsilon(g))(\mathcal{W}(yg) - \mathcal{W}(y)))dy\, dg$$

extends to a smooth function of $\delta \in M$ near δ_0.

Proof: Let $N^{op} = wNw^{-1}$ and let $y = mnk = m_1 n_1 k_1$, where $m, m_1 \in M$, $k, k_1 \in K$, $n \in N$, and $n_1 \in N^{op}$. Define $H^{op}(g)$ analogously to $H(g)$, but with respect to the opposite Borel subgroup. Then

$$\mathcal{W}(yg) - \mathcal{W}(y) = [H^{op}(y) - H^{op}(yg)] - [H(y) - H(yg)] = -H^{op}(k_1 g) - H(kg)\,.$$

and the function in the lemma is the sum of the integrals:

$$- |A_1 A_2 A_3| \int\limits_{\widetilde{Z}G\backslash\widetilde{G}} \int\limits_{N} \int\limits_{K} \phi(g^{-1}k^{-1}n^{-1}\delta nk\varepsilon(g))\; H(kg)\; dy\; dg$$

$$- |A_1 A_2 A_3| \int\limits_{\widetilde{Z}G\backslash\widetilde{G}} \int\limits_{N^{op}} \int\limits_{K} \phi(g^{-1}k^{-1}n^{-1}\delta nk\varepsilon(g))\; H^{op}(kg)\; dy\; dg$$

For δ and γ near 1, the ratio $|A_j(\gamma)/A_j(\delta)|$ is constant and hence, changing variables, we obtain a constant times the sum of the integrals:

$$- \int\limits_{ZG\backslash\widetilde{G}} \int\limits_{N} \int\limits_{K} \phi(g^{-1}k^{-1}\delta nk\varepsilon(g))\; H(kg)\; dndkdg$$

$$- \int\limits_{\widetilde{Z}G\backslash\widetilde{G}} \int\limits_{N^{op}} \int\limits_{K} \phi(g^{-1}k^{-1}\delta nk\varepsilon(g))H^{op}(kg)\; dndkdg.$$

The set of g such that $\phi(g^{-1}k^{-1}\delta nk\varepsilon(g)) \neq 0$ for some n, k, and δ near δ_0 is compact modulo $\widetilde{Z}G$ (cf. Lemma 4.12.1) and it follows that the integrals define smooth functions of δ in a neighborhood of δ_0.

By Proposition 9.5.2, the germ of $\mathfrak{I}(\delta, \phi)$ near δ_0 is the same as that of

$$|A_1 A_2 A_3| \int_{\widetilde{Z}G\backslash\widetilde{G}} \int_{M\backslash G} \phi(g^{-1}x^{-1}\delta x\varepsilon(g))\mathcal{W}(x)dx\, dg\,. \tag{9.5.3}$$

Let α be a smooth compactly supported function on $\widetilde{Z}\backslash\widetilde{G}$ such that

$$\overline{\alpha}(g) = \int_{Z\backslash G} \alpha(xg)dx$$

is equal to 1 for all $g \in \widetilde{G}$ such that $\phi(g^{-1}x\varepsilon(g)) \neq 0$ for some $x \in G$ near δ_0. Then (9.5.3) is equal to

$$|A_1 A_2 A_3| \int_{\widetilde{Z}G\backslash\widetilde{G}} \int_{M\backslash G} \overline{\alpha}(g)\phi(g^{-1}x^{-1}\delta x\varepsilon(g))\mathcal{W}(x)dxdg$$

$$= |A_1 A_2 A_3| \int_{M\backslash G} F_1(x^{-1}\delta x)\mathcal{W}(x)dx$$

where

$$F_1(x) = \int_{\widetilde{Z}\backslash\widetilde{G}} \alpha(g)\phi(g^{-1}x\varepsilon(g))\, dx$$

for $x \in G$ near 1. Define $F(x)$ near 1 by $F(x^2) = F_1(x)$. The germ of $\mathfrak{I}(\delta, \phi)$ is then the same as that of $\mathfrak{I}(\gamma, F)$. We may assume that ϕ is supported in a set of the form $\{g^{-1}x\varepsilon(g) : g \in \widetilde{G}, x \in \Omega\}$, where $\Omega \subset G$ is a small neighborhood of 1. By the construction of §4.12, we can take $f = F$ and $\phi^H(x) = \mu(\det_0(x'))f^H(x)$ for $x \in H$ near 1, where $x' \in H$ is such that $x'^2 = x$. By Lemma 9.1.1, the germ of $\mathfrak{I}(\gamma, \phi^H) - \mu(\gamma)^{-1}\mathfrak{I}(\gamma, f^H)$ near 1 is a constant times $|A_3(\gamma)|[\phi^H(\gamma_0) - \mu(\gamma_0)^{-1}f^H(\gamma_0)]$ and this vanishes. By construction, $\mathfrak{I}(\delta, \phi) - \mathfrak{I}(\gamma, f)$ extends to a smooth function near 1, and hence $R(\gamma, \phi, v)$ is smooth near γ_0. The arguments for the cases of non-central γ_0 and for $G = U(2)$ are similar. We omit the details.

CHAPTER 10

The stable trace formula

In this chapter, we put together the results of the previous chapters and carry out the comparisons between the ordinary and twisted trace formulas for $U(2)$ and $U(3)$. The main result is Theorem 10.3.1

10.1. Elliptic part of the fine $\mathcal{O}$-expansion. Let $G = U(2)$, $(U(2) \times U(1)$, or $U(3)$, and let $\mathcal{O}_{st}$ be a stable elliptic conjugacy class. Define $SJ_G(\mathcal{O}_{st}, f)$ inductively by the equation:

$$J_G(\mathcal{O}_{st}, f) = SJ_G(\mathcal{O}'_{st}, f) + i(G, H) \sum SJ_H(\mathcal{O}'_{st}, f)$$

where H is the unique proper elliptic endoscopic group for G and the sum is over the stable elliptic classes $\mathcal{O}'_{st}$ in H which transfer to $\mathcal{O}_{st}$. Here $J_G(\mathcal{O}_{st}, f)$ is the distribution defined in §7.6. This coincides with the definition given in §5.4 when $\mathcal{O}_{st}$ is regular by Proposition 5.4.1. According to the definitions of §5.2,

$$SJ_G(f) = \sum_{\mathcal{O}_{st}} SJ_G(\mathcal{O}_{st}, f) \ .$$

where $\mathcal{O}_{st}$ ranges over the stable elliptic classes modulo Z in G.

PROPOSITION 10.1.1: *$SJ_G(f)$ is a stable distribution.*

Proof: We have to show that $SJ_G(\mathcal{O}_{st}, f)$ is stable. If $\mathcal{O}_{st}$ is elliptic regular, this is a definition. For elliptic singular classes, it follows from the next proposition.

PROPOSITION 10.1.2: *Let $\mathcal{O}_{st} = \mathcal{O}_{st}(\gamma)$ where $\gamma \in G$ is elliptic and semisimple. Then $J_G(\mathcal{O}_{st}, f)$ is a stably invariant distribution.*

(a) *If $G = U(2) \times U(1)$ or $U(2)$ and γ is central, then*

$$SJ_G(\mathcal{O}_{st}, f) = m(\mathbf{ZG}\backslash\mathbf{G})\ f(\gamma) + C\Phi^M(\gamma, f) \ .$$

(b) *Let $G = U(3)$ and $H = U(2) \times U(1)$.*

(1) *If γ is singular but not central in G, then*

$$SJ_G(\mathcal{O}_{st}, f) = \frac{1}{2}m(\mathbf{Z}H\backslash\mathbf{H})\Phi^{st}(\gamma, f) + \frac{1}{2}C\Phi^M(\gamma, f)$$

(2) *If γ is central in G, then $SJ_G(\mathcal{O}_{st}, f)$ is equal to*

$$m(\mathbf{Z}G\backslash\mathbf{G})f(\gamma)+(D-\frac{1}{2}C)\Phi^M(\gamma, f)+\frac{1}{2}m(\mathbf{Z}S\backslash\mathbf{S})\int_{I_F} f^K(\gamma n(t))\omega_{E/F}(t)|t|^2 d^*t.$$

Proof: It follows from the definitions and Proposition 7.3.1 that the difference between the left and right-hand sides of the equality in (a) is equal to the difference of 7.3.1(b) and $\frac{1}{4}m(\mathbf{Z}C\backslash\mathbf{C})\, f^H(\gamma)$. This is zero by the results of [LL]. The term $m(\mathbf{Z}G\backslash\mathbf{G})f(\gamma)$ is stable by Proposition 8.4.1. Observe that $\Phi^M(\gamma, f)$ is stable for all $\gamma \in \mathbf{M}$ by continuity, since it is stable if γ is regular.

For part (b), suppose first that γ is singular but not central in G. We may assume that γ is central in H. By (a) and Proposition 7.2.2, $SJ_G(\mathcal{O}_{st}, f)$ is equal to the sum of the terms

(i) $\frac{1}{2}m(\mathbf{Z}H\backslash\mathbf{H})\ \Phi^{st}(\gamma, f)$

(ii) $\frac{1}{2}m(\mathbf{Z}H\backslash\mathbf{H})[\Phi^{\kappa}(\gamma, f) - f^H(\gamma)]$

(iii) $C(\Phi^M(\gamma, f) - \frac{1}{2}\Phi^M H(\gamma, f^H))$

(iv) 7.2.2(c) and $-\frac{1}{4}m(\mathbf{Z}T\backslash\mathbf{T})\Phi^{st}_H(\gamma_1, f^H)$

where γ_1 is a regular element of H which is stably conjugate to γ in G. Terms (ii) and (iv) vanish by Proposition 8.2.1(b) and (c). Part (1) follows since $\Phi^M(\gamma, f) = \Phi^M_H(\gamma, f^H)$ for all $\gamma \in M$. Term (i) is stable by Proposition 8.1.4 and $\Phi^M(\gamma, f)$ is stable, as before.

If γ is central in G, then, by (a) and Proposition 7.3.2, $SJ_G(\mathcal{O}_{st}, f)$ is the terms:

(i) $m(\mathbf{Z}G\backslash\mathbf{G})f(\gamma)$

(ii) 7.3.2(c)

(iii) 7.3.2(d) and $-\frac{1}{2}m(\mathbf{Z}H\backslash\mathbf{H})\ f^H(\gamma)$

(iv) $D\Phi^M(\gamma, f) - \frac{1}{2}C\Phi^M_H(\gamma, f^H)$

Term (iii) is zero by Proposition 8.3.3(a). Terms (i) and 7.3.2(c) are stable by Propositions 8.1.4 and 8.3.3(b), respectively. As before, $\Phi^M(\gamma, f) = \Phi^M_H(\gamma, f^H)$ for $\gamma \in M$ and is a stable distribution.

PROPOSITION 10.1.3: *Let $\widetilde{G} = \mathrm{Res}_{E/F}(G)$, where $G = U(2)$ or $U(3)$. Then*

$$J_{\widetilde{G}}(f) = \sum_{H \in \mathcal{E}(\widetilde{G})} i(G, H)\ SJ_H(f^H)$$

Proof: Let $\mathcal{O}_{\mathrm{st}}$ be an ε-elliptic semisimple class in $\widetilde{G}$. We have to show that

$$J_{\widetilde{G}}(\mathcal{O}_{\mathrm{st}}, f) = \sum_{H \in \mathcal{E}(G)} i(G, H) \sum SJ_H(\mathcal{O}'_{\mathrm{st}}, f^H)$$

where $\mathcal{O}'_{\mathrm{st}}$ runs over the semisimple classes which transfer to $\mathcal{O}_{\mathrm{st}}$. If δ is ε-regular, this is Proposition 5.5.2. Suppose that δ is singular and that $\gamma = N(\delta) \in M$.

If $G = U(2)$ and γ is central, then

$$J_{\widetilde{G}}(\mathcal{O}_{\mathrm{st}}, \phi) = m(\mathbf{Z}G \backslash \mathbf{G}) \sum_{\kappa} \Phi^{\kappa}_{\varepsilon}(\delta, \phi) + 2C\Phi^M_{\varepsilon}(\delta, \phi).$$

by Proposition 7.5.2. By Propositions 4.11.1(a) and 10.1.2(a), this is equal $J_G(\mathcal{O}_{\mathrm{st}}, f_1) + J_G(\mathcal{O}_{\mathrm{st}}, f_2)$, where $\mathcal{O}_{\mathrm{st}} = \mathcal{O}_{\mathrm{st}}(\gamma)$.

Let $G = U(3)$ and suppose that γ is singular but not central. By Proposition 7.4.1, $J_{\widetilde{G}}(\mathcal{O}_{\mathrm{st}}, \phi)$ is equal to the sum of the terms

(i) $\frac{1}{2} m(\mathbf{Z}H \backslash \mathbf{H})\ \Phi^{\mathrm{st}}_{\varepsilon}(\delta, \phi) = \frac{1}{2} m(\mathbf{Z}H \backslash \mathbf{H})\ \Phi^{\mathrm{st}}(\gamma, f)$

(ii) $C\Phi^M_{\varepsilon}(\delta, \phi) = \frac{1}{2} C\Phi^M(\gamma, f) + \frac{1}{2} C\Phi^M_H(\gamma, \phi^H)$

(iii) $\frac{1}{2} m(\mathbf{Z}H \backslash \mathbf{H})\Phi^{\kappa}_{\varepsilon}(\delta, \phi) = \frac{1}{2} m(\mathbf{Z}H \backslash \mathbf{H})\phi^H(\gamma)$

(iv) $7.4.1(\mathrm{c}) = \frac{1}{4} m(\mathbf{Z}T \backslash \mathbf{T})\Phi^{\mathrm{st}}(\gamma_1, \phi^H)$

The equalities (i) and (iii) (resp., (iv)) follow from Proposition 8.4.1(b) (resp., 8.4.2(a)). The sum of (i) and $\frac{1}{2} C\Phi^M(\gamma, f)$ is equal to $J_G(\mathcal{O}_{\mathrm{st}}(\gamma), f)$ by Proposition 10.1.2(b1), and, since $i(G, H) = 2$, it will suffice to check that the sum of (iii) , (iv), and $\frac{1}{2} C\Phi^M_H(\gamma, \phi^H)$ is equal to $\frac{1}{2} \sum SJ_H(\mathcal{O}'_{\mathrm{st}}, f^H)$. There are two stable conjugacy classes $\mathcal{O}'_{\mathrm{st}}$ in H which transfer to $\mathcal{O}_{\mathrm{st}}$, one central in H and the other regular. If $\mathcal{O}'_{\mathrm{st}}$ is the central class, then

$\frac{1}{2}SJ_H(\mathcal{O}'_{\rm st},\phi^H)$ is equal to (iii). If $\mathcal{O}'_{\rm st}$ is the regular class, then $SJ_H(\mathcal{O}'_{\rm st},\phi^H)$ is equal to $z_H(\gamma_1)\tau(H)\Phi^{\rm st}(\gamma_1,\phi^H)$. If we use a Tamagawa measure on T, then $\tau(H) = 4$ and since $T = H_{\gamma_1}$ is a Cartan subgroup of type (1), $z_H(\gamma_1) = \frac{1}{2}$, by the formula of §5.3. In this case, $\frac{1}{4}m(\mathbf{Z}T\backslash\mathbf{T}) = 1$. Hence $\frac{1}{2}SJ_H(\mathcal{O}'_{\rm st},\phi^H) = \Phi^{\rm st}(\gamma_1,\phi^H)$, which is equal to (iv).

If γ is central in G, then $J_{\tilde{G}}(\mathcal{O}_{\rm st},f)$ is the sum of

(i) $m(\mathbf{Z}G\backslash\mathbf{G})\Phi^{\rm st}_\varepsilon(\delta,\phi) = m(\mathbf{Z}G\backslash\mathbf{G})f(\gamma)$

(ii) $D\Phi^M_\varepsilon(\delta,\phi) = D\Phi^M(\gamma,f)$

(iii) $7.5.1(c) = 7.3.2(c)$

(iv) $7.5.1(d) = \frac{1}{2}m(\mathbf{Z}H\backslash\mathbf{H})\phi^H(\gamma)$

by Proposition 7.5.1. Equality (i) follows from Proposition 8.4.1(a) and (iii) and (iv) follow from Proposition 8.4.2(b). The sum of (i), (ii) and (iii) is equal to $SJ_G(\mathcal{O}_{\rm st},f) + \frac{1}{2}C\Phi^M(\gamma,f)$ where $\mathcal{O}_{\rm st} = \mathcal{O}_{\rm st}(\gamma)$, by Proposition 10.1.2(b2). By Proposition 10.1.2(a), (iv) is equal to

$$\frac{1}{2}SJ_H(\mathcal{O}_{\rm st},\phi^H) - \frac{1}{2}C\Phi^M_H(\gamma,\phi^H).$$

The assertion follows since $\Phi^M(\gamma,f)$ and $\Phi^M_H(\gamma,\phi^H)$ both coincide with $\Phi^M_\varepsilon(\delta,\phi)$ for $\gamma\in M$.

10.2 Stable trace formula. Let $G = U(2)$ or $U(3)$ and let H be the unique proper elliptic endoscopic group for G. At this point, by the results of §7.6, we can write the trace formulas for G and H as

$$\theta_G(f) + \theta_M(f) = J_G(f) + \frac{1}{2}m(\mathbf{Z}M\backslash\mathbf{M}^1)\sum J^T_M(\gamma,f) \tag{10.2.1}$$

$$\theta_H(f^H) + \theta_{M_H}(f^H) = J_H(f^H) + \frac{1}{2}m(\mathbf{Z}M\backslash\mathbf{M}^1)\sum J^T_{M_H}(\gamma,f) \tag{10.2.2}$$

where the sums are over $Z\backslash M$ (if $G = U(2)$, then $H = U(1)\times U(1)$ and M_H does not exist, hence the terms involving θ_{M_H} and $J^T_{M_H}$ do not occur). If $G = U(3)$, define

$$SJ^T_M(\gamma,f) = J^T_M(\gamma,f) - \frac{1}{2}\mu(\gamma)^{-1}J^T_M(\gamma,f^H)$$

for $\gamma\in\mathbf{M}$, where $H = U(2)\times U(1)$. Note that $\mu(\gamma) = 1$ if $\gamma\in M$. If $G = U(2)$ or $U(2)\times U(1)$, let $SJ^T_M(f,\gamma) = J^T_M(\gamma,f)$. We subtract $i(G,H)$

times (10.2.2) from (10.2.1) to obtain:

$$(10.2.3)\quad S\theta_G(f) + S\theta_M J(f) = SJ_G(f) + \frac{1}{2}m(\mathbf{Z}M\backslash\mathbf{M}^1)\sum SJ_M^T(\gamma, f)$$

Note that

$$SJ_M(f) = \frac{1}{2}m(\mathbf{Z}M\backslash\mathbf{M}^1)\sum SJ_M^T(\gamma, f).$$

according to the definitions of §5.2. In particular (5.2.2) holds. The next proposition is included even though it is not needed in §10.3. It can be used to prove that $S\theta_G$ is stable using the argument sketched in §5.2.

PROPOSITION 10.2.1: *Suppose that f is stably equivalent to 0. Then $SJ_M^T(\gamma, f$ defines a smooth function on* $\mathbf{M}$.

Proof: Suppose that $G = U(2)$. Then

$$SJ_M^T(\gamma, f) = 2T\Phi^M(\gamma, f) + \sum_{v\in S_0} \mathcal{I}(\gamma, f_v)\prod)_{w\neq v}\Phi^M(\gamma, f_w)$$

The function $\Phi^M(\gamma, f)$ is continuous on $\mathbf{M}$. If f is stably equivalent to 0, then $f(\gamma_0) = 0$ for all central γ_0 and $\mathcal{I}(\gamma, f_v)$ is continuous by Lemma 9.1.1 and Propositions 9.3.1 and 9.4.1. The continuity of $SJ_M^T(\gamma, f)$ follows. If $G = U(3)$, then $\Phi^M(\gamma, f_w) = \mu_w(\gamma)\Phi_H^M(\gamma, f_w^H)$ for all w, and

$$SJ_M^T(\gamma, f) = T\Phi^M(\gamma, f) + \sum_{v\in S_0} S\mathcal{I}(\gamma, f_v)\prod_{w\neq v}\Phi^M(\gamma, f_w)$$

This defines a continuous function on $\mathbf{M}$ by Proposition 9.2.5, 9.3.2, and 9.4.1.

10.3 The main equality. The proof of the next theorem follows the pattern described in §5.1 and §5.2.

THEOREM 10.3.1: *Let $G = U(2)$ or $U(3)$ and let $\widetilde{G} = Res_{E/F}(G)$.*

(a) *The following equality holds:*

$$\theta_{\widetilde{G}}(f) = \sum_{H\in\mathcal{E}(G)} i(G, H)\ S\theta_H(f^H)\ .$$

(b) *$S\theta_G$ is a stable distribution.*

Proof: We consider the case $G = U(3)$. The case $G = U(2)$ is similar. Let $H = U(2)\times U(1)$. By §7.6 and (10.2.3), we have

$$\theta_{\widetilde{G}}(\phi) + \theta_{\widetilde{M}}(\phi) = J_{\widetilde{G}}(\phi) + J_{\widetilde{M}}(\phi)$$

$$S\theta_G(f) + S\theta_M(f) = SJ_G(f) + \frac{1}{2}m(\mathbf{Z}M\backslash\mathbf{M}^1)\sum SJ^T_M(\gamma, f)$$

$$S\theta_H(\phi^H) + S\theta_{M_H}(\phi^H) = SJ_G(\phi^H) + \frac{1}{2}m(\mathbf{Z}M\backslash\mathbf{M}^1)\sum SJ^T_{M_H}(\gamma, f).$$

Proposition 10.1.3 implies that

$$\theta_{\widetilde{G}}(f) - S\theta_G(f) - \frac{1}{2}S\theta_H(f^H) \tag{10.3.1}$$

is equal to the sum of the two terms

$$\theta_{\widetilde{M}}(\phi) - S\theta_M(f) - \frac{1}{2}S\theta_{M_H}(\phi^H) \tag{10.3.2}$$

$$J_{\widetilde{M}}(\phi) - \frac{1}{2}m(\mathbf{Z}M\backslash\mathbf{M}^1)\sum\left[SJ^T_M(\gamma, f) + \frac{1}{2}SJ_{M_H}(\gamma, \phi^H)\right] \tag{10.3.3}$$

Recall that the distributions entering into (10.3.3) are defined in terms of a finite set S_0 of places of F which depends on ϕ. Fix a finite place w not in S_0. Then ϕ_w is the unit in $\mathcal{H}_w(\widetilde{G})$. Set

$$\phi^{\#} = \prod_{v \neq w} \phi_v, \quad \phi' = \phi^{\#} \times \varphi_w,$$

where φ_w ranges through $\mathcal{H}_w(\widetilde{G})$, and let $S' = S_0 \cup S(\varphi_w)$ be a finite set of places sufficiently large to define (10.3.3) for ϕ'. We now prove (a) for all φ_w. For convenience, we now denote ϕ' by ϕ. Let $f_w \in \mathcal{H}_w(G)$ be the image of φ_w under $\hat{\psi}_G$.

For $\gamma \in M$, let

$$\varepsilon(\gamma, f_w) = J^T_{\widetilde{M}}(\delta, \phi) - SJ^T_M(\gamma, f) - \frac{1}{2}SJ^T_{M_H}(\gamma, \phi^H) = \sum_{v \in S'} R(\gamma, \phi, v)$$

where $\delta \in \widetilde{\mathbf{M}}$, $N(\delta) = \gamma$. Then (10.3.3) is equal to

$$\frac{1}{2}m(\mathbf{Z}M\backslash\mathbf{M}^1) \sum_{\gamma \in Z\backslash M} \varepsilon(\gamma, f_w)\,. \tag{10.3.4}$$

We are interested in the varying φ_w, or equivalently, f_w. If $v \notin S_0 \cup \{w\}$, then ϕ_v is the unit in the Hecke algebra and $R(\gamma, \phi, v) = 0$ by Proposition 9.5.1. Hence

$$\varepsilon(\gamma, f_w) = \sum_{v \in S_0 \cup \{w\}} R(\gamma, \phi, v)$$

and again by Proposition 9.5.1, $\varepsilon(\gamma, f_w)$ defines a smooth function on $\mathbf{M}$.

Let $\varepsilon_1(\gamma, f_w)$ denote the restriction of $\varepsilon(\gamma, f_w)$ to $\mathbf{M}^1$. We apply the Poisson summation formula for $Z\backslash M$ and $\mathbf{Z}\backslash\mathbf{M}^1$ to write (10.3.4) as

$$\frac{1}{2}\sum \hat{\varepsilon}_1(\psi, f_w) \tag{10.3.5}$$

where the sum is over the set of characters ψ of $M\backslash\mathbf{M}^1$ whose restriction to $\mathbf{Z}$ is ω^{-1} and

$$\hat{\varepsilon}_1(\psi, f_w) = \int\limits_{\mathbf{Z}\backslash\mathbf{M}^1} \varepsilon_1(m)\psi(m)^{-1}dm\ .$$

Suppose that ϕ is a smooth compactly supported function on $\mathbf{M}$ of the form $\Pi\phi_v$ which transforms under $\mathbf{Z}$ via ω^{-1} and let ϕ_1 be the restriction of ϕ to $\mathbf{M}^1$. Let $\hat{\phi}$ denote the Fourier transform of ϕ:

$$\hat{\phi}(\psi') = \int\limits_{\mathbf{Z}\backslash\mathbf{M}} \phi(m)\psi'(m)^{-1}dm$$

for characters ψ' of $\mathbf{M}$ whose restriction to $\mathbf{Z}$ is ω^{-1}. If the restriction of ψ' to $\mathbf{M}^1$ is ψ, then, with a suitable normalization of measures,

$$\hat{\phi}_1(\psi) = \int_{-\infty}^{\infty} \hat{\phi}(\psi'\chi^{\mathrm{it}})dt = \int_{-\infty}^{\infty} \hat{\phi}_v(\psi'_w\chi_w^{\mathrm{it}})b(t)dt$$

where χ^{it} is defined by $\chi^{\mathrm{it}}(d(\alpha, \beta, \overline{\alpha}^{-1}) = |\alpha\overline{\alpha}|^{\mathrm{it}}$ and

$$b(t) = \prod_{v \neq w} \hat{\phi}_v(\psi'_v\chi_v^{\mathrm{it}}).$$

Now $\varepsilon(\gamma, f_w)$ is a sum of functions in product form and hence, as a function of f_w, each term in (10.3.5) is given by integration of the Satake transform $\hat{f}_w$ against a measure on the one-dimensional manifold of unitary unramified characters of M whose restrictions to Z_v is ω_v which is absolutely continuous with respect to Lebesgue measure. Hence this is true of (10.3.3). It is also true of (10.3.2), by §2.5.

We now apply an argument (decomposition of measures) due to Langlands ([L1], pg. 211]). We can assume that the place w remains prime in E, so that the Hecke algebra $\mathcal{H}_w(G)$ is isomorphic to the subalgebra of the algebra of Laurent polynomials in one variable $\mathbf{C}[z, z^{-1}]$ consisting of elements f such that $f(z) = f(z^{-1})$. Each term in (10.3.1) is defined as a discrete sum of traces of representations and, as a function of f_w, (10.3.1) is equal to a discrete sum of the form:

$$\sum c_j \hat{f}_w(z_j) . \tag{10.3.6}$$

The principal series representations corresponding to the z_j are all unitary and hence the z_j are contained in a compact subset X of $\mathbf{C}$. In fact, z_j lies either on the unit circle or in a compact sub-interval of $\mathbf{R}$. By the Stone-Weierstrass theorem, $\mathcal{H}_w(G)$ is dense in the space C of functions f on X such that $f(z) = f(z^{-1})$. The sum (10.3.6) is absolutely convergent for all $f \in \mathcal{H}_w(G)$ and hence (10.3.6) extends to a linear functional on C. Similarly, (10.3.2) and (10.3.3) extend to linear functionals on C, and their sum is equal to (10.3.6). This is an equality between a discrete measure and an absolutely continuous measure on X which, by the Riesz representation theorem, is impossible unless both sides are zero. This proves that (10.3.1) is zero and proves (a).

Part (b) is a consequence of (a). By (a),

$$\theta_{\tilde{G}}(\phi) = S\theta_G(f) + \frac{1}{2} S\theta_H(\phi^H) .$$

If f is a function on G all of whose stable orbital integrals vanish, then, this equality holds with $\phi = 0$ and $\phi^H = 0$. Hence $S\theta_G(f) = 0$.

CHAPTER 11

The Unitary group in two variables

This chapter contains local and global results on base change for $G = U(2)$. The results follow from the main equality by techniques which closely parallel but are more simple those used in §13. For this reason, we omit proofs of the results stated in §11.3 and §11.4 or for Theorem 11.5.1. Let $H = U(1) \times U(1)$ be the unique proper elliptic endoscopic group attached to G.

11.1 *L*-packets for $U(2)$. The group SL(2) and some related groups are considered in [LL]. Since the derived group of G is SL(2), the methods of [LL] apply to G with little modification, yielding the results of this section and Proposition 11.2.1.

Assume F is local. The group $\mathrm{PGL}_2(F)$ acts on G by conjugation and hence on $E(G)$. An L-packet on G is, by definition, a $\mathrm{PGL}_2(F)$-orbit in $E(G)$.

Let $\theta = \theta_1 \otimes \theta_2$ be a character of H. We call θ regular (resp., singular) if $\theta_1 \neq \theta_2$ (resp., $\theta_1 = \theta_2$). We will say $\theta' = \theta_1' \otimes \theta_2'$ is equivalent to θ if the sets $\{\theta_1', \theta_2'\}$ and $\{\theta_1, \theta_2\}$ coincide. If χ is a character of E^1, let $\tilde{\chi}$ be the character of $\mathbf{E}^*$, defined by $\tilde{\chi}(a) = \chi(a/\overline{a})$. If $\theta = \theta_1 \otimes \theta_1$ is singular, let $\tilde{\theta}$ denote the character $\tilde{\theta}_1$. We make analogous definitions in the global case.

We identify M with E^* and regard characters χ of M as characters of E^*. In the p-adic case, $i_G(\chi)$ is irreducible except in the following cases:

1) $\chi(\alpha) = \eta(\alpha)\|\alpha\|^{1/2}$ or $\eta(\alpha)\|\alpha\|^{-1/2}$ where $\eta|F^*$ is trivial.
2) $\chi|F^* = \omega_{E/F}$.

In both cases, $JH(i_G(\chi))$ has two elements. In case 1), $JH(i_G(\chi))$ consists of the one-dimensional representation $\xi(h) = \eta'(\det(h))$, where $\eta' \in \mathrm{Hom}(E^1, \mathbf{C})$ is defined by $\eta'(a/\overline{a}) = \eta(a)$, and of a square integrable (Steinberg) representation which we denote by $\mathrm{St}_G(\xi)$. In case 2), χ is unitary and $JH(i_G(\chi))$ is an l.d.s. L-packet.

To every character θ of H there is associated an L-packet $\rho(\theta)$ with two elements which depends only on the equivalence class of θ. This transfer corresponds to functoriality with respect to the embedding $\xi_H: {}^L H \to {}^L G$. In the global case, if θ is a regular character of $H\backslash\mathbf{H}$, then $\rho(\theta) = \otimes\rho(\theta_v)$ is a cuspidal L-packet which again depends only on the equivalence class of θ. If θ is singular, then $\rho(\theta)$ does not occur in the discrete spectrum.

PROPOSITION 11.1.1: (a) *If F is local (resp., global), an L-packet (resp., discrete L-packet) has more than one element if and only if it is of the form $\rho(\theta)$ for some θ (resp., θ regular).*
Assume that F is local.

(b) *Let θ be a character of H. Then the elements of $\rho(\theta)$ can be numbered as π_1, π_2 so that the character identity*

$$\mathrm{Tr}(\pi_1(f)) - \mathrm{Tr}(\pi_2(f)) = \mathrm{Tr}(\rho(\theta)(f^H))$$

holds.

(c) *If θ is singular, then $\rho(\theta)$ is the L-packet $JH(i_G(\tilde{\theta}\mu^{-1}))$.*

(d) *If F is archimedean, then $\rho(\theta)$ is square-integrable if and only if θ is regular and every square-integrable L-packet on H is of the form $\rho(\theta)$ with θ regular.*

(e) *If F is p-adic, then $\rho(\theta)$ is supercuspidal if and only if θ is regular.*

11.2 Trace formula. In the rest of the chapter, it will be tacitly assumed that, locally and globally, all representations of G and H transform under Z via ω and all representations of $\widetilde{\mathbf{G}}$ transform under $\widetilde{Z}$ via $\tilde{\omega}$. Let $\Pi(\mathbf{G})$ be the set of discrete L-packets on G.

The residual part of $\theta_G(f)$ is the sum:

$$\frac{1}{4}\sum_{\chi=w\chi} \mathrm{Tr}(M(\chi)I_\chi(f))$$

where χ ranges over the characters of $M\backslash\mathbf{M}$ such that $\chi = w\chi$ and $M(\chi)$ is the intertwining operator associated to w acting on I_χ. The relation $\chi = w\chi$ is satisfied if and only if $\chi\overline{\chi} = 1$.

PROPOSITION 11.2.1: (a) *If $\rho \in \Pi(\mathbf{G})$ is not of the form $\rho(\theta)$, then the multiplicities $m(\pi)$ for $\pi \in \rho$ all coincide. Denote their common value by $n(\rho)$.*

(b) *$S\theta_G(f)$ is equal to the sum of the following terms:*

(1) $\sum' n(\rho)\ \mathrm{Tr}(\rho(f))$

(2) $\frac{1}{2}\sum'\ \mathrm{Tr}(\rho(\theta)(f))$

(3) $-\frac{1}{4} \sum_{\substack{\chi = w\chi \\ \chi | C_F = 1}} \mathrm{Tr}(I_\chi(f))$

The first sum is over the discrete L-packets on G which are not of the form $\rho(\theta)$. The second sum is over a set of representatives for the equivalence classes of regular characters θ of $H\backslash\mathbf{H}$. The third sum is over the set of characters of $E^*\backslash I_E$ such that $\chi|C_F = 1$.

(b) *Let $n(\rho) = \frac{1}{2}$ if ρ is of the form $\rho(\theta)$. Then*

$$\mathrm{Tr}(\rho_d(f)) = \sum{}' n(\rho)\mathrm{Tr}(\rho(f)) + \frac{1}{2}\sum{}' \mathrm{Tr}(\theta(f^H)) .$$

Proof: Recall that $S\theta_G(f) = \theta_G(f) - \frac{1}{2} S\theta_H(f^H)$. Since H is anisotropic and abelian, $S\theta_H(f^H) = \sum \mathrm{Tr}(\theta(f^H))$, where θ ranges over the characters of $H\backslash\mathbf{H}$. This proposition follows by a minor modification of the methods used in [LL] for the case of SL(2). In particular, by [KyS], Theorem 5.1, if $\chi_1|C_F = 1$, then I_χ is irreducible and $M(\chi)$ acts as multiplication by -1. The terms involving $\mathrm{Tr}(M(\chi)I_\chi(f))$ where $\chi_1|C_F = \omega_{E/F}$ are cancelled by terms in $S\theta_H(f^H)$ associated to singular θ.

We regard a character χ of $\widetilde{M}$ as a pair of characters (χ_1, χ_2) of E^*. To a character $\theta = \theta_1 \otimes \theta_2$ of H we associate the following two characters of $\widetilde{M}$:

$$\mu(\theta) = (\tilde{\theta}_1\mu^{-1}, \tilde{\theta}_2\mu^{-1}) \ , \ \mu'(\theta) = (\tilde{\theta}_1, \tilde{\theta}_2) \ .$$

Using the multiplicity one theorem for GL(2), we obtain the following expression for $\theta_{\widetilde{G}}(\phi)$ (cf. §13.5).

PROPOSITION 11.2.2: *$\theta_{\widetilde{G}}(\phi)$ is equal to the sum of the following terms:*

(1) $\sum{}' \mathrm{Tr}(\pi(\phi)\pi(\varepsilon))$

(2) $\frac{1}{2}\sum \mathrm{Tr}(I_{\mu(\theta)}(\phi)I_{\mu(\theta)}(\varepsilon))$

(3) $\frac{1}{2}\sum \mathrm{Tr}(I_{\mu'(\theta)}(\phi)I_{\mu'(\theta)}(\varepsilon))$

(4) $-\frac{1}{4} \sum_{\substack{\chi = (\chi_1, \chi_1) \\ \chi_1\overline{\chi}_1 = 1}} \mathrm{Tr}(I_\chi(\phi)I_\chi(\varepsilon))$

The first sum is over the set of discrete representations of $\widetilde{G}$. The second and third sums are over a set of representatives for the equivalence classes of regular characters θ of $H\backslash\mathbf{H}$. Here $I_{\mu(\theta)}(\varepsilon)$ denotes the operator

$M_{\widetilde{B}|\widetilde{B}}(w)I^{\widetilde{B}}_{\mu(\theta)}(\varepsilon)$, *where* w *is the non-trivial element in the Weyl group. The fourth sum is over the set of characters* χ_1 *of* I_E *such that* $\chi_1\overline{\chi}_1 = 1$ *and* $I_\chi(\varepsilon)$ *denotes the standard action of* ε *on* I_χ.

11.3 The main equality. If $\chi' = (\chi_1, \chi_1)$ is a character of $\widetilde{\mathbf{M}}$ such that $\chi_1\overline{\chi}_1 = 1$, then either $\chi_1|C_F = 1$ and

$$\mathrm{Tr}(I_{\chi'}(\phi)I_{\chi'}(\varepsilon)) = \mathrm{Tr}(I_{\chi_1}(f_1)) ,$$

or $\chi_1 = \chi\mu|C_F$ where $\chi|C_F = 1$ and

$$\mathrm{Tr}(I_{\chi'}(\phi)I_{\chi'}(\varepsilon)) = \mathrm{Tr}(I_\chi(f_2)) ,$$

by Proposition 4.11.1(c). Every character such that $\chi_1\overline{\chi}_1 = 1$ is of the form χ or χ_μ where $\chi|C_F = 1$, and hence

$$-\frac{1}{4}\sum_{\substack{\chi'=(\chi_1,\chi_1)\\ \chi_1\overline{\chi}_1=1}} \mathrm{Tr}(I_{\chi'}(\phi)I_{\chi'}(\varepsilon)) + \frac{1}{4}\sum_{\substack{\chi=w\chi\\ \chi|C_F=1}} \mathrm{Tr}(I_\chi(f_1)) + \frac{1}{4}\sum_{\substack{\chi=w\chi\\ \chi|C_F=1}} \mathrm{Tr}(I_\chi(f_2))$$

is equal to zero. By Propositions 11.2.1 and 11.2.2,

$$\theta_{\widetilde{G}}(\phi) - \theta_G(f_1) - \theta_G(f_2)$$

is equal to the sum of the terms

$$\begin{aligned}
&(1) \quad \textstyle\sum' \mathrm{Tr}(\pi(\phi)\pi(\varepsilon)) - \sum' m(\rho)\,\mathrm{Tr}(\rho(f_1)) - \sum' m(\rho)\,\mathrm{Tr}(\rho(f_2))\\
&(2) \quad \frac{1}{2}\sum \mathrm{Tr}(I_{\mu(\theta)}(\phi)I_{\mu(\theta)}(\varepsilon)) - \frac{1}{2}\sum \mathrm{Tr}(\rho(\theta)(f_1))\\
&(3) \quad \frac{1}{2}\sum \mathrm{Tr}(I_{\mu'(\theta)}(\phi)I_{\mu'(\theta)}(\varepsilon)) - \frac{1}{2}\sum \mathrm{Tr}(\rho(\theta)(f_2)).
\end{aligned}$$

As shown in §13.7 for the case of $U(3)$, each of the lines (1)-(3) is equal to zero. The results of §11.4 and 11.5 are deduced from the resulting three equalities.

11.4 Local base change. Let $E_\varepsilon(\widetilde{G})$ be the set of irreducible ε-invariant representations of $\widetilde{G}$ whose central character is trivial on F^*. There are two base change maps from $\Pi(G)$ to $E_\varepsilon(G)$, reflecting the existence of the two maps ψ_G and ψ'_G (§4.7). Let $\rho \in \Pi(H)$. To define the base change lifts, assume first that ρ is tempered. We will say that $\tilde{\rho}$ is a lift of ρ with respect to ψ_G if the following relations hold:

$$\chi_{\pi\varepsilon}(\phi) = \begin{cases} \chi_\rho(f_1) & \text{if} \quad \pi = \tilde{\rho} \\ \chi_\rho(f_2) & \text{if} \quad \pi = \tilde{\rho}\otimes\mu. \end{cases} \tag{11.4.1}$$

If this is the case, then $\tilde{\rho}' = \tilde{\rho} \otimes \mu$ is the lift of ρ with respect to ψ'_G. These character identities determine $\chi_{\tilde{\rho}\varepsilon}$, which in turn determines $\tilde{\rho}$ ([L₁], Lemma 7.16), and hence $\tilde{\rho}$ is unique, if it exists. We denote the base change lifts $\psi_G(\rho)$ and $\psi'_G(\rho)$ by $\tilde{\rho}$ and $\tilde{\rho}'$, respectively.

For $\chi \in \mathrm{Hom}(M, \mathbf{C}^*)$, set

$$\tilde{\chi} = (\chi, \overline{\chi}^{-1}) = \chi \circ N, \ \tilde{\chi}' = (\chi_\mu, \overline{\chi}_\mu^{-1}) .$$

If χ is unitary and $\rho = JH(i_G(\chi))$, then $i_{\widetilde{G}}(\tilde{\chi}')$ and $i_{\widetilde{G}}(\tilde{\chi}')$ are irreducible and by Proposition 4.11.1(c), we have $\tilde{\rho} = i_G(\tilde{\chi})$ and $\tilde{\rho}' = i_{\widetilde{G}}(\tilde{\chi}')$.

If ρ is non-tempered, then ρ is a Langlands quotient of $i_B(\chi)$ for some χ which is positive with respect to B. The characters $\tilde{\chi}$ and $\tilde{\chi}'$ are then positive with respect to $\widetilde{B}$ and we define $\tilde{\rho}$ and $\tilde{\rho}'$ to be $i_{\widetilde{G}}(\tilde{\chi})$ and $i_{\widetilde{G}}(\tilde{\chi}')$, respectively. This completes the definition of the liftings. The next proposition gives its existence and some of properties.

PROPOSITION 11.4.1: *Let $\rho \in \Pi(G)$. There exist liftings of ρ with respect to ψ_G and ψ'_G.*

(a) *If $\rho = \rho(\theta)$, then $\tilde{\rho} = i_{\widetilde{G}}(\mu(\theta))$ and $\tilde{\rho}' = i_{\widetilde{G}}(\mu'(\theta))$.*

(b) *If $\dim(\rho) = 1$ and $\rho(h) = \eta(\det(h))$, where $\eta \in \mathrm{Hom}(E^1, \mathbf{C}^*)$, then $\tilde{\rho}(h) = \tilde{\eta}(\det(h))$ and $\tilde{\rho}'(h) = \tilde{\eta}\mu(\det(h))$. If $\pi \in E_\varepsilon(\widetilde{G})$ is one-dimensional $\pi = \tilde{\rho}$ or $\tilde{\rho}'$ for a unique one-dimensional representation ρ.*

(c) *Suppose that F is p-adic. Every square-integrable representation $\pi \in E_\varepsilon(\widetilde{G})$ is of the form $\tilde{\rho}$ or $\tilde{\rho}'$ for a unique square-integrable L-packet ρ which is not of the form $\rho(\theta)$. The sets $\psi_G(\Pi^2(G))$ and $\psi'_G(\Pi^2(G))$ are disjoint.*

The character identity (11.4.1) may not hold if ρ is not tempered. Suppose that ρ is a Langlands quotient of $i_G(\chi)$. If $i_G(\chi), i_{\widetilde{G}}(\tilde{\chi})$, and $i_G(\tilde{\chi}')$ are all irreducible, then (11.4.1) does hold by Proposition 4.11.1(c). The character identity breaks down if $i_G(\chi)$ is irreducible and $i_G(\tilde{\chi})$ or $i_G(\tilde{\chi}')$ are reducible, e.g., if $\chi(\alpha) = \psi(\alpha)\|\alpha\|^{1/2}$, where $\psi|F^* = \omega_{E/F}$. Suppose that ρ is non-tempered and unitary. If ρ is one-dimensional, then so are $\tilde{\rho}$ and $\tilde{\rho}'$ and (4.11.1) holds (cf. the argument for Proposition 12.4.11). If ρ is infinite-dimensional, then $i_G(\chi)$ is an irreducible complementary series representation and $i_{\widetilde{G}}(\tilde{\chi})$ and $i_{\widetilde{G}}(\tilde{\chi}')$ are also irreducible. Hence the character identity holds for unitary ρ and, in particular, for ρ which occur as local components of discrete automorphic representations.

11.5. Global base change: Let $\Pi_\varepsilon(\widetilde{\mathbf{G}})$ be the set of ε-invariant discrete representations of $\widetilde{G}$ such that the restriction of ω_π to I_F is trivial. We call a global L-packet cuspidal if all of its discrete members are cuspidal. Let $\Pi(\mathbf{G})$ be the set of discrete L-packets and let $\Pi_0(\mathbf{G})$ be the set of cuspidal L-packets on G. Then $\Pi(\mathbf{G})$ is the union of $\Pi_0(\mathbf{G})$ and the one-dimensional automorphic representations of G. Let $\Pi_s(\mathbf{G})$ be the set of cuspidal L-packets on G which are not of the form $\rho(\theta)$. For $\rho = \otimes\rho_v \in \Pi(\mathbf{G})$, define global base change maps $\psi_G(\rho) = \otimes\psi_G(\rho_v)$ and $\psi'_G(\rho) = \otimes\psi'_G(\rho_v)$.

THEOREM 11.5.1: (a) *The maps ψ_G and ψ'_G are injective. The sets $\psi_G(\Pi(\mathbf{G}))$ and $\psi'_G(\Pi(\mathbf{G}))$ are disjoint and every element of $\Pi_\varepsilon(\widetilde{\mathbf{G}})$ is in the image of ψ_G or ψ'_G.*

(b) *If $\rho \in \Pi_0(\mathbf{G})$, then $\psi_G(\rho)$ (or $\psi'_G(\rho)$) is cuspidal if and only if $\rho \in \Pi_s(\mathbf{G})$.*

(c) *$n(\rho) = 1$ if $\rho \in \Pi_s(\mathbf{G})$.*

THEOREM 11.5.2: *If π is an ε-invariant discrete representation of $\widetilde{\mathbf{G}}$, then the restriction of ω_π to I_F is trivial.*

Proof of Theorem 11.5.2: Let $\tilde{\omega}$ be a character of $\widetilde{\mathbf{Z}}$ whose restriction to I_F is $\omega_{E/F}$. Let ρ_d be the representation of $\widetilde{\mathbf{G}}$ on the space $L_d(\widetilde{G})$ defined with respect to $\tilde{\omega}$. Then $\rho_d(\varepsilon)$ is defined since $\tilde{\omega}$ is ε-invariant. The assertion of the theorem is that the space $L_d(\widetilde{G})$ is trivial. It will suffice to show that $\mathrm{Tr}(\rho_d(\phi)\rho_d(\varepsilon)) = 0$ for all ϕ such that $\phi(zg) = \tilde{\omega}(z)^{-1}\phi(g)$. To simplify the notation and without loss of generality, we assume that $\tilde{\omega} = \mu$.

Let $\delta \in \widetilde{G}$ be an ε-semisimple element. Observe that if δ is ε-conjugate to $z\delta$ for some $z \in I_F$ such that $\omega_{E/F}(z) = -1$, then $\Phi_\varepsilon(\delta,\phi) = 0$ by a change of variables. Let γ be a norm of δ. Suppose that v is a place of F that remains prime in E. Then δ is ε-conjugate to $z\delta$ in $\widetilde{G}_v$ for all $z \in F_v^*$ unless γ is a regular element in a Cartan subgroup of G_v of type (1). Suppose that γ is regular and let $T = G_\gamma$. It is immediate from the classification of Cartan subgroups of G (§3.6) that if T_v is of type (1) for all places v which do not split in E, then T is of type (1). Indeed, if $T = T_K$ where $K \neq E$, then the set of places of F which split in E contains the set of places of F which split in K, and this is impossible. This shows that $\Phi_\varepsilon(\delta,\phi) = 0$ unless T is of type (1). In addition, $\Phi_\varepsilon^M(\delta,\phi_v)$ is identically zero for v which remain prime in E and hence the global weighted orbital integrals for ε-regular $\delta \in \widetilde{M}$ vanish.

The twisted trace formula therefore becomes very simple. We will simplify it still further by making the following strong assumption: for all places v which remain prime in E, either ϕ_v belongs to the Hecke algebra or $\mathrm{supp}(\phi_v)$ is contained in the ε-regular subset of $\widetilde{G}_v$. We also assume that the latter condition holds for at least one v. Let H be a Cartan subgroup of type (1) and let $\widetilde{H}$ be the centralizer of H in $\widetilde{G}$. By our assumption, all that remains of the $\mathcal{O}$-expansion in the twisted trace formula is:

$$\sum_{\delta \in \mathcal{C}'} \varepsilon(\delta)^{-1} m(\mathbf{Z}G_\gamma \backslash \mathbf{G}_\gamma) \Phi_\varepsilon(\delta, \phi) \tag{11.5.1}$$

where $\mathcal{C}'$ is a set of representatives modulo $\widetilde{Z}$ for the ε-conjugacy classes of elements in $\widetilde{H}$ such that $\gamma = N(\delta)$ is regular.

Observe that H is an elliptic endoscopic group for $\widetilde{G}$. We now re-write (11.5.1) in terms of κ-orbital integrals and then as the trace formula applied to a function on H.

Let $\delta \in \mathcal{C}'$ and let $\widetilde{\mathcal{O}} = \mathcal{O}_{\varepsilon-\mathrm{st}}(\delta)$. Locally and globally, the ε-conjugacy classes within $\widetilde{\mathcal{O}}$ are parametrized by $H^1(F, H)$. We identify $H^1(F, H)$ with $(F^*/NE^*)^2$ in such a way that $H^1(F, Z) = F^*/NE^*$ is embedded diagonally.

Suppose that F a local field. If E/F is a quadratic extension, let κ_1 and κ_2 be the two characters of $(F^*/NE^*)^2$ whose restriction to $H^1(F, Z)$ is non-trivial and define

$$\Phi_\varepsilon^{\kappa_j}(\delta, \phi_v) = \sum \kappa_j(\mathrm{inv}(\delta, \delta'))\Phi_\varepsilon(\delta', \phi_v) \ ,$$

where the sum is over a set $\{\delta'\}$ of representatives for the ε-conjugacy classes within $\widetilde{\mathcal{O}}$ modulo F^*. The terms in the sum depend only on δ' modulo F^*. Fix identifications of H and $\widetilde{H}$ with $E^1 \times E^1$ and $E^* \times E^*$, respectively. The norm of $(\delta_1, \delta_2) \in H$ is $(\delta_1/\overline{\delta}_1, \delta_2/\overline{\delta}_2)$. By §3.11, the elements $(\delta_1, \delta_2), (\nu\delta_1, \delta_2), (\delta_1, \nu\delta_2), (\nu\delta_1, \nu\delta_2)$, where $\nu \in F^* - NE^*$, make up a set of representatives for the ε-conjugacy classes within $\widetilde{\mathcal{O}}_{\mathrm{st}}$. We number the κ_j numbered so that:

$$\Phi_\varepsilon^{\kappa_1}(\delta, \phi_v) = \Phi_\varepsilon((\delta_1, \delta_2), \phi_v) - \Phi_\varepsilon((\nu\delta_1, \delta_2), \phi_v)$$
$$\Phi_\varepsilon^{\kappa_2}(\delta, \phi_v) = \Phi_\varepsilon((\delta_1, \delta_2), \phi_v) - \Phi_\varepsilon((\delta_1, \nu\delta_2), \phi_v).$$

If E is not a field, set $\Phi_\varepsilon^{\kappa_j}(\delta, \phi_v) = \Phi(\delta, \phi_v)$. Then

$$\Phi_\varepsilon^{\kappa_j}((\xi_1\delta_1, \xi_2\delta_2), \phi_v) = \mu^{-1}(\xi_j)\Phi_\varepsilon^{\kappa_j}((\delta_1, \delta_2), \phi_v)$$

for $\xi_1, \xi_2 \in F^*$. Define

$$\widetilde{\Delta}_1(\delta) = \mu^{-1}(\delta_1)\mu(\gamma_1 - \gamma_2)D_G(\gamma), \ \widetilde{\Delta}_2(\delta) = \mu^{-1}(\delta_2)\mu(\gamma_1 - \gamma_2)D_G(\gamma).$$

and let

$$\phi_{vj}^H(\gamma) = \widetilde{\Delta}_j(\delta)\Phi_\varepsilon^{\kappa_j}(\delta, \phi_v)$$

where $\gamma = N(\delta)$. Then $\widetilde{\Delta}_j(\delta)\Phi_\varepsilon^{\kappa_j}(\delta, \phi_v)$ depends only on γ and $\phi_{vj}^H(\gamma)$ is well-defined function on H. If $z \in \widetilde{Z}$, then

$$\widetilde{\Delta}_j(z\delta) = \mu^{-1}(z\delta_2)\mu((z/\overline{z})(\gamma_1 - \gamma_2))\ D_G((z/\overline{z})\gamma) = \mu(z)\widetilde{\Delta}_j(z\delta)$$

and $\phi_{vj}^H(\gamma)$ in fact depends only on the image of γ in $Z\backslash T$.

Globally, let $\Phi_\varepsilon^{\kappa_j}(\delta, \phi)$ be the product of the local integrals $\Pi\Phi_\varepsilon^{\kappa_j}(\delta, \phi_v)$ and let $\phi_j^H(\gamma) = \Pi\phi_{vj}^H(\gamma)$. Then $\phi_j^H(\gamma) = \Phi_\varepsilon^{\kappa_j}(\delta, \phi)$, by the product formula and, as in §5.5, we can write (11.5.1) as

$$\frac{1}{2}\sum_{j=1,2}\sum_{\delta\in\mathfrak{C}''} \varepsilon_{\mathrm{st}}(\delta)^{-1} m(\mathbf{Z}G_\gamma\backslash\mathbf{G}_\gamma)\Phi_\varepsilon^{\kappa_j}(\delta, \phi)$$

where $\mathfrak{C}''$ is a set of representatives modulo $\widetilde{Z}$ for the stable ε-conjugacy classes of elements in $\widetilde{H}$ such that $\gamma = N(\delta)$ is regular.

The endoscopic embeddings η_1, η_2 of ${}^L\widetilde{G}$ were defined in §4.7. The next result is the fundamental lemma for the η_j. We refer to [BR$_1$] for a proof.

LEMMA 11.5.3: *Suppose that $\phi_v \in \mathcal{H}(\widetilde{G}, \mu_v)$. Then $\phi_{vj}^H(\gamma) = \hat{\eta}_j(\phi_v)$.*

As noted in §4.7, η_1 and η_2 are equivalent and hence $\hat{\eta}_1$ and $\hat{\eta}_2$ coincide. Explicitly, if E/F is a quadratic extension, then the lemma asserts that $\phi_{vj}^H(\gamma) = \mathrm{Tr}(I_{\chi_j}(\phi)I_{\chi_j}(\varepsilon))$ for $\phi_v \in \mathcal{H}(\widetilde{G}, \mu_v)$, where $\chi_1 = (\mu, 1)$ and $\chi_2 = (1, \mu)$, and $I_{\chi_j}(\varepsilon)$ is normalized by the requirement that it fix the space of $\widetilde{K}$-fixed vectors. The equality $\hat{\eta}_1 = \hat{\eta}_2$ is clear since I_{χ_1} is isomorphic to I_{χ_2}.

The functions ϕ_{vj}^H extend to smooth functions on $Z_v\backslash T_v$. This is immediate if v splits in E, but requires proof if v remains prime. In view of our assumption on ϕ and Lemma 11.5.3, however, it automatically holds, and (11.5.1) can be written as:

$$\frac{1}{4} m(\mathbf{Z}T\backslash\mathbf{T}) \sum_{j=1,2}\sum_{\gamma\in Z\backslash T} \phi_j^H(\gamma) \tag{11.5.2}$$

Here we use that an ε-conjugacy class modulo $\widetilde{Z}$ occurs $2/\varepsilon_{st}(\delta)$ times as γ ranges through a set of representative for $Z\backslash T$. By the trace formula (Poisson summation formula) for $Z\backslash T$, (11.5.2) is equal to

$$\frac{1}{4}\sum \mathrm{Tr}(\theta(\phi_j^H))$$

where $\theta = (\theta_1, \theta_2)$ ranges over the characters of $Z\backslash T$.

The χ-expansion for $\widetilde{G}$ is the sum of $\theta_{\widetilde{G}}(\phi)$ and $\theta_{\widetilde{M}}(\phi)$. If a quintuple $(\widetilde{M}, \widetilde{M}, \mathcal{A}, \sigma, s)$ (cf. §2.3) contributes to $\theta_{\widetilde{M}}(\phi)$, then $s = 1$ and hence $\varepsilon(\sigma) = \sigma$. However, an ε-invariant character is trivial on I_F and the term $\theta_{\widetilde{M}}(\phi)$ is therefore empty. On the other hand, $\theta_{\widetilde{G}}(\phi)$ is equal to the sum of two terms:

$$\sum{}' \mathrm{Tr}(\pi(\phi)\pi(\varepsilon))$$

$$\frac{1}{2}\sum \mathrm{Tr}(I_{\theta'}(\phi)I_{\theta'}(\varepsilon))$$

where the first sum is over the set of ε-invariant discrete representations of $\widetilde{G}$ with central character μ and θ' ranges over a set of equivalence classes of characters $\theta' = (\chi_1, \chi_2)$ such that $\chi_j\overline{\chi}_j = 1$ and $\chi_1\chi_2 = \mu$. For example we can take $\theta' = (\tilde{\theta}_1\mu, \tilde{\theta}_2)$, where $\theta = (\theta_1, \theta_2)$ runs over the set of characters of $Z\backslash T$. In writing down the first term, we are using the multiplicity one theorem for GL(2). The twisted trace formula becomes the equality.

$$\sum{}' \mathrm{Tr}(\pi(\phi)\pi(\varepsilon)) = \frac{1}{4}\sum_{j=1}^{2}\sum_{\theta} \mathrm{Tr}(\theta(\phi_j^H) - \frac{1}{2}\sum_{\theta'} \mathrm{Tr}(I_{\theta'}(\phi)I_{\theta'}(\varepsilon)) \tag{11.5.3}$$

To complete the proof, we use the terminology of §13.6. If $\theta = (\theta_1, \theta_2)$, the distribution $\phi \to \mathrm{Tr}(\theta(\phi_j^H))$ defines the same e.v.p. on $\widetilde{G}$ as $I_{\theta'}$, where $\theta' = (\tilde{\theta}_1, \mu, \theta_2)$ by Lemma 11.5.3. Hence each term on the right-hand side of (11.5.3) defines an e.v.p. of the form $t_S(I_{\theta'})$. Suppose that π occurs on the left-hand side. Let S be a finite set of places such that π_v is unramified for $v \notin S$ and assume that $\phi_v \in \mathcal{H}(\widetilde{G}, \mu_v)$ for all $v \notin S$. For $v \in S$, we can choose ϕ_v with support in the ε-regular set if v remains prime in F (and we assume that S contains at least one such v) such that $\mathrm{Tr}(\pi_v(\phi_v)\pi_v(\varepsilon)) \neq 0$. By the strong multiplicity theorem for GL(2), π is the only term on the left-hand side with e.v.p. $t_S(\pi)$ and hence $t_S(\pi)$ must appear on the left-hand side. However, π is cuspidal (there are no ε-invariant one-dimensional representations of G with central character μ) and therefore, by the Jacquet-

Shalika theorem ([JS], cf. §13.6), $t_S(\pi)$ does not coincide with $t_S(I_{\theta'})$ for any character θ of $\widetilde{M}\backslash\widetilde{M}$. The sum on the left-hand side is therefore empty.

CHAPTER 12

Representation theory

Let F denote a local field. Unless otherwise stated, we set $G = U(3)$, $H = U(2) \times U(1)$, and $C = U(1) \times U(1) \times U(1)$. The purpose of this chapter is to describe the L-packets of representations for H and G. Certain L-packets on G can be defined directly. Their existence in general, for the p-adic case, is proved in §13. In addition, some results needed in §13 are assembled.

12.1 *L*-packets for *H*. Let χ_1 be a character of E^* and let χ_2 be a character of E^1. We denote by $\chi = (\chi_1, \chi_2)$ the character of M defined by

$$\chi(d(\alpha, \beta, \overline{\alpha}^1)) = \chi_1(\alpha)\chi_2(\alpha\overline{\alpha}^{-1}\beta)$$

By the results quoted in §11.1, in the p-adic case $i_H(\chi)$ is irreducible except in the following cases:

1) $\chi_1(\alpha) = \eta(\alpha)\|\alpha\|^{1/2}$ or $\eta(\alpha)\|\alpha\|^{1/2}$ where $\eta|F^*$ is trivial.
2) $\chi_1|F^* = \omega_{E/F}$.

In case 1), define $\eta' \in \mathrm{Hom}(E^1, \mathbf{C}^*)$ by $\eta'(a/\overline{a}) = \eta(a)$ and let ξ be the character of H defined by:

$$\xi(h) = \eta'(\det_0(h))\ \chi_2(\det(h))\ .$$

Here $\det_0$ denotes the determinant on the $U(2)$-factor of H. Then $\xi \in JH(i_H(\chi))$. The remaining constituent is a square-integrable (Steinberg) representation of H which we denote by $St_H(\xi)$. In case 2), χ is unitary and $JH(i_H(\chi))$ is an l.d.s. L-packet.

To each character θ of C is associated an L-packet $\rho(\theta)$ on H. Suppose that $\theta = \theta_1 \otimes \theta_2 \otimes \theta_3$. Then $\rho(\theta_1 \otimes \theta_3)$ is an L-packet on $U(2)$ (§11.1) and we define $\rho(\theta)$ to be the L-packet $\rho(\theta_1 \otimes \theta_3) \otimes \theta_2$ on $U(2) \times U(1)$. We call θ regular (resp., semi-regular) if the θ_j are distinct (resp., $\theta_1 \neq \theta_3$ and $\theta_2 = \theta_1$ or θ_3) and singular if $\theta_1 = \theta_2 = \theta_3$. A semi-regular character θ can be uniquely written in the form $\psi \otimes \varphi\psi \otimes \varphi\psi$ or $\varphi\psi \otimes \varphi\psi \otimes \psi$ for $\varphi, \psi \in \mathrm{Hom}(E^1, \mathbf{C}^*)$ where $\varphi \neq 1$. Call θ H-singular if $\theta_1 = \theta_3$, and H-

regular otherwise. An H-singular character can be uniquely written in the form $\theta = \varphi\psi \otimes \psi \otimes \varphi\psi$. Recall that if χ is a character of E^1, then $\tilde{\chi}$ is the character of E^* defined by $\tilde{\chi}(a) = \chi(a/\bar{a})$. If θ is semi-regular or H-singular, let $\tilde{\theta}$ denote the character $(\tilde{\varphi}, \psi)$ of M. If θ is H-singular, the L-packet $\rho(\theta)$ is $JH(i_H(\tilde{\theta}\mu^{-1}))$.

Let $\theta' = \theta'_1 \otimes \theta'_2 \otimes \theta'_3$. We will say that θ' and θ are G-equivalent (resp., H-equivalent) if the sets $\{\theta'_1, \theta'_2, \theta'_3\}$ and $\{\theta_1, \theta_2, \theta_3\}$ coincide (resp., if $\theta'_2 = \theta_2$ and the sets $\{\theta'_1, \theta'_3\}$ and $\{\theta_1, \theta_3\}$ coincide). The same notation will be used in the global case.

By Jacquet's theorem ([Ca], Theorem 2.5), in the p-adic case an irreducible representation of H is either supercuspidal or a constituent of an induced representation $i_H(\chi)$. We obtain the following list of L-packets on H for F p-adic; the first four types are the L-packets of cardinality one.

(1) $i_H(\chi)$, irreducible principal series
(2) $St_H(\xi)$, ξ = one-dimensional representation of H
(3) one-dimensional representation of H
(4) ρ supercuspidal, not in any L-packet $\rho(\theta)$.
(5) $\rho(\theta)$, where θ is H-regular; $\mathrm{Card}(\rho(\theta)) = 2$
(6) $\rho = JH(i_H(\tilde{\theta}\mu^{-1}))$, where θ is H-singular; $\mathrm{Card}(\rho) = 2$.

We consider the two base change maps ψ_H and ψ'_H from H to $\widetilde{H}$. We will denote the transfer $\psi_H(\rho)$ of ρ with respect to ψ_H by $\tilde{\rho}$ and the transfer with respect to ψ'_H by $\tilde{\rho}'$. If $\rho = \rho_1 \otimes \chi$, where ρ_1 is a representation of $U(2)$ and χ is a character of $U(1)$, then $\tilde{\rho} = \tilde{\rho_1} \otimes \tilde{\chi}$ and $\widetilde{\rho'} = \widetilde{\rho'_1} \otimes \tilde{\chi}$, where $\tilde{\rho}_1$ and $\tilde{\rho}'_1$ are as in §11.4.

The maps are again given explicitly in certain cases as follows. If χ_1, χ_2, χ_3 are characters of E^*, we denote by (χ_1, χ_2, χ_3) the character of $\widetilde{M}$ sending $d((a_1, a_2, a_3))$ to $\Pi\chi_j(a_j)$. For a character $\chi = (\chi_1, \chi_2)$ of M, set

$$\tilde{\chi} = (\chi_1, \chi_2, \overline{\chi_1}^{-1}) = \chi \circ N \quad , \quad \tilde{\chi}' = (\chi_1\mu, \chi_2, \overline{\chi_1}^{-1}\mu) \, .$$

If χ is unitary and $\rho = JH(i_H(\chi))$, then $\tilde{\rho} = i_{\widetilde{H}}(\tilde{\chi})$ and $\tilde{\rho}' = i_{\widetilde{H}}(\tilde{\chi}')$.

Let $\theta = \theta_1, \otimes\theta_2 \otimes \theta_3 \in \mathrm{Hom}(C, \mathbf{C}^*)$ and set $\rho = \rho(\theta)$. Define:

$$\mu(\theta) = (\tilde{\theta}_1\mu^{-1}, \tilde{\theta}_2, \tilde{\theta}_3\mu^{-1}) \, , \; \mu'(\theta) = (\tilde{\theta}_1, \tilde{\theta}_2, \tilde{\theta}_3)$$

Then $\tilde{\rho} = i_{\widetilde{H}}(\mu(\theta))$ and $\tilde{\rho}' = i_{\widetilde{H}}(\mu'(\theta))$, as follows from Proposition 11.4.1. Note that this is what is expected from the relation $\psi'_H \circ \xi_C = \xi_{\tilde{C}} \circ \psi_C$, remarked in §4.8

Let $\eta_1, \eta_2 \in \mathrm{Hom}(E^1, \mathbf{C}^*)$ and let ξ be the character of H defined by:

$$\xi(h) = \eta_1(\det_0(h))\ \eta_2(\det(h)).$$

Then

$$\tilde{\xi}(h) = \tilde{\eta_1}(\det_0(h))\tilde{\eta_2}(\det(h)),\ \tilde{\xi}'(h) = \tilde{\eta_1}\mu(\det_0(h))\tilde{\eta_2}(\det(h))$$

Every ε-invariant one-dimensional representation of $\widetilde{H}$ is either of the form $\tilde{\xi}$ or $\tilde{\xi}'$ for a unique one-dimensional ξ. If ψ is a one-dimensional representation of $\widetilde{H}$, let $St_{\widetilde{H}}(\psi)$ denote the square-integrable representation such that $\{\psi, St_{\widetilde{H}}(\psi)\}$ is the set of constituents of an induced representation $i_{\widetilde{H}}(\chi)$. If $\rho = St_H(\xi)$, then $\tilde{\rho} = St_{\widetilde{H}}(\tilde{\xi})$ and $\tilde{\rho}' = St_{\widetilde{H}}(\tilde{\xi}')$.

In the p-adic case, if ρ is square-integrable but not of the form $\rho(\theta)$, then $\tilde{\rho}$ and $\tilde{\rho}'$ are distinct square-integrable representations by the results of §11. Every ε-invariant square-integrable representation of $\widetilde{H}$ is either of the form $\tilde{\rho}$ or $\tilde{\rho}'$ for a unique L-packet ρ.

12.2 U(3) over a p-adic field. In this section, E/F will denote a quadratic extension of p-adic fields. Let $E(G)$ be the set of irreducible admissible representations of G. An element of $E(G)$ is either supercuspidal or a constituent of an induced representation $i_G(\chi)$, by Jacquet's theorem. If $i_G(\chi)$ is reducible, it contains exactly two irreducible constituents ([BZ]). Furthermore $i_G(\chi)$ and $i_G(w\chi)$ have the same sets of constituents. If $\chi = (\chi_1, \chi_2)$, then $w(\chi_1, \chi_2) = (\chi_1^{-1}, \chi_2)$. According to results of Keys ([Ky]), $i_G(\chi)$ is irreducible except in the following cases:

(1) $\chi_1(\alpha) = \|\alpha\|$ or $\|\alpha\|^{-1}$
(2) $\chi_1(\alpha) = \eta(\alpha)\|\alpha\|^{1/2}$ or $\eta(\alpha)\|\alpha\|^{-1/2}$, where $\eta\big|F^* = \omega_{E/F}$
(3) χ_1 is non-trivial and $\chi_1\big|F^*$ is trivial.

In case (1), $JH(i_G(\chi))$ consists of the one-dimensional representation $\psi = \chi_2 \circ \det_G$ and of a square-integrable Steinberg representation, which we denote by $St_G(\psi)$. Each of these representations is declared to lie in an L-packet by itself.

In case (2), $i_G(\chi)$ has a unique square-integrable constituent. Define $\eta' \in \mathrm{Hom}(E^1, \mathbf{C}^*)$ by $\eta'(a/\overline{a}) = \eta\mu^{-1}(a)$ and define a character ξ of H by:

$$\xi(h) = \eta'(\det_0(h))\chi_2(\det(h)).$$

Then ξ determines χ and to each character ξ of H there corresponds a character χ in this way. Denote the square-integrable constituent of $i_G(\chi)$

by $\pi^2(\xi)$ and let $\pi^n(\xi)$ be the remaining constituent. It is known that $\pi^n(\xi)$ is unitary. This will be proved in §13, where $\pi^n(\xi)$ will be exhibited as the local component of a cuspidal automorphic representation. It follows from the theory of matrix coefficients that $\pi^n(\xi)$ is non-tempered.

In case (3), χ is of the form $\tilde{\theta}$ for some semi-regular character θ of C, and the two elements of $JH(i_B(\tilde{\theta}))$ make up an l.d.s. L-packet, which we denote by $\Pi(\theta)$. Since an induced representation of G has a unique Whittaker functional, there is a unique element $\pi_1(\theta)$ of $\Pi(\theta)$ which possesses a Whittaker model. Denote the remaining element by $\pi_2(\theta)$. If θ is a singular character of C, then $i_B(\tilde{\theta})$ is irreducible and we set $\Pi(\theta) = i_B(\tilde{\theta})$.

In §13, an L-packet $\Pi(\rho)$ on G will be defined for all $\rho \in \Pi(H)$. We can define $\Pi(\rho)$ directly in the following cases. If $\rho = \rho(\theta)$ where θ is semi-regular or singular, set $\Pi(\rho) = \Pi(\theta)$. If $\rho = JH(i_H(\chi))$, where $\chi \in \mathrm{Hom}(M, \mathbf{C}^*)$ is unitary, then $\Pi(\rho) = JH(i_G(\chi\mu))$. Assuming the results of §13.1, we obtain the following list of L-packets on G:

(1) $\{i_G(\chi)\}$, irreducible principal series
(2) $\Pi(St_H(\xi)), \xi =$ one-dimensional representation of H; $\mathrm{Card}(\Pi) = 2$
(3) $\{\pi^n(\xi)\}, \xi =$ one-dimensional representation of H
(4) $\Pi(\rho), \rho$ supercuspidal and not of the form $\rho(\theta)$; $\mathrm{Card}(\Pi) = 2$
(5) $\Pi(\rho), \rho = \rho(\theta)$ with θ regular; $\mathrm{Card}(\Pi) = 4$
(6) $\Pi(\theta)$, where θ is semi-regular (l.d.s. L-packet); $\mathrm{Card}(\Pi) = 2$
(7) $\{\chi\}$, $\chi =$ one-dimensional representation of G
(8) $\{\mathrm{St}_G(\chi)\}$, $\chi =$ one-dimensional representation of G.
(9) $\{\pi\}, \pi$ supercuspidal, not in any L-packet of the form $\Pi(\rho)$.

12.3 The case $E/F = \mathbf{C}/\mathbf{R}$. In this section, let E/F be $\mathbf{C}/\mathbf{R}$. If $\chi \in \mathrm{Hom}(M, \mathbf{C}^*)$, there are unique $m, n \in \mathbf{C}^*$ such that $m - n \in \mathbf{Z}$ and $r \in \mathbf{Z}$ such that:

$$\chi\left(\begin{pmatrix} \alpha & & \\ & \beta & \\ & & \overline{\alpha}^{-1} \end{pmatrix}\right) = (\alpha\overline{\alpha})^{1/2(m+n)}(\alpha/\overline{\alpha})^{1/2(m-n)}((\alpha/\overline{\alpha})\beta)^r$$

where, for $s \in \frac{1}{2}\mathbf{Z}$, $(\alpha/\overline{\alpha})^s = (\alpha/|\alpha|)^{2s}$. We write $\chi = (m, n, r)$. There is an integer t such that

$$\mu(z) = (\alpha/\overline{\alpha})^{t+(1/2)} .$$

Recall that $W_{\mathbf{C}/\mathbf{R}} = \mathbf{C}^* \cup \mathbf{C}^*\tau$ where $\tau \in W_{\mathbf{C}/\mathbf{R}}$ is an element such that $\tau^2 = -1$ and $\tau z \tau^{-1} = \overline{z}$. The Langlands classification gives a bijection between $\Pi(G)$ and the set of $\widehat{G}$-conjugacy classes of L-maps $W_{\mathbf{C}/\mathbf{R}} \to {}^LG$. Such maps are called L-parameters. The symbol σ (resp., φ) will denote an L-parameter on H (resp. G) and $\rho(\sigma)$ (resp., $\Pi(\varphi)$) will denote the corresponding L-packet. By abuse of notation, we write $\varphi = \xi_H \circ \sigma$ if φ is $\widehat{G}$-conjugate to $\xi_H \circ \sigma$. Functoriality for the embedding ξ_H associates to $\rho = \rho(\sigma)$ the L-packet $\Pi(\rho) = \Pi(\varphi)$ where $\varphi = \xi_H \circ \sigma$.

Let $\sigma = \sigma(a, b, c)$ be the L-parameter defined by:

$$\sigma(z) = \begin{pmatrix} \sigma_1(z) & & \\ & \sigma_2(z) & \\ & & \sigma_3(z) \end{pmatrix} \times z \qquad z \in \mathbf{C}^*$$

$$\sigma(\tau) = \begin{pmatrix} & & 1 \\ & 1 & \\ 1 & & \end{pmatrix} \times \tau$$

where

$$\sigma_1(z) = (z/\overline{z})^{a-t-(1/2)}, \sigma_3(z) = (z/\overline{z})^{c-t-(1/2)}, \sigma_2(z) = (z/\overline{z})^b .$$

The L-packet $\rho(\sigma) = \rho(a, b, c)$ is determined by b and the set $\{a, c\}$. It is square-integrable if and only if $a \neq c$. In the notation of §12.1, $\rho(\sigma)$ coincides with the L-packet $\rho(\theta)$ where $\theta \in \operatorname{Hom}(C, \mathbf{C}^*)$ is defined by $\theta((\alpha, \beta, \gamma)) = \alpha^a \beta^b \gamma^c$.

Suppose that $a \neq c$ and let $\xi = \xi(a, b, c)$ be the irreducible representation of H of dimension $|a - c|$ on which the center of H acts by the character $(\alpha, \lambda) \to \alpha^{a+c-2t-1} \lambda^b$. Then $\rho = \rho(a, b, c)$ consists of the two square-integrable representations of H on which the center of the enveloping algebra of H acts by the same character by which it acts on ξ. We also denote ρ by $\rho(\xi)$.

PROPOSITION 12.3.1: *Let $\chi = (m, n, r)$. Then $i_G(\chi)$ is reducible if and only if $(m, n) \in \mathbf{Z}$ and $(m, n) \neq (0, 0)$.*

Proof: [W].

Let $\varphi = \varphi(a, b, c)$ be the L-parameter on G defined by:

$$\varphi(z) = \begin{pmatrix} (z/\overline{z})^a & & \\ & (z/\overline{z})^b & \\ & & (z/\overline{z})^c \end{pmatrix} \times z$$

for $z \in \mathbf{C}^*$, and

$$\varphi(\tau) = \begin{pmatrix} & & 1 \\ & -1 & \\ 1 & & \end{pmatrix} \times \tau .$$

Then $\varphi = \xi_H \circ \sigma(a,b,c)$. The L-packet $\Pi(\varphi)$ depends only on the set $\{a,b,c\}$ and is square-integrable if and only if a, b, and c are distinct. Let $\Pi = \Pi(\varphi)$.

We now review some results from an article of Wallach ([W]). Assume that $a \geq b \geq c$. Set $m = a - b$ and $n = b - c$ and put:

$$\chi_\varphi = (m, n, b), \quad \chi_\varphi^+ = (-m, m+n, a), \quad \chi_\varphi^- = (m+n, -n, c).$$

If $n > 0$ (resp. $m > 0$), then the representation $i_G(\chi_\varphi^+)$ (resp. $i_G(\chi_\varphi^-)$) has a unique irreducible (infinite-dimensional) quotient which we denote by J_φ^+ (resp. J_φ^-). According to [W], J_φ^+ (resp., J_φ^-) is unitary if and only if $n = 1$ (resp. $m = 1$).

Suppose that $m > 0$ and $n > 0$, Then $i_G(\chi_\varphi)$ has a unique finite-dimensional quotient which we denote by F_φ. The highest weight of F_φ with respect to B is the character

$$\begin{pmatrix} \alpha & & \\ & \beta & \\ & & \gamma \end{pmatrix} \to \alpha^{a-1}\beta^b\gamma^{c+1} .$$

The L-packet Π consists of the 3 square-integrable representations of G on which the center of the enveloping algebra of G acts by the same character by which it acts on F_φ. By [W], there is a unique labelling of the members of Π such that:

$$\begin{aligned} \Pi &= \{D_\varphi, D_\varphi^+, D_\varphi^-\} \\ JH(i_G(\chi_\varphi)) &= \{F_\varphi, J_\varphi^+, J_\varphi^-, D_\varphi\} \\ JH(i_G(\chi_\varphi^+)) &= \{J_\varphi^+, D_\varphi^+, D_\varphi\} \\ JH(i_G(\chi_\varphi^-)) &= \{J_\varphi^-, D_\varphi^-, D_\varphi\} . \end{aligned}$$

If $m > 0$ and $n = 0$, then χ_φ^+ is unitary and φ is $\widehat{G}$-conjugate to the L-parameter associated to the L-packet $JH(i_G(\chi_\varphi^+))$. In this case, $\Pi = JH(i_G(\chi_\varphi^+))$. Similarly, if $n > 0$ and $m = 0$, then $\Pi = JH(i_G(\chi_\varphi^-))$. By

[W], Π has two elements π_φ^1 and π_φ^2 which can be labelled so that

$$JH(i_G(\chi_\varphi^+)) = \{J_\varphi^+, \pi_\varphi^1\} \quad \text{if} \quad n > 0,\ m = 0$$
$$JH(i_G(\chi_\varphi^-)) = \{J_\varphi^-, \pi_\varphi^1\} \quad \text{if} \quad m > 0,\ n = 0$$

In both cases, π_φ^1 is the unique member of π possessing a Whittaker model. Furthermore, $\chi_\varphi^+ = \tilde{\theta}$ (resp., $\chi_\varphi^- = \tilde{\theta}$) if $m > 0$ and $n = 0$ (resp., $n > 0$ and $m = 0$), where θ is a semi-regular character of C. Consistent with the p-adic case, we will also write $\Pi(\theta)$, $\pi^1(\theta)$, $\pi^2(\theta)$ for $\pi, \pi_\varphi^1, \pi_\varphi^2$, respectively. Finally, if $m = n = 0$, then $i_G(\chi_\varphi)$ is irreducible and $\Pi = \{i_G(\chi_\varphi)\}$

Set

$$\rho = \rho(a,b,c), \quad \rho^+ = \rho(b,a,c), \quad \rho^- = \rho(a,c,b).$$

Then $\{\rho, \rho^+, \rho^-\}$ is the set of L-packets $\rho' \in \Pi(H)$ such that $\Pi = \Pi(\rho')$. Furthermore, ρ, ρ^+ and ρ^- are distinct if and only if $mn \neq 0$ and if and only if Π is square-integrable. Every square-integrable L-packet on H is of the form ρ, ρ^+, or ρ^- for unique $a \geq b \geq c$ such that m and n are not both zero. If $m, n > 0$, define functions $\langle \rho,\ \rangle$, $\langle \rho^+,\ \rangle$, $\langle \rho^-,\ \rangle$ on Π by:

$$\begin{array}{lll} \langle \rho, D_\varphi \rangle = 1, & \langle \rho, D_\varphi^+ \rangle = -1, & \langle \rho, D_\varphi^- \rangle = -1 \\ \langle \rho^+, D_\varphi \rangle = 1, & \langle \rho^+, D_\varphi^+ \rangle = +1, & \langle \rho^+, D_\varphi^- \rangle = -1 \\ \langle \rho^-, D_\varphi \rangle = 1, & \langle \rho^-, D_\varphi^+ \rangle = -1, & \langle \rho^-, D_\varphi^- \rangle = +1 . \end{array}$$

If $n = 0$, define $\langle \rho, \pi_\varphi^1 \rangle = 1$, $\langle \rho, \pi_\varphi^2 \rangle = -1$ and $\langle \rho^+, \pi_\varphi^1 \rangle = \langle \rho^+, \pi_\varphi^2 \rangle = 1$ (in this case, $\rho = \rho^-$). If $m = 0$, define $\langle \rho, \pi_\varphi^1 \rangle = 1, \langle \rho, \pi_\varphi^2 \rangle = -1$ and $\langle \rho^-, \pi_\varphi^1 \rangle = \langle \rho^-, \pi_\varphi^2 \rangle = 1$ (in this case $\rho = \rho^+$). The next proposition is a special case of results of Shelstad ([S$_1$]).

PROPOSITION 12.3.2: *In the above notation, the following relation holds:*

$$\chi_{\rho'}(f^H) = \sum_{\pi \in \Pi} \langle \rho', \pi \rangle \chi_\pi(f)$$

for $\rho' = \rho,\ \rho^+$, *or* ρ^-.

We now introduce the following notation in order to mesh with the notation in the p-adic case in §12.2. Let ξ be a one-dimensional representation of H. Then there are unique integers $a \geq b \geq c$ such that either $n = b - c = 1$ and $\xi = \xi(b,a,c)$ or $m = a - b = 1$ and $\xi = \xi(a,c,b)$. Let $\varphi = \varphi(a,b,c)$ and set:

$$\pi^n(\xi) = \begin{cases} J_\varphi^+ & \text{if} \quad \xi = \xi(b,a,c) \\ J_\varphi^- & \text{if} \quad \xi = \xi(a,c,b) \end{cases}$$

$$\pi^s(\xi) = \begin{cases} D_\varphi^- & \text{if} \quad \xi = \xi(b,a,c) \quad \text{and} \quad m \neq 0 \\ D_\varphi^+ & \text{if} \quad \xi = \xi(a,c,b) \quad \text{and} \quad n \neq 0 \\ \pi_\varphi^2 & \text{if} \quad mn = 0 \end{cases}$$

PROPOSITION 12.3.3: *Let ξ be a one-dimensional representation of H.*
(a) $\mathrm{Tr}(\xi(f^H)) = \mathrm{Tr}(\pi^n(\xi)(f)) + \mathrm{Tr}(\pi^s(\xi)(f))$.
(b) $\mathrm{Tr}(\pi^n(\xi)(f)) - \mathrm{Tr}(\pi^s(\xi)(f))$ *defines a stable distribution.*

Proof: In the above notation, set

$$\chi = \begin{cases} \chi_\varphi^+ & \text{if} \quad \xi = \xi(b,a,c) \\ \chi_\varphi^- & \text{if} \quad \xi = \xi(a,c,b) \end{cases}$$

and let $\pi = \pi(\varphi)$. Then

$$\mathrm{Tr}(i_G(\chi)(f)) - \mathrm{Tr}(\Pi(f)) = \mathrm{Tr}(\pi^n(\xi)(f)) - \mathrm{Tr}(\pi^2(\xi)(f))$$

and (b) follows, since $i_G(\chi)$ and Π define stable distributions. We have

$$\chi\mu^{-1}(d(\alpha,\beta,\overline{\alpha}^{-1})) = \begin{cases} (\alpha\overline{\alpha})^{1/2} \left(\frac{\alpha}{\overline{\alpha}}\right)^{-m-t-1} \left(\frac{\alpha}{\overline{\alpha}}\beta\right)^a & \text{if} \quad \xi = \xi(b,a,c) \\ (\alpha\overline{\alpha})^{1/2} \left(\frac{\alpha}{\overline{\alpha}}\right)^{n-t} \left(\frac{\alpha}{\overline{\alpha}}\beta\right)^c & \text{if} \quad \xi = \xi(a,c,b) \end{cases}$$

and

$$\xi(h) = \begin{cases} (\det_0(h))^{-m-t-1}(\det(h))^a & \text{if} \quad \xi = \xi(b,a,c) \\ (\det_0(h))^{n-t}(\det(h))^c & \text{if} \quad \xi = \xi(a,c,b). \end{cases}$$

By the well known structure of induced representations of $U(2)$, we have $JH(i_H(\chi\mu^{-1})) = \{\xi, \rho(\xi)\}$ and since $\mathrm{Tr}(i_G(\xi)(f)) = \mathrm{Tr}(i_H(\chi\mu^{-1})(f^H))$, we obtain

$$\mathrm{Tr}(\xi(f^H)) = \mathrm{Tr}(i_G(\chi)(f)) - \mathrm{Tr}(\rho(\xi)(f^H)) = \mathrm{Tr}(\pi^n(\xi)(f)) + \mathrm{Tr}(\pi^s(\xi)(f))$$

by Proposition 12.3.2.

In the present situation, $\widetilde{G} = \mathrm{GL}_2(\mathbf{C})$ and $\widetilde{H} = \mathrm{GL}_2(\mathbf{C}) \times \mathrm{GL}_1(\mathbf{C})$. We identify $\widetilde{H}$ with the Levi factor of the parabolic subgroup P as before. The next proposition is a special case of results of Clozel on stable archimedean base change ([C$_1$]).

PROPOSITION 12.3.4: *Let $\Pi \in \Pi(G)$ be a tempered L-packet and let $\tilde{\pi}$ be the base change of π to $\widetilde{G}$ with respect to ψ_G. Then*

$$\mathrm{Tr}(\tilde{\pi}(\phi)\tilde{\pi}(\varepsilon)) = \mathrm{Tr}(\Pi(f))$$

for a suitable choice of $\tilde{\pi}(\varepsilon)$. In particular, if $\rho \in \Pi(H)$ is tempered, then

$$\mathrm{Tr}(I_{\tilde{\rho}'}(\phi)I_{\tilde{\rho}'}(\varepsilon)) = \mathrm{Tr}(\Pi(\rho)(f))$$

for a suitable choice of $I_{\tilde{\rho}'}(\varepsilon)$.

COROLLARY 12.3.5: *If $\xi \in \Pi(H)$ is one-dimensional, then*

$$\mathrm{Tr}(\pi^n(\xi)(f)) - \mathrm{Tr}(\pi^s(\xi)(f)) = \mathrm{Tr}(I_{\widetilde{\xi}'}(\phi)I_{\widetilde{\xi}'}(\varepsilon))$$

for some choice of $I_{\tilde{\xi}'}(\varepsilon)$.

Proof: Let $\pi = i_G(\chi)$, where χ is as in the proof of Proposition 12.3.3, and let $\tilde{\pi} = i_{\widetilde{G}}(\tilde{\chi})$, where $\tilde{\chi} = \chi \circ N$. We will check that $JH(\tilde{\pi}) = \{I_{\tilde{\rho}'}, I_{\tilde{\xi}'}\}$. Assuming this, and defining $I_{\tilde{\xi}'}(\varepsilon)$ and $I_{\tilde{\rho}'}(\varepsilon)$ via the canonical choice of $\tilde{\pi}(\varepsilon)$, we have

$$\mathrm{Tr}(I_{\tilde{\xi}'}(\phi)I_{\tilde{\xi}'}(\varepsilon)) = \mathrm{Tr}(\pi(f)) - \mathrm{Tr}(\Pi(f)) = \mathrm{Tr}(\pi^n(\xi)(f)) - \mathrm{Tr}(\pi^s(\xi)(f))$$

since

$$\mathrm{Tr}(\tilde{\pi}(\phi)\tilde{\pi}(\varepsilon)) = \mathrm{Tr}(\pi(f)) \quad \text{and} \quad \mathrm{Tr}(\Pi(f)) = \mathrm{Tr}(I_{\tilde{\rho}'}(\phi)I_{\tilde{\rho}'}(\varepsilon)),$$

by Proposition 12.3.4, and the corollary follows. By the definitions,

$$\chi(d(\alpha, \beta, \overline{\alpha}^{-1})) = \begin{cases} (\alpha\overline{\alpha})^{1/2} \left(\frac{\alpha}{\overline{\alpha}}\right)^{-m-\frac{1}{2}} \left(\frac{\alpha}{\overline{\alpha}}\beta\right)^a & \text{if} \quad \xi = \xi(b, a, c) \\ (\alpha\overline{\alpha})^{1/2} \left(\frac{\alpha}{\overline{\alpha}}\right)^{n+\frac{1}{2}} \left(\frac{\alpha}{\overline{\alpha}}\beta\right)^c & \text{if} \quad \xi = \xi(a, c, b) \end{cases}$$

and it follows that $\tilde{\chi}(d(x, y, z)) = \|\frac{x}{z}\|^{1/2}\tilde{\xi}'(d(x, y, z))$. From the well-known structure of induced representations of $\mathrm{GL}_2(\mathbf{C})$, we have $JH(i_{\widetilde{H}}(\tilde{\chi})) = \{\tilde{\xi}', \rho_1\}$, where $\rho_1 = i_{\widetilde{H}}(\psi)$ and $\psi(\gamma) = \xi'(\gamma)\left(\frac{x\overline{z}}{\overline{x}z}\right)^{1/2}$ for $\gamma = d(x, y, z)$. We have

$$\psi(\gamma) = \begin{cases} \left(\frac{x}{\overline{x}}\right)^b \left(\frac{y}{\overline{y}}\right)^a \left(\frac{z}{\overline{z}}\right)^c & \text{if} \quad \xi = \xi(b, a, c) \\ \left(\frac{x}{\overline{x}}\right)^a \left(\frac{y}{\overline{y}}\right)^c \left(\frac{z}{\overline{z}}\right)^b & \text{if} \quad \xi = \xi(a, c, b) \end{cases}$$

and hence $\rho_1 = \tilde{\rho}'$. Induction by stages gives $JH(\tilde{\pi}) = \{I_{\tilde{\rho}'}, I_{\tilde{\xi}'}\}$.

Suppose that $\varphi : W_{\mathbf{C}} \to {}^L\widetilde{G}$ is the L-parameter associated to a representation $\tilde{\pi}$ of $\widetilde{G}$ which is ε-invariant. Here we regard $\widetilde{G}$ as a complex group and take ${}^L\widetilde{G} = \mathrm{GL}_3(\mathbf{C})$. We can assume that φ is of the form $\varphi(z) = d(\xi_1(z), \xi_2(z), \xi_3(z))$ and the ε-invariance implies that, up to conjugacy, φ satisfies one of the following two conditions

(1) $\xi_1(z) = \xi_3(\overline{z})^{-1}$ and $\xi_2(z\overline{z}) = 1$

(2) $\xi_j(z\overline{z}) = 1$ for $j = 1,2,3$, and the ξ_j are distinct.

Assuming, as we are, that the central character is trivial on F^*, we have in addition that $\xi_2|\mathbf{R}^* = 1$ in case (1) and $\xi_j|\mathbf{R}^*$ is trivial either for $j = 1,2,3$ or for precisely one index j in case (2). In case (1), φ is the restriction to $W_{\mathbf{C}}$ of a parameter $\varphi_{\mathbf{R}} : W_{\mathbf{R}} \to {}^LG$, where we take the Galois form ${}^LG = \mathrm{GL}_3(\mathbf{C}) \rtimes \mathrm{Gal}(\mathbf{C}/\mathbf{R})$. The representation $\tilde{\pi}$ is a quotient of $i_{\widetilde{G}}(\chi)$ for some character χ of the form $\chi = \chi_0 \circ N$ where $\chi_0 \in \mathrm{Hom}(M, \mathbf{C}^*)$. If χ_0 is unitary, then $i_{\widetilde{G}}(\chi)$ is irreducible and $\chi_{\tilde{\pi}\varepsilon}(\phi) = \chi_\pi(f)$ where π is the L-packet $JH(i_G(\chi_0))$. Suppose that χ is not unitary, but that $\tilde{\pi}$ is unitary and generic. This is the case of interest to us in §13 since the local component of a cuspidal representation is unitary and generic. As noted in [C$_5$], Lemma 4.9, we then have $\xi_1(z) = \psi(z))\|z\|^s$ where $0 < s < \frac{1}{2}$ and ψ satisfies $\psi(z) = \psi(\overline{z})^{-1}$. Both $i_{\widetilde{G}}(\chi)$ and $i_G(\chi_0)$ are irreducible in this case and the identity $\chi_{\tilde{\pi}\varepsilon}(\phi) = \chi_\pi(f)$ again holds, with $\Pi = \{i_G(\chi_0)\}$. In case (2), $\tilde{\pi}$ is tempered and φ factors through ψ_G if and only if $\xi_j|\mathbf{R}^* = 1$ for $j = 1,2,3$. If $\xi_j|\mathbf{R}^* = 1$ for a unique index j, then φ corresponds to a parameter that factors through the endoscopic embedding η_1. In this case, $\tilde{\pi} = I_{\tilde{\rho}}$ where $\rho \in \Pi^2(H)$ and $\tilde{\rho}$ is the standard base change of ρ to $\widetilde{H}$.

12.4 Representations of $\widetilde{G}$. For the rest of this chapter, assume that F is a p-adic field. Let $E_\varepsilon(\widetilde{G})$ be the set of irreducible admissible representations π of $\widetilde{G}$ such that $\varepsilon(\pi) = \pi$ and the restriction of ω_π to F^* is trivial. Define $E_\varepsilon(\widetilde{H})$ similarly. Let P' be a standard parabolic subgroup of $\widetilde{G}$ and let M' be its standard Levi subgroup. Suppose that (M'', P'') is another such pair and let σ', σ'' be supercuspidal representations of M' and M'', respectively. By a basic result of Bernstein-Zelevinsky ([BZ]), either $JH(i_{p'}(\sigma'))$ and $JH(i_{P''}(\sigma''))$ are disjoint or (M', σ') is conjugate to (M'', σ''), i.e., for some $g \in \widetilde{G}$, $gM'g^{-1} = M''$ and $\sigma''(m)$ is equivalent to $\sigma'(g^{-1}mg)$. The automorphism ε maps $JH(i_{P'}(\sigma))$ to $JH(i_{\varepsilon(P')}(\varepsilon(\sigma)))$ and hence if σ is supercuspidal and $JH(i_{P'}(\sigma))$ contains an ε-invariant representation, then the pairs (M', σ) and $(\varepsilon(M'), \varepsilon(\sigma))$ are conjugate, i.e., $s\varepsilon(M') = M'$ and $s\varepsilon(\sigma) = \sigma$ for some $s \in \Omega(\widetilde{M})$. By Jacquet's theorem, every element of $E_\varepsilon(\widetilde{G})$ lies in $JH(i_{P'}(\sigma))$ for some (P',σ). From this and the results of [Z], we deduce the following rough classification for $E_\varepsilon(\widetilde{G})$.

Let $\chi = (\chi_1,\chi_2,\chi_3) \in \mathrm{Hom}(M,\mathbf{C}^*)$ and suppose that $i_{\widetilde{G}}(\chi)$ has an ε-invariant constituent. Up to the action of $\Omega(\widetilde{M})$, we can assume that χ satisfies one of the following two conditions:

$$\begin{array}{ll}(1) & \chi_3 = \overline{\chi_1}^{-1} \quad \text{and} \quad \chi_2\big|F^* = 1\\ (2) & \chi_j\chi_j = 1 \quad \text{for} \quad j=1,2,3 \quad \text{and} \quad \chi_1\chi_2\chi_3\big|F^* = 1\end{array}$$

In case (2), $i_{\widetilde{G}}(\chi)$ is irreducible by the irreducibility criterion of [Z]. In case (1), $\chi = \chi_0 \circ N$ for some $\chi_0 \in \mathrm{Hom}(M,\mathbf{C}^*)$ and $i_{\widetilde{G}}(\chi)$ is irreducible unless $\chi_0 = (\|\ \|,\ \psi)$, where $\psi \in \mathrm{Hom}(E^1,\mathbf{C}^*)$, again by [Z]. In this latter case, $\chi = (\tilde{\psi}\|\ \|, \tilde{\psi}, \tilde{\psi}\|\ \|^{-1})$, where $\tilde{\psi}(a) = \psi(a/\overline{a})$, and $JH(i_{\widetilde{G}}(\chi))$ consists of the one-dimensional representation $\tilde{\psi} \circ \det \widetilde{G}$, a square-integrable (Steinberg) representation which we denote by $St_{\widetilde{G}}(\tilde{\psi})$, and two additional distinct irreducible representations π_1 and π_2. It is not hard to check that $\varepsilon(\pi_1) = \pi_2$. For example, the Jacquet modules π_{1N} and π_{2N} of π_1 and π_2 are known explicitly ([Z]) and, up to labelling, are isomorphic to $s_2\chi \oplus s_1s_2\chi$ and $s_1\chi \oplus s_2s_1\chi$, respectively. The assertion follows since ε interchanges π_{1N} and π_{2N}. Let $\tilde{\pi} = St_{\widetilde{G}}(\tilde{\psi})$, $\tilde{\pi}' = i_{\widetilde{G}}(\chi)$, and $\pi' = i_G(\chi_0)$. The canonical choice of $\tilde{\pi}'(\varepsilon)$ (cf. §4.10) induces a choice of $\tilde{\pi}(\varepsilon)$ and also an action of ε on $\tilde{\psi} \circ \det$ which is, in fact, the trivial action. It follows that

$$\mathrm{Tr}(\tilde{\pi}'(\phi)\pi'(\varepsilon)) = \mathrm{Tr}(\tilde{\pi}(\phi)\tilde{\pi}(\varepsilon)) + \mathrm{Tr}(\tilde{\psi} \circ \det(\phi)).$$

By Proposition 4.10.2,

$$\mathrm{Tr}(\pi'(f)) = \mathrm{Tr}(\tilde{\pi}'(\phi)\pi'(\varepsilon))$$

and

$$\mathrm{Tr}(\pi'(f)) = \mathrm{Tr}(St_G(\psi)(f)) + \mathrm{Tr}(\psi \circ \det(f))$$

since $JH(\pi') = \{\pi, \psi \circ \det\}$. $\tilde{\pi} = St_{\widetilde{G}}(\tilde{\psi})$. The relation

$$\mathrm{Tr}(\tilde{\psi} \circ \det(\phi)) = \mathrm{Tr}(\psi \circ \det(f))$$

follows immediately from the Weyl integration formulas (cf. §12.5), and the next proposition follows.

PROPOSITION 12.4.1: *Let $\psi \in \mathrm{Hom}(E^1,\mathbf{C}^*)$. Then $\chi_\Pi(f) = \chi_{\pi\varepsilon}(\phi)$ in the following two cases*

(a) $\Pi = \{St_G(\psi)\}$ *and* $\pi = St_{\widetilde{G}}(\tilde{\psi})$
(b) $\Pi = \{\psi \circ \det\}$ *and* $\pi = \tilde{\psi} \circ \det$

if the twisted character is defined relative to the action of ε *on* π *induced by its canonical action on* $i_{\widetilde{G}}(\chi)$, *with* χ *as above.*

Regard $\widetilde{H}$ as a subgroup of $\widetilde{G}$ and let P be a parabolic subgroup of $\widetilde{G}$ with standard Levi subgroup $\widetilde{H}$. For $\sigma \in E_\varepsilon(\widetilde{H})$, we denote $i_P(\sigma)$ by I_σ. By [Z], I_σ is irreducible if σ is supercuspidal. If $\mu = (\mu_1, \mu_2, \mu_3) \in \mathrm{Hom}(\widetilde{M}, \mathbf{C}^*)$ satisfies $\mu_j \overline{\mu}_j = 1$ and $\mu_1\mu_2\mu_3 \big| F^* = 1$, denote $i_{\widetilde{G}}(\mu)$ by I_μ. Then I_μ and I_σ lie in $E_\varepsilon(\widetilde{G})$. We obtain the following list of possibilities for an element π of $E_\varepsilon(\widetilde{G})$:

(1) $i_{\widetilde{G}}(\chi \circ N)$, irreducible principal series ($\chi \in \mathrm{Hom}(M, \mathbf{C}^*)$)
(2) I_σ, $\sigma \in E_\varepsilon(\widetilde{H})$, σ square-integrable or one-dimensional
(3) I_μ, where $\mu_j\overline{\mu}_j = 1$ and $\mu_1\mu_2\mu_3|F^* = 1$
(4) $\tilde{\psi} \circ \det_{\widetilde{G}}$, where $\psi \in \mathrm{Hom}(E^1, \mathbf{C}^*)$
(5) $\mathrm{St}_{\widetilde{G}}(\tilde{\psi})$, where $\psi \in \mathrm{Hom}(E^1, \mathbf{C}^*)$
(6) π supercuspidal, ε-invariant.

12.5 Weyl integration formula. Assume that f transforms under Z by ω^{-1} and that α is a class function on G which transforms under Z by ω. The Weyl integration formula is:

$$\int_{Z\backslash G} f(g)\alpha(g)dg = \sum |\Omega(T,G)|^{-1} \int_{Z\backslash T} D_G(\gamma)^2 \Phi(\gamma, f)\alpha(\gamma)d\gamma$$

where the sum is over a set of representatives for the conjugacy classes of Cartan subgroups of G. Here, $dg = \frac{dg'}{dz}$ where dg' and dz are measures on G and Z, respectively, and $d\gamma = \frac{dt}{dz}$ where $\frac{dg'}{dt}$ is the measure on $T\backslash G$ used to define $\Phi(\gamma, f)$. The number of $\nu \in \mathcal{D}(T/F)$ such that T^ν is conjugate to T is equal to $|\Omega(T,G)|^{-1}|\Omega_F(T,G)|$. Hence, if α is stable (i.e., takes the same value at stably conjugate points)

$$\int_{Z\backslash G} f(g)\alpha(g)\, dg = \sum |\Omega_F(T,G)|^{-1} \int_{Z\backslash T} D_G(\gamma)^2 \Phi^{\mathrm{st}}(\gamma, f)\alpha(\gamma)\, d\gamma$$

where the sum is over a set of representatives for the stable conjugacy classes of Cartan subgroups of G.

Suppose that $\delta \in \mathcal{D}(T/F)$ is represented by a cocycle $\{\sigma(g)g^{-1}\}$ for some $g \in G(\overline{F})$. For $w \in \Omega_F(T,G)$, let $w(\delta)$ be the element represented by the cocycle $\{\sigma(wg)g^{-1}w^{-1}\}$. Then $w(\delta)$ is independent of the choice of g and of the implicit choice of representative for w in $G(\overline{F})$. Let T^δ denote the Cartan subgroup $g^{-1}Tg$ and for $t \in T$, let $t^\delta = g^{-1}tg$ (T^δ and t^δ are well-defined up to conjugacy).

Let α be a stable class function on H^r. The transfer $f \to f^H$ allows us to pull back α to a distribution α^G on G defined by:

$$f \to \int_{Z\backslash H} f^H(h)\alpha(h)dh$$

It is given by integration against a class function on G described in the next lemma.

LEMMA 12.5.1: *Suppose that $T \subset H$. For $\delta \in \mathcal{D}(T/F)$ and $\gamma \in T^r$,*

$$D_G(\gamma)\alpha^G(\gamma^\delta) = \sum_{w\in\Omega_F(T,H)\backslash\Omega_F(T,G)} \tau(w\gamma w^{-1})\ D_H(w\gamma w^{-1})\kappa(w(\delta))\alpha(w\gamma w^{-1})$$

where $\kappa \in \mathcal{R}(T/F)$ is defined by the endoscopic group H

Proof: Let $\{T'\}$ be a set of representatives for the conjugacy classes within the stable class of T in G. By the Weyl integration formula, we clearly have

$$\sum |\Omega(T',G)|^{-1} \int_{Z\backslash T'} D_G(\gamma)^2\Phi(\gamma,f)\ \alpha^G(\gamma)\ d\gamma$$
$$= |\Omega_F(T,H)|^{-1} \int_{Z\backslash T} D_H(\gamma)^2\ \Phi_H^{st}(\gamma,f^H)\alpha(\gamma)\ d\gamma\ .$$

Recall that $\tau(\gamma)D_G(\gamma)\sum\kappa(\delta)\Phi(\gamma^\delta,f) = D_H(\gamma)\Phi_H^{st}(\gamma,f^H)$. Let $\delta \in \mathcal{D}(T/F)$ and set $T_1 = g^{-1}Tg$ where δ is represented by $\{\sigma(g)g^{-1}\}$. Then:

(12.5.1)

$$\int_{Z\backslash T_1} D_G(\gamma)^2\Phi(\gamma,f)\alpha^G(\gamma)d\gamma$$
$$= \frac{|\Omega(T_1,G)|}{|\Omega_F(T,H)|}\int_{Z\backslash T} D_G(\gamma)\tau(\gamma)\ D_H(\gamma)\left(\sum_{\delta'\in X}\kappa(\delta')\Phi(\gamma^{\delta'},f)\right)\alpha(\gamma)\ d\gamma$$

where X is the set of $\delta' \in \mathfrak{D}(T/F)$ such that $T^{\delta'}$ is conjugate to T_1. We have

$$X = \{w(\delta) : w \in \Omega_F(T,G)/g\Omega(T_1,G)g^{-1}\}.$$

Note that if $w \in \Omega_F(T,H)$, then δ and $w(\delta)$ have the same image in $H^1(F,H)$. In the p-adic case, $H^1(F,H)$ has a natural structure of abelian group and the map $H^1(F,T) \to H^1(F,H)$ is a homomorphism (cf. §3.2). Hence δ and $w(\delta)$ differ by an element in $\mathfrak{D}_H(T/F)$ and $\kappa(\delta) = \kappa(w(\delta))$, since κ is trivial on the image of $\mathfrak{D}_H(T/F)$ in $\mathfrak{D}(T/F)$. Therefore (12.5.1) is equal to

$$\int_{Z\backslash T} D_G(\gamma)\tau(\gamma)D_H(\gamma)\left(\sum \kappa(w(\delta))\Phi((w^{-1}\gamma w)^\delta, f)\right)\alpha(\gamma)\, d\gamma$$

$$= \int_{Z\backslash T} D_G(\gamma)\left(\sum \tau(w\gamma w^{-1})D_H(w\gamma w^{-1})\kappa(w(\delta))\alpha(w\gamma w^{-1})\right)\Phi(\gamma^\delta, f)\, d\gamma$$

where the sums are over $w \in \Omega_F(T,H)\backslash\Omega_F(T,G)$ and the lemma follows.

Let G^e be the set of elliptic regular elements in G and let α_1 and α_2 be class functions supported on G^e. Let $\mathfrak{C}$ be a set of representatives for the stable conjugacy classes of elliptic Cartan subgroups in G. Define the elliptic inner product $\langle\alpha_1,\alpha_2\rangle_e = \langle\alpha_1,\alpha_2\rangle_{G,e}$ by:

$$\langle\alpha_1,\alpha_2\rangle_{G,e} = \sum |\Omega(T,G)|^{-1} \int_{Z\backslash T} D_G(\gamma)^2\alpha_1(\gamma)\overline{\alpha_2(\gamma)}\, d\gamma$$

$$= \sum_{T\in\mathfrak{C}} |\Omega_F(T,G)|^{-1} \sum_{\delta\in\mathfrak{D}(T/F)} \int_{Z\backslash T} D_G(\gamma)^2\alpha_1(\gamma^\delta)\,\overline{\alpha_2(\gamma^\delta)\, d\gamma}$$

where the first sum is over a set of representatives $\mathfrak{C}$ for the conjugacy classes of elliptic Cartan subgroups in G, and the measures $d\gamma$ are normalized so that $\text{meas}(Z\backslash T) = 1$. If α_1 and α_2 are stable class functions, then

$$\sum_{T\in\mathfrak{C}} |\mathfrak{D}(T/F)|\cdot|\Omega_F(T,G)|^{-1} \int_{Z\backslash T} D_G(\gamma)^2\alpha_1(\gamma)\overline{\alpha_2(\gamma)}\, d\gamma$$

Define $\langle\alpha_1,\alpha_2\rangle_{H,e}$ similarly. If α_1 and α_2 are stable class functions on H^e, then

$$\langle\alpha_1,\alpha_2\rangle_{H,e} = \sum_{T\in\mathfrak{C}_{G/H}} |\mathfrak{D}_H(T/F)|\cdot|\Omega_F(T,H)|^{-1} \int_{Z\backslash T} D_G(\gamma)^2\alpha_1(\gamma)\overline{\alpha_2(\gamma)}\, d\gamma$$

where $\mathcal{C}_{G/H}$ is a set of representatives for the stable conjugacy classes of elliptic Cartan subgroups T of G with a representative in H. We assume that $T \subset H$. We let κ denote the element of $\mathcal{R}(T/F)$ defined by the endoscopic group H.

PROPOSITION 12.5.2: *Let α_1 and α_2 be stable class functions on H^e. Then*

$$\langle \alpha_1^G, \alpha_2^G \rangle_{G,e} = 2\langle \alpha_1, \alpha_2 \rangle_{H,e} .$$

Proof: By Proposition 12.5.1, $\langle \alpha_1^G, \alpha_2^G \rangle_{G,e}$ is equal to the sum over $T \in \mathcal{C}_{G/H}$ and $\delta \in \mathfrak{D}(T/F)$ of $|\Omega_F(T,G)|^{-1}$ times the integral over $Z\backslash T$ of:

$$\sum \tau(\gamma_1)\overline{\tau(\gamma_2)}^{-1} D_H(\gamma_1)D_H(\gamma_2)\kappa(w_1(\delta))\overline{(w_2(\delta))}\alpha_1(\gamma_1)\overline{\alpha_2(\gamma_2)}$$

where the sum is over $w_1, w_2 \in \Omega_F(T,H)\backslash\Omega_F(T,G)$ and $\gamma_j = w_j\gamma w_j^{-1}$. The sum over δ is zero unless the characters $\kappa(w_1(\delta))$ and $\kappa(w_2(\delta))$ of $\mathfrak{D}(T/F)$ coincide, by the orthogonality relations. They coincide if and only if $w_1 w_2^{-1} \in \Omega_F(T,H)$ (this assertion is trivial for T of type (2) and is easy to check for T of type (1), cf. Lemma 3.7.1). A change of variables gives:

$$|\mathfrak{D}(T/F)| \cdot |\Omega_F(T,H)|^{-1} \int_{Z\backslash T} D_H(\gamma)^2 \alpha_1(\gamma)\overline{\alpha_2(\gamma)}\, d\gamma .$$

The proposition follows from the equality $|\mathfrak{D}_H(T/F)| = 2|\mathfrak{D}(T/F)|$ (Lemma 3.6.1).

Let α be a class function on G^e, and for $T \in \mathcal{C}_{G/H}$ and $\gamma \in T^r$, set:

$$\alpha^\kappa(\gamma) = D_G(\gamma)D_H(\gamma)^{-1}\overline{\tau(\gamma)} \sum_{\delta \in \mathfrak{D}(T/F)} \overline{\kappa(\delta)}\, \alpha(\gamma^\delta) .$$

Extend α^κ to a stably invariant class function on H^e.

PROPOSITION 12.5.3: *A class function α on G^e is stable if and only if $\alpha^\kappa = 0$. If β is a stable class function on H^e, then:*

$$\langle \alpha, \beta^G \rangle_e = \langle \alpha^\kappa, \beta \rangle_{H,e} .$$

Proof: For the first assertion, we have to check that for all Cartan subgroups T, $\alpha(\gamma^\delta) = \alpha(\gamma)$ for all regular elliptic $\gamma \in T$ and $\delta \in \mathfrak{D}(T/F)$ if and only if $\alpha^\kappa(\gamma) = 0$ for all regular elliptic $\gamma \in T$. We can assume that T is of type (1) or (2). It is clear for T of type (2) since $|\mathfrak{D}(T/F)| = 2$ in this case. For T of type (1), note that $\Omega_F(T,G)$ acts transitively on the set of non-trivial characters of $\mathfrak{D}(T/F)$. Since $\alpha^{\kappa}(w^{-1}\gamma w) = 0$ for all $w \in \Omega_F(T,G)$, $\alpha^{\kappa'}(\gamma) = 0$ for all non-trivial character κ' of $\mathfrak{D}(T/F)$ and the assertion follows. By Proposition 12.5.1 and a change of variables, $\langle \alpha, \beta^G \rangle_e$ is the sum over $T \in \mathfrak{C}_{G/H}$ of

$$\sum_{\delta \in \mathfrak{D}(T/F)} |\Omega_F(T,G)|^{-1}$$

$$\int_{Z\backslash T} D_G(\gamma) D_H(\gamma) \overline{\tau(\gamma)} \left(\sum_{w \in \Omega_F(T,H)\backslash \Omega_F(T,G)} \overline{\kappa(w(\delta))} \alpha((w^{-1}\gamma w)^{\delta}) \right) \overline{\beta(\gamma)} d\gamma.$$

$$= |\Omega_F(T,H)|^{-1} \int_{Z\backslash T} D_G(\gamma) D_H(\gamma) \overline{\tau(\gamma)} \left(\sum_{\delta \in \mathfrak{D}(T/F)} \overline{\kappa(\delta)} \alpha(\gamma^{\delta}) \right) \overline{\beta(\gamma)} d\gamma.$$

and the proposition follows.

COROLLARY 12.5.4: *Let α be a class function on G^e. Then α is stable if and only if $\langle \alpha, \chi_\rho^G \rangle_{G,e} = 0$ for all $\rho \in \Pi^2(H)$.*

Proof: The set of restrictions to the elliptic set of the square-integrable representations (with fixed unitary central character) form an orthonormal basis with respect to the elliptic norm. As is well-known, this follows immediately from the local Jacquet-Langlands correspondence, since the corresponding fact is true for the multiplicative group of a quaternion algebra by the representation theory of compact groups. From this, it is easy to show that the restrictions of the functions χ_ρ , $\rho \in \Pi^2(H)$, give an orthogonal basis for the stably invariant class functions on H^e with respect to the elliptic norm. One uses that the derived group of H is $\mathrm{SL}_2(F)$. By Proposition 12.5.3, α is stable if and only if $\alpha^\kappa = 0$. The corollary follows from the relation, $\langle \alpha, \chi_\rho^G \rangle_{G,e} = \langle \alpha^\kappa, \chi_\rho \rangle_{H,e}$.

Let T be a Cartan subgroup of G and let $\widetilde{T}$ be the centralizer of T in $\widetilde{G}$. Then $\widetilde{T}$ is a Cartan subgroup of $\widetilde{G}$. By Proposition 3.11.1(a), every stable ε-semisimple regular conjugacy class intersects $\widetilde{T}$ for some T. Let $\widetilde{T}^N = \{\delta \in T : N(\delta) = 1\}$. Then the sequence:

$$1 \to \widetilde{Z}\widetilde{T}^N \to \widetilde{T} \xrightarrow{N} Z\backslash T \to 1$$

is exact. If α is a stable ε-class function which transforms under $\widetilde{Z}$ by $\tilde{\omega}$, then the following Weyl integration formula holds.:

$$\int_{\widetilde{Z}\backslash\widetilde{G}} \phi(g)\alpha(g)dg = \sum_{T\in\mathfrak{C}} |\Omega_F(T,G)|^{-1} \int_{\widetilde{Z}\widetilde{T}^N\backslash\widetilde{T}} D_G(N(\delta))^2 \Phi_\varepsilon^{\mathrm{st}}(\delta,\phi)\alpha(\delta)d\delta \ .$$

The measure dg on $\widetilde{Z}\backslash\widetilde{G}$ and the measure on $\widetilde{Z}T\backslash\widetilde{G}$ used to define $\Phi_\varepsilon^{\mathrm{st}}(\delta,\phi)$ determine a measure on $\widetilde{Z}\backslash\widetilde{Z}T = Z\backslash T$. By the above sequence, this defines a measure on $\widetilde{Z}\widetilde{T}^N\backslash\widetilde{T}$ which we take to be $d\delta$ above.

Let α_1 and α_2 be ε-class functions on $G^{\varepsilon r}$ which transforms by $\tilde{\omega}$ and assume that α_2 is stable. Define an ε-elliptic inner product:

$$\langle\alpha_1,\alpha_2\rangle_\varepsilon = \sum_{T\in\mathfrak{C}} |\Omega_F(T,G)|^{-1} \int_{Z\widetilde{T}^N\backslash\widetilde{T}} D_G(N(\gamma))^2 \left(\sum_{\delta\in\mathfrak{D}(T/F)} \alpha_1(\gamma^\delta)\right) \overline{\alpha_2(\gamma)}d\gamma \, .$$

where $d\gamma$ is the measure on $Z\widetilde{T}^N\backslash\widetilde{T}$ with total mass one.

If α is a stable class function on G, then the function $\tilde{\alpha}$ on $\widetilde{G}$ defined by $\tilde{\alpha}(\delta) = \alpha(\gamma)$ if $\gamma \in \mathcal{N}(\delta)$ is an ε-stable class function on $\widetilde{G}$. Similarly, if $\tilde{\alpha}$ is a stable class function on $\widetilde{G}$, we can define α by $\alpha(\gamma) = \tilde{\alpha}(\delta)$. The Weyl integration formula implies

$$\int_{\widetilde{Z}\backslash\widetilde{G}} \phi(g)\tilde{\alpha}(g)dg = \int_{Z\backslash G} f(g)\alpha(g)dg \ .$$

PROPOSITION 12.5.5: *If α_1, α_2 are stable class functions on G which transform by ω, then $\langle\alpha_1,\alpha_2\rangle_e = \langle\tilde{\alpha}_1,\tilde{\alpha}_2\rangle_\varepsilon$.*

12.6 Orthogonality relations. An irreducible representation π will be called elliptic if the character χ_π does not vanish identically on G^e. Let $\pi \in E(G)$ be an elliptic representation. A function f_π is called a pseudo-coefficient for π if:

$$\Phi(\gamma, f_\pi) = \begin{cases} 0 & \text{if} \quad \gamma \in G^r - G^e \\ \overline{\chi_\pi(\gamma)} & \text{if} \quad \gamma \in G^e \end{cases}$$

By the Weyl integration formula, $\mathrm{Tr}(\pi'(f_\pi)) = \langle\chi_{\pi'},\chi_\pi\rangle_e$. The existence of pseudo-coefficients follows from [K], Theorem 4.1. The character of a principal series representation is supported on the conjugacy classes that meet M. It follows from the results of §12.2 that if π is elliptic, then π is either supercuspidal, or belongs to one of the following three types of sets:

$\{St_G(\psi), \psi \circ \det_G\}$ for some $\psi \in \mathrm{Hom}(E^1, \mathbf{C}^*)$, $\{\pi^2(\xi), \pi^{nt}(\xi)\}$ for some one-dimensional $\xi \in E(H)$, an l.d.s. L-packet. All representations of this type are in fact elliptic since all representations of G which are not of the form $i_G(\chi)$ are elliptic by an application of the results of [K]. The next proposition will be referred to as the orthogonality relations.

PROPOSITION 12.6.1: *Let $\pi, \pi' \in E(G)$ be elliptic representations. Then*

(a) $\langle \chi_\pi, \chi_\pi \rangle_e \neq 0$. *If π is square-integrable, then $\langle \chi_\pi, \chi_\pi \rangle_e = 1$.*

(b) *If $\langle \chi_\pi, \chi_{\pi'} \rangle_e \neq 0$ and $\pi \neq \pi'$, then either $\{\pi, \pi'\}$ is an l.d.s. L-packet, $\{\pi, \pi'\} = \{St_G(\psi), \psi\}$ for some $\psi \in \mathrm{Hom}(E^1, \mathbf{C}^*)$, or $\{\pi, \pi'\} = \{\pi^2(\xi), \pi^{nt}(\xi)\}$ for some one-dimensional $\xi \in E(H)$.*

(c) *For $\{\pi, \pi'\}$ as in* (b), $\langle \chi_\pi, \chi_{\pi'} \rangle_e = -1$.

Proof: Applying [K], Theorem F, we see that $\langle \chi_\pi, \chi_{\pi'} \rangle = 0$ unless $\pi, \pi' \in JH(i_G(\chi))$ for some χ, and (b) follows. The first part of (a) is clear. If π is supercuspidal, then $\langle \chi_\pi, \chi_\pi \rangle = 1$ by [H$_4$], Theorem 17. As shown by Harish-Chandra, the corresponding fact for square-integrable representations follows from the Howe conjecture, which has been proved by Clozel ([C$_4$]). For (c), observe that $\chi_\pi = -\chi_{\pi'}$ on G^e.

If $\Pi = \{\pi^1, \pi^2\}$ is an l.d.s. L-packet, let f_Π denote f_π for some $\pi \in \Pi$. Note that $\chi_{\pi^1} = -\chi_{\pi^2}$ on G^e since $\chi_{\pi^1} + \chi_{\pi^2}$ is the character of a principal series representation, and hence $\chi_{\pi^1}(f_\Pi) = -\chi_{\pi^2}(f_\Pi)$.

We will call a representation π in $E_\varepsilon(\widetilde{G})$ ε-elliptic if $\chi_{\pi\varepsilon}$ is not identically zero on $G^{\varepsilon e}$. By [KR], π is ε-elliptic if and only if $\chi_{\pi\varepsilon}$ is not of the form $i_{\widetilde{G}}(\chi \circ N)$ for any $\chi \in \mathrm{Hom}(M, \mathbf{C}^*)$. A function ϕ_π is called an ε-pseudo-coefficient for π if:

$$\Phi_\varepsilon(\gamma, \phi_\pi) = \begin{cases} 0 & \text{if} \quad \gamma \in G^{\varepsilon r} - G^{\varepsilon e} \\ \overline{\chi_{\pi\varepsilon}(\gamma)} & \text{if} \quad \gamma \in G^{\varepsilon e}\,. \end{cases}$$

The existence of ε-pseudo-coefficients follows from [KR]. If π is supercuspidal, then $\chi_{\pi\varepsilon}(\phi_\pi) = 1$. Suppose that $\pi = \mathrm{St}_{\widetilde{G}}(\tilde{\psi})$ for some $\psi \in \mathrm{Hom}(E^1, \mathbf{C}^*)$. Then $\chi_{\pi\varepsilon}(\delta) = \tilde{\psi}(\det(\delta))$ for $\delta \in \widetilde{G}^{\varepsilon r}$ by the discussion of §12.4, since the twisted character of $i_{\widetilde{G}}(\chi)$ vanishes on $G^{\varepsilon r}$. By the Weyl integration formula, $\chi_{\pi\varepsilon}(\phi_\pi) = \chi_{\pi'}(f_{\pi'})$ where $\pi' = \mathrm{St}_G(\psi)$ and $\chi_{\pi'}(f_{\pi'}) = 1$ by Proposition 12.6.1. Hence $\chi_{\pi\varepsilon}(\phi_\pi) = 1$ for all square-integrable π. The following twisted orthogonality relations follow from [KR].

PROPOSITION 12.6.2: *Let* $\pi, \pi' \in E_\varepsilon(\widetilde{G})$ *be elliptic and assume that* $\chi_{\pi\varepsilon}$ *is stably* ε*-invariant. Then:*

(a) $\langle \chi_{\pi\varepsilon}, \chi_{\pi\varepsilon} \rangle_\varepsilon \neq 0$. *If* π *is square-integrable, then* $\langle \chi_{\pi\varepsilon}, \chi_{\pi\varepsilon} \rangle_\varepsilon = 1$.

(b) *If* $\pi \neq \pi'$ *and* $\langle \chi_{\pi\varepsilon}, \chi_{\pi'\varepsilon} \rangle_\varepsilon \neq 0$, *then* $\{\pi, \pi'\} = \{St_{\widetilde{G}}(\psi), \psi \circ \det_{\widetilde{G}}\}$.

12.7 Some lemmas. In this section, some results that will be used in §14 are proved. We will consider sums of the form:

$$\sum_{\pi \in X} a(\pi) \operatorname{Tr}(\pi(f))$$

where the $a(\pi)$ are non-zero complex numbers. Here and below, it is assumed that X is a countable set of irreducible unitary representations of G with central character ω and that the sum is absolutely convergent for all $f \in C(G, \omega)$. If the sum vanishes for all f, then X is empty (cf. Proposition 13.8.1).

If π is square-integrable and not supercuspidal, there is a unique representation π^{nt} such that $\{\pi, \pi^{nt}\} = JH(i_G(\chi))$ for some character χ of M by §12.2. Then:

$$\begin{aligned}
&a(\pi)\operatorname{Tr}(\pi(f)) + a(\pi^{nt})\operatorname{Tr}(\pi^{nt}(f)) \\
&= (a(\pi) - a(\pi^{nt}))\operatorname{Tr}(\pi(f)) + a(\pi^{nt})\operatorname{Tr}(i_G(\chi)(f)) \\
&= (a(\pi^{nt}) - a(\pi))\operatorname{Tr}(\pi^{nt}(f)) + a(\pi)\operatorname{Tr}(i_G(\chi)(f)) \ .
\end{aligned}$$

If $a(\pi) = a(\pi^{nt})$ for almost all square-integrable π, then the sum can be re-written in the form:

$$\sum_{\pi \in X'} b(\pi)\operatorname{Tr}(\pi(f))$$

where the $b(\pi)$ are integers and, for every square-integrable representation π which is not supercuspidal, at most one of π, π^{nt} occurs in X'. However X' may contain reducible p.s. representations. The sum is again absolutely convergent.

LEMMA 12.7.1: *Let* $\Pi = \{\pi', \pi''\}$ *be an l.d.s. L-packet and let* $\rho \in \Pi^2(H)$ *be a supercuspidal L-packet* ρ *such that* $\operatorname{Card}(\rho) = 2$. *Suppose that:*

$$\pm\frac{1}{4}(\operatorname{Tr}(\pi'(f)) - \operatorname{Tr}(\pi''(f)) + \sum_{\pi \in X} a(\pi)\operatorname{Tr}(\pi(f)) = \frac{1}{4}\operatorname{Tr}(\rho(f^H))$$

where $a(\pi) \in \mathbf{Z}$. *Then* X *is empty.*

Proof: Suppose that X contains a representation which belongs to a set $\{\pi, \pi^{nt}\}$ where π is square-integrable, and let $b(\pi) = a(\pi) - a(\pi^{nt})$. The equality, applied to a pseudo-coefficient f_π, yields $b(\pi) = \frac{1}{4}\chi_\rho^G(f_\pi)$. We have:

$$\langle \chi_\rho^G, \chi_\rho^G \rangle_e = 2\langle \chi_\rho, \chi_\rho \rangle_{H,e} = 2\mathrm{Card}(\rho) = 4$$

by Proposition 12.5.2, and the orthogonality relations give $|\chi_\rho^G(f_\pi)| \leq 4$. Since $b(\pi)$ is an integer, if $b(\pi) \neq 0$, then $b(\pi) = \pm 1$ and $\chi_\pi = \pm\chi_\rho^G$ on the elliptic set. This leads to a contradiction as follows. Let f_Π be a pseudo-coefficient for Π and apply the equality to f_Π. Then $\chi_\rho^G(f_\Pi) = 0$ by the orthogonality relations and the relation $\chi_\pi = \pm\chi_\rho^G$, and hence:

$$\frac{1}{2}\,\mathrm{Tr}(\pi'(f_\Pi)) \pm [a(\pi') - a(\pi'')]\;\mathrm{Tr}(\pi'(f_\Pi)) = 0\;.$$

This is impossible since $a(\pi')$, $a(\pi'') \in \mathbf{Z}$ and $\mathrm{Tr}(\pi'(f_\Pi)) \neq 0$. Therefore $b(\pi) = 0$. A similar argument shows that X contains no supercuspidal representations and we can write the equality as

$$\pm\frac{1}{4}(\mathrm{Tr}(\pi'(f)) - \mathrm{Tr}(\pi''(f)) + \sum_{\pi\in X'} b(\pi)\mathrm{Tr}(\pi(f)) = \frac{1}{4}\mathrm{Tr}(\rho(f^H))$$

where X' consist entirely of p.s. and l.d.s. representations.

Let M'' be the subset of M of elements of the form $d(\alpha, \beta, \overline{\alpha}^{-1})$ such that $|\alpha\overline{\alpha}| \neq 1$ and let S be the space of functions $f \in C(G, \omega)$ which are supported in the set of conjugacy classes which meet M''. By Casselman's theorem ([C]), $\chi_{\pi'}$ and $\chi_{\pi''}$ coincide on M'' and χ_ρ vanishes on M''. Hence

$$\mathrm{Tr}(\rho(f^H)) = \mathrm{Tr}(\pi'(f)) - \mathrm{Tr}(\pi''(f)) = 0$$

for all $f \in S$.

Assume that $f \in S$ and let $F_f(\gamma) = D_G(\gamma)\Phi(\gamma, f)$ for $\gamma \in M^r$. Set

$$\widehat{F}_f(\chi) = \int_{Z\backslash M} F_f(\gamma)\chi(\gamma)d\gamma$$

for characters χ of M whose restriction to Z is ω. For all $\pi \in X'$,

$$b(\pi)\mathrm{Tr}(\pi(f)) = c(\pi)\widehat{F}_f(\chi^\pi)$$

for some character χ^π, where $c(\pi) = b(\pi)$ if π is a p.s. representation and $c(\pi) = \frac{1}{2}b(\pi)$ if π is an l.d.s. representation. We obtain:

$$\sum_{\pi\in X'} c(\pi)\widehat{F}_f(\chi^\pi) = 0.$$

Identify E^* with the subgroup of M of elements of the form $d(\alpha, 1, \overline{\alpha}^{-1})$. Since the central character ω is fixed, F_f is determined by its restriction to E^*. A smooth, compactly-supported function φ on E^* is of the form F_f if and only if $\varphi(\alpha) = \varphi(\overline{\alpha}^{-1})$. Fix a prime element η of E and let $\varepsilon = \overline{\eta}/\eta$. Fix a character χ_0 of $\mathcal{O}_{E^*}$ a non-zero positive integer m. Then there exists $f \in S$ such that F_f has support in $\eta^m \mathcal{O}_E^* \cup \eta^{-m}\mathcal{O}_E^*$ and for $u \in \mathcal{O}_E^*$:

$$F_f(\eta^n u) = \begin{cases} 0 & \text{if} \quad m \neq \pm n \\ \chi_0(u)^{-1} & \text{if} \quad n = m \\ \chi_0(\varepsilon^{-m}\overline{u}) & \text{if} \quad n = -m. \end{cases}$$

Then

$$\widehat{F}_f(\chi) = \chi(\eta^m) \int_{\mathcal{O}_E^*} \chi(u)\chi_0(u)^{-1} du + \chi(\eta^{-m})\chi_0(\varepsilon^{-m}) \int_{\mathcal{O}_E^*} \chi(u)\chi_0(\overline{u}) du \ .$$

This is zero unless the restriction of χ to $\mathcal{O}_E^*$ is χ_0 or $\overline{\chi_0}^{-1}$. If $\chi|\mathcal{O}_E^* = \chi_0$, then, up to a constant,

$$\widehat{F}_f(\chi) = \begin{cases} \chi(\eta^m) & \text{if} \quad \chi_0 \neq \overline{\chi_0}^{-1} \\ \chi(\eta^m) + \chi(\overline{\eta}^{-m}) & \text{if} \quad \chi_0 = \overline{\chi_0}^{-1}, \end{cases}$$

and $\widehat{F}_f(\overline{\chi}^{-1}) = \widehat{F}_f(\chi)$. If $\chi_0 \neq \overline{\chi_0}^{-1}$, we obtain

$$\sum c(\pi) z(\pi)^m = 0$$

for all non-zero integers m, where

$$z(\pi) = \begin{cases} \chi^\pi(\eta^m) & \text{if} \quad \chi|\mathcal{O}_E^* = \chi_0, \\ \chi^\pi(\overline{\eta}^{-m}) & \text{if} \quad \chi|\mathcal{O}_E^* = \overline{\chi_0}^{-1} \\ 0 & \text{otherwise}. \end{cases}$$

If $\chi_0 = \overline{\chi_0}^{-1}$, we obtain

$$\sum c(\pi)(\chi^\pi(\eta^m) + \chi^\pi(\overline{\eta}^{-m})) = 0$$

where the sums is over $\pi \in X$ such that the restriction of χ^π to $\mathcal{O}_E^*$ is χ_0. Since the π are unitary, there is a positive constant c such that $|\chi^\pi(\eta)|^{\pm 1} \geq c$. The constants $c(\pi)$ belong to $\frac{1}{2}\mathbf{Z}$ and since the above sums are absolutely convergent, they must be finite. Given finitely many pairs c_j, z_j of complex numbers such that $z_j \neq 0$, the sum $\sum c_j z_j^m = 0$ is zero for all non-zero integers m if and only if $c_j = 0$ for all j. It follows that X' is empty.

In §13, the preceding lemma will be used to prove that

$$\chi_{\pi^1(\theta)} - \chi_{\pi^2(\theta)} = \pm\chi^G_{\rho(\theta)}$$

for any semi-regular character θ of C. We assume this result for the proof of Lemmas 12.7.2 and 12.7.5.

LEMMA 12.7.2: *Let $\rho \in \Pi^2(H)$ and assume that ρ is not of the form $\rho(\theta)$ for θ semi-regular. Suppose that*

$$\sum_{\pi\in X} a(\pi)\mathrm{Tr}(\pi(f) = \mathrm{Tr}(\rho(f^H))$$

where $a(\pi) \in \mathbf{Z}$. Then Card(X) = 2Card(ρ) *and X consists entirely of square-integrable representations. The $a(\pi)$ are equal to ± 1 and do not all have the same sign. Furthermore:*

$$\sum_{\pi\in X} a(\pi)\ d(\pi) = 0$$

where $d(\pi)$ is the formal degree of π. If ρ is supercuspidal, then all elements of X are supercuspidal.

Proof: If X contains elements from an l.d.s. L-packet $\Pi = \Pi(\theta) = \{\pi', \pi''\}$, then

$$a(\pi')\mathrm{Tr}(\pi'(f)) + a(\pi'')\mathrm{Tr}(\pi''(f))$$
$$= \frac{1}{2}(a(\pi') + a(\pi''))\mathrm{Tr}(i_G(\widetilde{\theta})(f)) \pm \frac{1}{2}(a(\pi') - a(\pi''))\mathrm{Tr}(\rho(\theta)(f^H))\ .$$

If f_Π is a pseudo-coefficient for Π, then $\mathrm{Tr}(i_G(\widetilde{\theta})(f)) = 0$ and $\mathrm{Tr}(\rho(f^H_\Pi)) = 0$ by the orthogonality relations and the hypothesis $\rho \neq \rho(\theta)$. However,

$$\mathrm{Tr}(\rho(\theta)(f^H_\Pi)) = \pm(\mathrm{Tr}(\pi'(f_\Pi)) - \mathrm{Tr}(\pi''(f_\Pi))) = \pm 2(\mathrm{Tr}(\pi'(f_\Pi)) \neq 0$$

since the characters of π' and π'' differ by a sign on the elliptic set, and hence $a(\pi') = a(\pi'')$. We can write the equality of the lemma as

$$(12.7.1) \qquad \sum_{\pi\in X''} b(\pi)\ \mathrm{Tr}(\pi(f)) + \sum_{\pi\in X'''} b(\pi)\ \mathrm{Tr}(\pi(f)) = \mathrm{Tr}(\rho(f^H))$$

where $b(\pi) \in \mathbf{Z}, X'''$ is a countable set of p.s. representations, and X'' is the set of representations in X which are either supercuspidal or belong to sets of the form $\{\pi, \pi^{nt}\}$ for π square-integrable. If π is square-integrable but not supercuspidal and f_π is a pseudo-coefficient, then $b(\pi) - b(\pi^{nt}) =$

$\pm\chi_\rho^G(f_\pi)$. Since χ_ρ^G has bounded elliptic norm and the $b(\pi)$ are integers, the orthogonality relations imply that $b(\pi) - b(\pi^{nt})$ is zero for all but finitely many square-integrable π. If $b(\pi) = b(\pi^{nt})$, we can re-write $b(\pi)\,\mathrm{Tr}(\pi(f)) + b(\pi^{nt})\,\mathrm{Tr}(\pi(f^{nt}))$ as $\mathrm{Tr}(i_G(\chi)(f^{nt}))$ for some character χ. We can therefore assume, in (12.7.1), that X'' consists entirely of square-integrable representations. The same argument applied with π supercuspidal implies that X'' is finite.

Let $\gamma = d(\alpha, \beta, \overline{\alpha}^1)$, where $\|\alpha\| < 1$. Recall that by Casselman's theorem, $\chi_\pi(\gamma) = \chi_{\pi_N}(\gamma)$, where π_N is the Jacquet module of π with respect to N. As is well-known ([BZ]), the semi-simplification of the Jacquet module of $i_B(\varphi)$ is $\delta^{1/2}\varphi \oplus \delta^{1/2}(w\varphi)$, where $\delta^{1/2} = \|\alpha\|$ is the modulus function for B. If $i_B(\varphi)$ is reducible and π is an irreducible constituent of $i_B(\varphi)$, then π_N is one-dimensional. Suppose that $\pi_N = \delta^{1/2}\varphi$. By the theory of the constant term ([Si]), §3), if ψ is a matrix coefficient, then $\psi(\gamma) = \delta^{1/2}(\gamma)\varphi(\gamma)$ for $\|\alpha\|$ sufficiently small. If $\pi = \mathrm{St}_G(\xi)$, where ξ is a one-dimensional representation of G, then $\varphi(\gamma) = \xi(\det(\gamma))\|\alpha\|^{\pm 1}$ and if $\pi_N = \delta^{1/2}\varphi$, we must have $\varphi(\gamma) = \xi(\det(\gamma))\|\alpha\|$ since the matrix coefficient ψ is square integrable. Similarly, if $\pi = \pi^2(\xi)$, where ξ is a one-dimensional representation of H, then π is a constituent of $i_B(\varphi)$ where $\varphi(\gamma) = \xi(\gamma)\mu(\alpha)\|\alpha\|^{1/2}$ ($\xi(\gamma)$ is defined by regarding γ as an element of H) and $\pi_N = \delta^{1/2}(\gamma)\xi(\gamma)\mu(\alpha)\|\alpha\|^{1/2}$. We have $D_G(\gamma) = \|\alpha\|^{-1} = \delta^{1/2}(\gamma)^{-1}$, and hence

$$D_G(\gamma)\chi_\pi(\gamma) = \begin{cases} \xi(\det(\gamma))\|\alpha\| & \text{if} \quad \pi = St_G(\xi) \\ \xi(\gamma)\mu(\alpha)\|\alpha\|^{1/2} & \text{if} \quad \pi = \pi^2(\xi) \end{cases}$$

for $\|\alpha\| < 1$. Since $\chi_\pi(\gamma) = \chi_\pi(w\gamma w^{-1})$, we see for all square-integrable π, the restriction of $D_G(\gamma)\chi_\pi(\gamma)$ to M is an integrable function. Similarly, if ρ is a square-integrable representation of H which is not supercuspidal, then $\rho = St_H(\xi)$, where ξ is a one-dimensional representation of H and ρ is a constituent of $i_{B_H}(\varphi)$ where $\varphi(\gamma) = \xi(\gamma)\|\alpha\|^{1/2}$. In this case, $D_H(\gamma) = \delta_H^{1/2}(\gamma) = \|\alpha\|^{-1/2}$ and $D_H(\gamma)\chi_\rho(\gamma) = \xi(\gamma)\|\alpha\|^{1/2}$ where δ_H is the modulus function for B_H. Again, the restriction of $D_H(\gamma)\chi_\rho$ to M is integrable.

We now use the notation of the previous proof. The equality (12.7.1) and the results of the previous paragraph show that

$$\sum_{\pi \in X'''} b(\pi)\mathrm{Tr}(\pi(f)) = \int_{E^*} F_f(\alpha)g(\alpha)d^*\alpha$$

for some function $g(\alpha)$ which is integrable on E^* with respect to $d^*\alpha$. Assume that $d^*\alpha$ is such that $\text{meas}(\mathcal{O}_E) = 1$.

Fix a unitary character χ_0 of E^* and let ω be a prime element in E^*. Up to replacing χ_0 by $\chi_0\| \, \|^s$ for some s, we may assume that $\chi_0(\omega) = 1$. Let φ be a compactly supported function on E^* such that $\varphi(u\alpha) = \chi_0(u)^{-1}\varphi(\alpha)$ for all $u \in \mathcal{O}_E^*$. Then there exists a function f such that $F_f(\alpha) = \varphi(\alpha) + \varphi(\overline{\alpha}^{-1})$. Define

$$\hat{\varphi}(\chi) = \int_{E^*} \varphi(\alpha)\chi(\alpha)d^*\alpha = \begin{cases} \sum_{n=-\infty}^{\infty} \varphi(\omega^n)\chi(\omega)^n & \text{if} \quad \chi|\mathcal{O}_E^* = \chi_0|\mathcal{O}_E^* \\ 0 & \text{otherwise.} \end{cases}$$

Then $\widehat{F}_f(\chi) = \hat{\varphi}(\chi) + \hat{\varphi}(\overline{\chi}^{-1})$. Let Y be the set of $\pi \in X'''$ such that the restriction of χ^π to $\mathcal{O}_E^*$ is either χ_0 or $\overline{\chi_0}^{-1}$. As before, Y is a finite set and

$$\sum_{\pi \in X'''} b(\pi)\text{Tr}(\pi(f)) = \sum_{n=-\infty}^{\infty} \varphi(\omega^n)(\sum_j z_j^n)$$

for some finite collection $\{z_j\}$ of complex numbers. On the other hand, we have

$$\int_{E^*} F_f(\alpha)g(\alpha)d^*\alpha = \sum_{n=-\infty}^{\infty} \varphi(\omega^n)g_n$$

for some sequence $\{g_n\}$ such that $\sum |g_n| < \infty$. This is a contradiction if Y is non-empty since $|\sum z_j^n|$ cannot converge to zero as $n \to -\infty$ and as $n \to \infty$.

This shows that X consists of finitely-many square-integrable representations. As follows from the classification of §12.1 and §12.2, a representation of G or H is supercuspidal if and only if the restriction of its character to M is compactly supported. If ρ is supercuspidal, then the restriction of $\sum a(\pi)\chi_\pi$ to M is compactly supported and it follows easily that all of the π are supercuspidal. The orthogonality relations and Proposition 12.5.2 give

$$\sum a(\pi)^2 = \langle \chi_\rho^G, \chi_\rho^G \rangle_e = 2\langle \chi_\rho, \chi_\rho \rangle_{H,e} = 2\text{Card}(\rho) \ .$$

To prove that the $a(\pi)$ do not all have the same sign, let f_π be a pseudo-coefficient for π and let T be a Cartan subgroup of type (3). The constant term in the germ expansion of $\Phi(\gamma, f_\pi)$ for $\gamma \in T^r$ near 1 is of the form $cf_\pi(1)$ where c is a non-zero real constant independent of π. By the Plancherel formula, $f_\pi(1) = d(\pi)$. However, $\Phi(\gamma, f_\pi) = \overline{\chi_\pi(\gamma)}$ and hence

$$\sum a(\pi)\chi_\pi(\gamma) = c\sum a(\pi)d(\pi) + \alpha(\gamma)$$

where $\alpha(\gamma)$ is either zero for γ near 1 or $\alpha(\gamma)$ tends to infinity as $\gamma \to 1$ (cf. §8.1). Since χ_ρ^G vanishes on T (Cartan subgroups of type 3 do not occur in H), we obtain $\sum a(\pi)d(\pi) = 0$. Formal degrees are positive and hence the $a(\pi)$ do not all have the same sign. In particular, $\mathrm{Card}(X) > 1$ and since $\mathrm{Card}(\rho) = 1$ or 2, $a(\pi) = \pm 1$.

LEMMA 12.7.3: *If the hypothesis of Lemma 12.7.2 holds for $\rho = St_H(\xi)$, where ξ is a character of H, then there exists a unique supercuspidal representation $\pi^s(\xi)$ such that:*

$$\mathrm{Tr}(\pi^2(\xi)(f)) - \mathrm{Tr}(\pi^s(\xi)(f)) = \mathrm{Tr}(St_H(\xi)(f^H)) .$$

Proof: If the assumptions of Lemma 12.7.2 hold for ρ, then

$$\sum_{j=1,2} \varepsilon_j \mathrm{Tr}(\pi_j(f)) = \mathrm{Tr}(\rho(f^H))$$

where $\varepsilon_j = \pm 1$ and the π_j are square-integrable, since $\mathrm{Card}(\rho) = 1$. Let $\gamma = d(\alpha, \beta, \overline{\alpha}^{-1})$, where $\|\alpha\| < 1$. Then $\tau(\gamma) = \mu(\alpha)$ by §4.9, and by the proof of the previous lemma,

$$\mu(\gamma) D_H(\gamma) \chi_\rho(\gamma) = \xi(\gamma)\mu(\alpha)\|\alpha\|^{1/2}$$

On the other hand, by Proposition 12.5.1,

$$D_G(\gamma)\chi_\rho^G(\gamma) = \sum \varepsilon_j \ D_G(\gamma)\chi_{\pi_j}(\gamma) = \xi(\gamma)\mu(\alpha)\|\alpha\|^{1/2} .$$

This is only possible if, up to ordering, $\pi_1 = \pi^2(\xi)$ and $\varepsilon_1 = 1$ (cf. the proof of Lemma 12.7.2). The character of π_2 must then be compactly supported on M and π_2 is thus supercuspidal. The previous lemma gives $\varepsilon_2 = -1$.

COROLLARY 12.7.4: *Suppose that the hypothesis of Lemma 12.7.3 holds. Then:*

$$\mathrm{Tr}(\pi^n(\xi)(f)) + \mathrm{Tr}(\pi^s(\xi)(f)) = \mathrm{Tr}(\xi(f^H)) .$$

Proof: Choose $\chi \in \mathrm{Hom}(M, \mathbf{C}^*)$ so that $JH(i_H(\chi)) = \{St_H(\xi),\ \xi\}$. Then $JH(i_G(\chi\mu)) = \{\pi^2(\xi), \pi^n(\xi)\}$ and the result follows from Lemma 4.9.2.

PROPOSITION 12.7.5: *Let $\tilde{\pi} \in E_\varepsilon(\widetilde{G})$ be a supercuspidal representation. Suppose that*

$$\chi_{\tilde{\pi}_\varepsilon}(\phi) = \sum_{\pi \in X} a(\pi)\mathrm{Tr}(\pi(f)) \qquad (12.7.2)$$

where $a(\pi) \in \mathbf{Z}$. *Then* X *consists of a single supercuspidal representation* π *and* $a(\pi) = \pm 1$.

Proof: Let $\Pi = \Pi(\theta) = \{\pi', \pi''\}$ be an l.d.s. L-packet on G and let f_Π be a pseudo-coefficient for Π. Then $\chi_\pi = -\chi_{\pi'}$ on the elliptic regular set. Since we are assuming that $\chi_\pi - \chi_{\pi'} = \pm\chi^G_{\rho(\theta)}$, the characters χ_π and $\chi_{\pi'}$ are orthogonal to all stable class functions under the $\langle\ ,\ \rangle_e$ by Proposition 12.5.3. It follows (cf. §12.5) that the stable orbital integrals of f_π vanish and the equality of the proposition holds with $\phi = 0$. Since $\mathrm{Tr}(\pi'(f_\Pi)) = -\mathrm{Tr}(\pi''(f_\Pi))$, $a(\pi') = a(\pi'')$ and the contribution of Π to the sum can be replaced by $a(\pi')\mathrm{Tr}(i_G(\widetilde{\theta})(f))$. Arguing as in the proof of Lemma 12.7.2, we can write the sum over X as

$$\sum_{\pi \in X'} b(\pi)\mathrm{Tr}(\pi(f)) \ .$$

where $b(\pi) \in \mathbf{Z}$ and X' consists of (possibly reducible) p.s. representations and at most finitely many square-integrable representations.

The equality (12.7.2) implies that the right-hand side depends only on the stable orbital integrals of f, i.e., that $f \to \chi_{\tilde{\pi}\varepsilon}(\phi)$ is a well-defined stable distribution. However, since $\tilde{\pi}$ is ε-elliptic, the subset Y of elliptic representations in X' is non-empty and consists of square-integrable representations. By the orthogonality relations, card(Y) must equal 1 and the sum over X' can therefore be written as:

$$\pm\mathrm{Tr}(\pi(f)) + \sum_{\chi \in X'''} b(\chi)\ \mathrm{Tr}(i_G(\chi)(f)$$

where χ ranges over a countable set of characters of M and π is supercuspidal. The restriction to M of $\chi_{\varepsilon\pi}(\phi) - \chi_\pi(f)$, viewed as a distribution on G, is integrable, and therefore X'' is empty by the argument of the proof of Lemma 12.7.2. Applying the twisted version of Casselman's theorem ([R3], Proposition 7.4), we see that the restriction of χ_π to M is compactly supported and hence π is supercuspidal.

LEMMA 12.7.6: *Let* $\rho = St_H(\xi)$. *Suppose that the hypothesis of Lemma 12.7.3 is satisfied and that*

$$\mathrm{Tr}(I_{\widetilde{\rho'}}(\phi)|_{\widetilde{\rho'}})) = \mathrm{Tr}(\pi^2(\xi)(f)) + \mathrm{Tr}(\pi^s(\xi)(f)).$$

Then

$$\mathrm{Tr}(I_{\widetilde{\xi'}}(\phi)I_{\widetilde{\xi'}}(\varepsilon)) = \mathrm{Tr}(\pi^n(\xi)(f)) - \mathrm{Tr}(\pi^s(\xi)(f)) \ .$$

Proof: Let $\chi \in \mathrm{Hom}(M, \mathbf{C}^*)$ be such that $JH(i_H(\chi)) = \{\xi, \rho\}$. Then $JH\{i_G(\chi\mu)\} = \{\pi^2(\xi),\ \pi^n(\xi)\}$ and $JH\{i_{\widetilde{G}}(\chi\mu \circ N)\} = \{I_{\widetilde{\rho'}}, I_{\widetilde{\xi'}}\}$. The conclusion follows from Proposition 4.10.2.

LEMMA 12.7.7: *Let $\rho \in \Pi^2(H)$ and suppose that ρ is not of the form $\rho(\theta)$ for θ semi-singular. Let $\pi = I_{\widetilde{\rho'}}$ and define $\chi(\gamma) = \chi_{\pi\varepsilon}(\delta)$ where $\delta \in \widetilde{M}$ satisfies $N(\delta) = \gamma$. Then $D_G(\gamma)\chi(\gamma)$ is integrable as a function on M.*

Proof: Let $\gamma = d(\alpha, \beta, \overline{\alpha}^{-1})$, where $\|\alpha\| < 1$. Let σ be semisimplification of the Jacquet module of π with respect to $\widetilde{N}$. By the twisted version of Casselman's theorem ([R₃], Proposition 7.4), $\chi(\gamma) = \chi_{\sigma\varepsilon}(\delta)$. If $\rho = \rho(\theta)$ where $\theta = \theta_1 \otimes \theta_2 \otimes \theta_3$ is a regular character, then $\pi = I_{\mu'}$ where $\mu' = (\tilde{\theta}_1, \tilde{\theta}_2, \tilde{\theta}_3)$, and $\sigma = \oplus \delta^{1/2}(w\mu')$, where δ is the modulus function for $\widetilde{B}$ and the sum is over the Weyl group $\Omega(\widetilde{M})$. The eigenspaces $\delta^{1/2}\mu$, for $\mu = w\mu'$, are distinct since θ is regular and $\sigma(\varepsilon)$ permutes them without fixed points, taking μ to $\varepsilon\mu$. Hence $\chi(\gamma) = \mathrm{Tr}(\sigma(\delta)\sigma(\varepsilon)) = 0$. Suppose that ρ is not of the form $\rho(\theta)$. If ρ is supercuspidal, then $\tilde{\rho}'$ is also supercuspidal, in which case $\sigma = \{0\}$ by [BZ]. If ρ is not supercuspidal, then $\rho = St_H(\xi)$ and $\widetilde{\rho'} = St_{\widetilde{H}}(\tilde{\xi}')$ is as in §12.1. Let $\phi(d(a, b, c)) = \tilde{\xi}'(\delta)\|\frac{a}{c}\|^{1/2}$. By [BZ], §2.11, there exist distinct non-trivial elements $w_1,\ w_2 \in \Omega(\widetilde{M})$ such that

$$\sigma = \delta^{1/2}\phi \oplus \delta^{1/2}w_1\phi \oplus \delta^{1/2}w_2\phi\ .$$

As before, $\sigma(\varepsilon)$ permutes the eigenspaces. Since $\delta^{1/2}\phi$ is regular and ε-invariant, ε has two orbits of length one and two orbits of length two in $\{\delta^{1/2}w\phi : w \in \Omega(\widetilde{M})\}$, hence $\sigma(\varepsilon)$ must interchange the $\delta^{1/2}w_1\phi$ and $\delta^{1/2}w_2\phi$-eigenspaces. It follows that $\chi(\gamma) = \delta^{1/2}\phi(\delta)$ and

$$D_G(\gamma)\chi(\gamma) = D_G(\gamma)\delta^{1/2}\phi(\delta) = \xi(\gamma)\mu(\alpha)\|\alpha\|^{1/2}$$

for $\|\alpha\| < 1$. Now γ is conjugate to $\gamma' = d(\overline{\alpha}^{-1}, \beta, \alpha)$ and $\chi(\gamma)$ is a class function. Hence if $\|\alpha\| > 1$, then

$$D_G(\gamma)\chi(\gamma) = D_G(\gamma')\chi(\gamma') = \xi(\gamma)\mu(\alpha)\|\alpha\|^{-1/2}$$

and the lemma follows.

CHAPTER 13

Automorphic representations

The first three sections of this chapter contain statements of the main local and global results. In the succeeding sections, these results are derived from the results of §10.

13.1 Classification of local L-packets for $U(3)$. In this section and the next, let E/F be a quadratic extension of p-adic fields.

THEOREM 13.1.1: *There exists a unique partition of $E(G)$ into L-packets and a map $\xi_H : \Pi(H) \to \Pi(G)$, where $\Pi(G)$ is the set of L-packets on G, such that:*

(1) $\mathrm{Card}(\Pi) > 1$ *if and only if* $\Pi = \xi_H(\rho)$ *for some* $\rho \in \Pi^2(H)$.

(2) *Let* $\rho \in \Pi(H)$. *Assume that* $\dim(\rho) \neq 1$ *and that* ρ *is not of the form* $i_H(\chi\mu^{-1})$, *where* $\chi_1(\alpha) = \|\alpha\|$ *or* $\|\alpha\|^{-1}$. *Then there is a unique map* $\pi \to \langle\rho, \pi\rangle$ *from* $\xi_H(\rho)$ *to* $\{\pm 1\}$ *such that*

$$\chi_\rho(f^H) = \sum_{\pi \in \xi_H(\rho)} \langle\rho, \pi\rangle \chi_\pi(f) \ .$$

(3) *If* $\dim(\rho) = 1$, *then* $\xi_H(\rho) = \pi^n(\rho)$.

A list of the types of L-packets is given in §12.2. The next two propositions list important properties of L-packets.

PROPOSITION 13.1.2: *Let* $\Pi \in \Pi(G)$.

(a) $\mathrm{Card}(\Pi) = 1, 2,$ *or* 4. *If* $\Pi = \xi_H(\rho)$, *where* $\rho \in \Pi(H)$ *and* Π *is square-integrable, then* $\mathrm{Card}(\Pi) = 2\mathrm{Card}(\rho)$.

(b) $\mathrm{Card}(\Pi) = 4$ *if and only if there is a set* $\{\theta_1, \theta_2, \theta_3\}$ *of distinct G-equivalent regular characters such that* $\Pi = \xi_H(\rho(\theta_j))$ *for* $j = 1, 2, 3$. *There is a bijection between* Π *and* $(\mathbf{Z}/2)^2$ *under which the set of functions* $\{\langle\rho_j, \rangle\}$ *is identified with the set of non-trivial characters of* $(\mathbf{Z}/2)^2$.

(c) *Let* ρ_1, ρ_2 *be distinct elements in* $\Pi(H)$. *Then* $\xi_H(\rho_1) = \xi_H(\rho_2)$ *if and only if there exist G-equivalent characters* θ_1 *and* θ_2 *such that* $\rho_1 = \rho(\theta_1)$ *and* $\rho_2 = \rho(\theta_2)$.

PROPOSITION 13.1.3: *Let $\rho \in \Pi(H)$.*

(a) *If $\rho \in \Pi^2(H)$, then $\langle \rho, \ \rangle$ is surjective.*

(b) *The L-packet $\xi_H(\rho)$ is square-integrable (resp. supercuspidal) if and only if ρ is square-integrable (resp. supercuspidal) and is not of the form $\rho(\theta)$ with θ semi-regular. If $\xi_H(\rho)$ is square-integrable, then the members of $\xi_H(\rho)$ have the same formal degree.*

(c) *If $\rho = \rho(\theta)$, where θ is semi-regular, then $\xi_H(\rho) = \Pi(\theta) = \{\pi^1(\theta), \pi^2(\theta)\}$ and $\langle \rho, \pi^1(\theta)\rangle = 1$, $\langle \rho, \pi^2(\xi)\rangle = -1$.*

(d) *If $\rho = St_H(\xi)$, where $\xi \in \Pi(H)$ is one-dimensional, there is a supercuspidal representation $\pi^s(\xi)$ such that $\xi_H(\rho) = \{\pi^2(\xi), \pi^s(\xi)\}$ and $\langle \rho, \pi^2(\xi)\rangle = 1$, $\langle \rho, \pi^s(\xi)\rangle = -1$.*

If $\dim(\rho) \neq 1$, the image of ρ under ξ_H will be denoted by $\Pi(\rho)$. If $\xi \in \Pi(H)$ is one-dimensional, define $\Pi(\xi) = \{\pi^n(\xi), \pi^s(\xi)\}$. We refer to $\Pi(\xi)$ as an A-packet. Define $\langle \xi, \ \rangle$ to be the function with constant value 1 on $\Pi(\xi)$. Conditions (2) and (3) of Proposition 13.1.1 determine $\xi_H(\rho)$ uniquely unless ρ is of the form $i_H(\chi\mu^{-1})$, where $\chi_1(\alpha) = \|\alpha\|$ or $\|\alpha\|^{-1}$. In this latter case, the problem is that $\chi_\rho(f^H) = \chi_\pi(f)$, where $\pi = i_G(\chi)$, and $i_G(\chi)$ has two constituents, a one-dimensional representation π' and a Steinberg representation, each of which makes up an L-packet on its own. We define $\xi_H(\rho) = \pi'$. However ρ plays no role in global questions since no twist of ρ is unitary and hence ρ does not occur as a local component of an automorphic representation.

PROPOSITION 13.1.4: *If $\xi \in \Pi(H)$ and $\dim(\xi) = 1$, the relation:*

$$\chi_\xi(f^H) = \sum_{\pi \in \Pi(\xi)} \langle \xi, \pi \rangle \chi_\pi(f)$$

holds.

Let $\Pi_e(G)$ be the set of L-packets of the form $\Pi(\rho)$ where $\rho \in \Pi^2(H)$ and let $\Pi_s(G)$ be the set of L-packets of cardinality 1 which are not of the form $\{\pi^n(\xi)\}$ for any one-dimensional representation ξ of H. Let $\Pi_a(G)$ be the set of A-packets $\Pi(\xi)$. Let $\Pi'(G)$ be the union of $\Pi_e(G)$, $\Pi_s(G)$, and $\Pi_a(G)$. Each representation π of G lies in at least one packet in $\Pi'(G)$ and in at most one unless $\pi = \pi^s(\xi)$ for some ξ.

13.2 Local base change for $U(3)$. The main results on local base change are contained in the following two propositions. Recall that $E_\varepsilon(\widetilde{G})$

is the set of ε-invariant representations π of $\widetilde{G}$ such that the restriction of ω_π to F^* is trivial. In this section, E/F is again p-adic.

Let $\Pi \in \Pi'(G)$. If $\Pi \notin \Pi_a(G)$, let $\langle 1, \pi \rangle = 1$ for all $\pi \in \Pi$ and if $\Pi = \Pi(\xi)$ where $\dim(\xi) = 1$, let $\langle 1, \pi^n(\xi) \rangle = 1$ and $\langle 1, \pi^s(\xi) \rangle = -1$. Define

$$\mathrm{Tr}(\Pi(f)) = \chi_\Pi(f) = \sum_{\pi \in \Pi} \langle 1, \pi \rangle \chi_\pi(f)$$

We define the notion of a lift with respect to ψ_G in two steps. If Π is tempered or if $\Pi = \Pi(\xi)$ for some $\xi \in \Pi(H)$ of dimension one, we will say that $\tilde{\pi} \in E_\varepsilon(\widetilde{G})$ is a lift of Π if there is a choice of $\tilde{\pi}(\varepsilon)$ such that $\chi_{\tilde{\pi}\varepsilon}(\phi) = \chi_\pi(f)$ whenever $\phi \to f$. This condition determines $\tilde{\pi}$ uniquely by the linear independence of twisted characters. If Π is non-tempered but not of the form $\Pi(\xi)$, then Π consists of a Langlands quotient of $i_G(\chi)$ for some $\chi \in \mathrm{Hom}(M, \mathbf{C}^*)$ which is positive with respect to B and we define the lift $\tilde{\pi}$ of Π to be the Langlands quotient of $i_{\widetilde{G}}(\chi \circ N)$ (note that $\tilde{\chi} \circ N$ is positive with respect to $\widetilde{B}$, so that $i_{\widetilde{G}}(\chi \circ N)$ has a Langlands quotient). In this case, either $i_G(\chi)$ is irreducible or Π consists of the one-dimensional representation $\psi \circ \det$. In the latter case, $\tilde{\pi} = \tilde{\psi} \circ \det$ and the character identity $\chi_{\tilde{\pi}\varepsilon}(\phi) = \chi_\pi(f)$ holds by Proposition 12.4.1. It also holds if $i_{\widetilde{G}}(\tilde{\chi})$ is irreducible. It does not hold, however, if $i_{\widetilde{G}}(\tilde{\chi})$ is reducible. This happens when $\chi = (\chi_1, \chi_2)$, where $\chi_1(\alpha) = \eta(\alpha)\|\alpha\|^{1/2}$ and $\eta|F^* = 1$, for then, $i_G(\chi)$ is irreducible but $i_{\widetilde{G}}(\tilde{\chi})$ is reducible by the results of §12.

THEOREM 13.2.1: *For all $\Pi \in \Pi'(G)$, there exists a unique lifting $\tilde{\pi} \in E_\varepsilon(\widetilde{G})$. The map $\Pi \to \tilde{\pi}$ is injective.*

We denote the lift of $\Pi \in \Pi(G)$ by $\psi_G(\Pi)$.

PROPOSITION 13.2.2: *Let $\Pi \in \Pi(G)$ and suppose that $\pi = \psi_G(\Pi)$.*

(a) *π is square-integrable (resp. supercuspidal) if and only if Π is square-integrable (resp. supercuspidal) and $\Pi \in \Pi_s(G)$. The map ψ_G defines a bijection between the set of square-integrable (resp. supercuspidal) L-packets $\Pi \in \Pi_s(G)$ and the set of square-integrable (resp. supercuspidal) representations in $E_\varepsilon(\widetilde{G})$.*

(b) *If $\psi \in \mathrm{Hom}(E^1, \mathbf{C}^*)$, then the lift of $\mathrm{St}_G(\psi)$ (resp., $\psi \circ \det_G$) is $\mathrm{St}_{\widetilde{G}}(\widetilde{\psi})$ (resp., $\widetilde{\psi} \circ \det \widetilde{G}$).*

(c) *Let $\rho \in \Pi^2(H)$. Then $\psi_G(\Pi(\rho)) = I_{\tilde{\rho}'}$.*

(d) *If $\xi \in \Pi(H)$ is one-dimensional, then $\psi_G(\Pi(\xi)) = I_{\widetilde{\xi}'}$.*

13.3 Global results.

THEOREM 13.3.1: *Let π be a discrete automorphic representation of G. Then the multiplicity $m(\pi)$ of π in the discrete spectrum of G is equal to* 1.

To define global L-packets on G, choose $\Pi_v \in \Pi'(G_v)$ for all v, such that for almost all finite v, Π_v contains an unramified representation π_v^0. Note that if v splits in E, then G_v is isomorphic to $\mathrm{GL}_3(F_v)$ and an L-packet consists of a single irreducible representation. We define a global L-packet $\Pi = \otimes\Pi_v$ on G to be the set of $\pi = \otimes\pi_v$ such that $\pi_v \in \Pi_v$ for all v and $\pi_v = \pi_v^0$ for almost all v. A global L-packet Π will be called discrete if some member of Π occurs in the discrete spectrum and will be called cuspidal if each member of Π which occurs discretely occurs only in the space of cusp forms. Let $\Pi(\mathbf{G})$ and $\Pi(\mathbf{H})$ be the set of discrete L-packets on G and H, respectively, and let $\Pi_0(\mathbf{H})$ be the set of cuspidal L-packets on H. If ρ is a discrete L-packet on H, let $\Pi(\rho) = \otimes\Pi(\rho_v)$.

THEOREM 13.3.2: *Let $\rho \in \Pi(\mathbf{H})$. Then $\Pi(\rho)$ is discrete if and only if ρ is not of the form $\rho(\theta)$ with θ semi-regular. If $\Pi(\rho)$ is discrete, and* $\dim(\rho) \neq 1$, *then it is cuspidal.*

Define:

$$\Pi_a(\mathbf{G}) = \{\Pi(\xi) : \xi \in \Pi(\mathbf{H}) \quad \text{and } \dim(\xi) = 1\}$$
$$\Pi_e(\mathbf{G}) = \{\Pi(\rho) : \rho \in \Pi_0(\mathbf{H}), \rho \neq \rho(\theta) \quad \text{for } \theta \text{ semi-regular}\}$$

It is clear that $\Pi_a(\mathbf{G})$ and $\Pi_e(\mathbf{G})$ are disjoint. Let $\Pi_s(\mathbf{G})$ be the set of discrete L-packets $\Pi = \otimes\Pi_v$ on G such that there is no $\rho \in \Pi(\mathbf{H})$ with the property that $\Pi_v = \Pi(\rho_v)$ for a.a.v. L-packets in $\Pi_e(\mathbf{G})$ and $\Pi_s(\mathbf{G})$ will be called endoscopic and stable, respectively.

Let $\Pi_\varepsilon(\widetilde{\mathbf{G}})$ be the set of ε-invariant discrete representations of $\widetilde{\mathbf{G}}$ whose restriction to the subgroup I_F of the center of $\widetilde{\mathbf{G}}$ is trivial. Then $\Pi_\varepsilon(\widetilde{\mathbf{G}})$ is the union of the subset $\Pi_{0\varepsilon}(\mathbf{G})$ of its cuspidal elements and the set of one-dimensional ε-invariant automorphic representations of $\widetilde{\mathbf{G}}$ ([MW]). If $\Pi = \otimes\Pi_v$ is an L-packet on $\mathbf{G}$, the base change $\psi_G(\Pi)$ of Π is, by definition, the representation $\pi = \otimes\pi_v$ where $\psi_G(\Pi_v) = \pi_v$.

THEOREM 13.3.3: *Let* $\Pi \in \Pi(\mathbf{G})$.

(a) $\psi_G(\Pi) \in \Pi_\varepsilon(\widetilde{\mathbf{G}})$ *if and only if* $\Pi \in \Pi_s(\mathbf{G})$. *The map ψ_G defines a bijection between $\Pi_s(\mathbf{G})$ and $\Pi_\varepsilon(\widetilde{\mathbf{G}})$.*

(b) *If* $\Pi \in \Pi_s(\mathbf{G})$, *then* Π *is cuspidal if and only if* $\psi_G(\Pi)$ *is cuspidal. In particular,* Π *is cuspidal unless* $\Pi = \{\psi\}$, *where* ψ *is a one-dimensional automorphic representation of* G.

(c) *If* $\Pi \in \Pi_s(\mathbf{G})$, *then* $m(\pi) = 1$ *for all* $\pi \in \Pi$.

THEOREM 13.3.4: *Let* $\rho, \rho' \in \Pi(\mathbf{H})$. *Then* $\Pi(\rho) = \Pi(\rho')$ *if and only if there exists* G*-conjugate characters* θ *and* θ' *such that* $\rho = \rho(\theta)$ *and* $\rho' = \rho(\theta')$.

If Π is of the form $\Pi(\rho)$, where $\rho = \rho(\theta)$, we write $\Pi = \Pi(\theta)$.

THEOREM 13.3.5: *Let* $\Pi, \Pi' \in \Pi(\mathbf{G})$. *If* $\Pi_v = \Pi_v$ *for almost all* v, *then* $\Pi = \Pi'$.

By Theorem 13.3.5, $\Pi(\mathbf{G})$ is the disjoint union of $\Pi_s(\mathbf{G})$, $\Pi_e(\mathbf{G})$, and $\Pi_a(\mathbf{G})$.

Let $\xi \in \Pi(\mathbf{H})$ be one-dimensional. There exist Hecke characters η and ψ of I_E^1 such that $\xi(h) = \eta(\det_0(h))\psi(\det(h))$. Let φ be the Hecke character of I_E defined by $\varphi(a) = \mu(a)\tilde{\eta}(a)$, where $\tilde{\eta}(a) = \eta(a/\overline{a})$.

THEOREM 13.3.6: *Let* $\pi \in \Pi(\xi)$.

(a) *If* $m(\pi) \neq 0$ *and* $\pi \neq \otimes\pi^n(\xi_v)$, *then* π *is cuspidal.*

(b) *If* $\pi = \otimes\pi^n(\xi_v)$, *then* $m(\pi) = 1$. *Furthermore,* π *is cuspidal if and only if* $L\left(\frac{1}{2}, \varphi\right) = 0$.

(c) *If* π' *is a discrete automorphic representation of* G *such that* π'_v *is of the form* $\pi^n(\xi_v)$ *for some place* v *of* F *which does not split in* E, *then* $\pi' \in \Pi(\xi)$ *for some one-dimensional* $\xi \in \Pi(\mathbf{H})$

Let $\Pi = \Pi(\rho)$ where $\rho \in \Pi_0(\mathbf{H})$. If v splits in E, define $\langle \rho_v, \pi_v \rangle = \langle 1, \pi_v \rangle = 1$. For $\pi \in \Pi$, define

$$\langle \rho, \pi \rangle = \Pi \langle \rho_v, \pi_v \rangle$$

and set $\langle 1, \pi \rangle = \Pi \langle 1, \pi_v \rangle$. The products are well-defined, since $\langle \rho_v, \pi_v \rangle = \langle 1, \pi_v \rangle = 1$ if π_v is unramified. Let

$$\widehat{\Pi} = \{\rho \in \Pi(\mathbf{H}) : \Pi = \Pi(\rho)\} \cup \{1\}\ .$$

We regard $\langle\ ,\ \rangle$ as a pairing between $\widehat{\Pi}$ and Π. Observe that $\mathrm{Card}(\widehat{\Pi})$ is 1 if $\Pi \in \Pi_s(\mathbf{G})$, and if $\Pi \in \Pi_e(\mathbf{G}) \cup \Pi_a(\mathbf{G})$, it is 4 or 2 according as Π is or is not of the form $\Pi(\theta)$ for some regular θ.

THEOREM 13.3.7: *Let* $\Pi \in \Pi_e(\mathbf{G}) \cup \Pi_a(\mathbf{G})$. *For all* $\pi \in \Pi$,

$$m(\pi) = \mathrm{Card}(\widehat{\Pi})^{-1} \sum_{s \in \widehat{\Pi}} \langle s, \pi \rangle \ .$$

Let $n(\Pi) = \mathrm{Card}(\widehat{\Pi})^{-1}$. Define

$$\mathrm{Tr}(\Pi(f)) = \prod_v \mathrm{Tr}(\Pi_v(f_v)) = \sum_{\pi \in \Pi} \langle 1, \pi \rangle \mathrm{Tr}(\pi(f))$$

The sum on the right is finite since $\mathrm{Tr}(\pi(f)) = 0$ unless π_v is the unique unramified member of Π_v for v such that $f_v \in \mathcal{H}_v$. Recall that ρ_d is the representation of $\mathbf{G}$ on the space $L_d(G)$ defined with respect to the central character ω.

THEOREM 13.3.8: $\mathrm{Tr}(\rho_d(f)) = \sum n(\Pi)\mathrm{Tr}(\Pi(f)) + \frac{1}{2} \sum n(\rho)\mathrm{Tr}(\rho(f^H))$ *where* Π *ranges over the subset of* $\Pi(\mathbf{G})$ *of packets with central character* ω *and* ρ *ranges over the subset of* $\Pi(\mathbf{H})$ *of packets with central character* $\omega\mu^{-1}$ *which are not of the form* $\rho(\theta)$ *for some semi-regular* θ.

Proof: We have to check that if Π is a discrete L-packet on G, then

$$\sum_{\pi \in \Pi} m(\pi)\mathrm{Tr}(\pi(f)) = n(\Pi)\mathrm{Tr}(\Pi(f)) + \frac{1}{2} \sum_{\substack{\rho \in \Pi(\mathbf{H}) \\ \Pi = \Pi(\rho)}} n(\rho)\mathrm{Tr}(\rho(f^H))$$

If $\Pi \in \Pi_s(\mathbf{G})$, then $n(\Pi) = 1$, the sum on the right is empty, and the equality follows from Theorem 13.3.1(c). If $\Pi = \Pi(\rho)$, then $n(\Pi) = \frac{1}{2}n(\rho)$ and $\mathrm{Tr}(\rho(f^H)) = \sum \langle \rho, \pi \rangle \mathrm{Tr}(\pi(f))$ by Propositions 13.1.1 and 13.1.4. Hence the right-hand side is equal to

$$\mathrm{Card}(\widehat{\Pi})^{-1} \sum_{\pi \in \Pi} \sum_{s \in \widehat{\Pi}} \langle s, \pi \rangle \mathrm{Tr}(\pi(f)) \ .$$

and the result follows from Theorem 13.3.7.

13.4 In the next lemma and corollary, the pair (G, ε) can be arbitrary. We use the notation of §2.3.

LEMMA 13.4.1: *Let* M_1, M_2 *be Levi subgroups of* G *and let* $s \in \Omega(M_1, M_2)$. *Let* $P_j \in \wp(M_j)$. *Then*

$$I^{P_2}_{s(\sigma), s(\lambda)}(\varepsilon) \circ M_{P_2|P_1}(s, \lambda) = M_{\varepsilon(P_2)|\varepsilon(P_1)}(\varepsilon(s), \varepsilon(\lambda)) \circ I^{P_1}_{\sigma, \lambda}(\varepsilon) \ ,$$

where σ is a discrete automorphic representation of M_1.

Proof: This follows from a change of variables.

COROLLARY 13.4.2: *Let $P_1, P_2 \in \mathcal{P}(M)$, where M is a Levi subgroup, and let $s \in \Omega(\varepsilon(M), M)$. Let σ be a discrete automorphic representation of M such that $s\varepsilon(\sigma) \sim \sigma$. Then:*

$$\mathrm{Tr}(I_\sigma^{P_1}(\phi) \circ M_{P_1|\varepsilon(P_1)}(s) \circ I_\sigma^{P_1}(\varepsilon)) = \mathrm{Tr}(I_\sigma^{P_2}(\phi) \circ M_{P_2|\varepsilon(P_2)}(s) \circ I_\sigma^{P_2}(\varepsilon))$$

Proof: By the functional equations from the theory of Eisenstein series,

$$M_{P_1|P_2}(1) = M_{P_2|P_1}(1)^{-1}$$
$$M_{P_1|P_2}(1) \circ M_{P_2|\varepsilon(P_2)}(s) \circ M_{\varepsilon(P_2)|\varepsilon(P_1)}(1) = M_{P_1|\varepsilon(P_1)}(s) \ .$$

It follows from Lemma 13.4.1 that the operator on the left is conjugate to the one on the right via $M_{P_1|P_2}(1)$ and hence they have the same trace.

In view of Corollary 13.4.2, the contribution of a quintuple $\{M, G, \{0\}, \sigma, s\}$ to the fine χ-expansion can be written as

$$|\Omega_0|^{-1} \cdot |\Omega(M)| \cdot \Delta(s)^{-1} \mathrm{Tr}(I_\sigma^P(\phi) \circ M_{P|\varepsilon(P)}(s) \circ I_\sigma^P(\varepsilon))$$

for any $P \in \mathcal{P}(M)$. On the other hand, if $w \in \Omega(M, M')$, $P \in \mathcal{P}(M)$, and $P' \in \mathcal{P}(M')$, then

$$M_{P'|P}(w) \circ I_\sigma^P(\phi) \circ M_{P|\varepsilon(P)}(s) \circ I_\sigma^P(\varepsilon) \circ M_{P|P'}(w^{-1})$$
$$= I_{w(\sigma)}^{P'}(\phi) \circ M_{P'|\varepsilon(P')}(ws\varepsilon(w)^{-1}) I_{w(\sigma)}^{P'}(\phi) \ .$$

The contributions of $\{M, G, \{0\}, \sigma, s\}$ and $\{M, G, \{0\}, w(\sigma), ws\varepsilon(w)^{-1}\}$ therefore coincide. This holds, in particular, if $M = M'$.

13.5 For the rest of the chapter, let $G = U(3)$, $H = U(2) \times U(1)$, and $\widetilde{G} = \mathrm{Res}_{E/F}(U(3))$. We will tacitly assume, locally and globally, that all representations of G, H and related subgroups have central character ω and that all representations of $\widetilde{G}$ and related subgroups have central character $\tilde{\omega}$, but will not make this assumption explicit to avoid repetition.

We now determine the quintuples $(M, \widetilde{G}, \{0\}, \sigma, s)$ which correspond to discrete term $\theta_{\widetilde{G}}$ in the trace formula for $\widetilde{G}$. Let s_j be the reflection with respect to the root α_j. Let M_j be the centralizer in $\widetilde{G}$ of the kernel of α_j in $\widetilde{M}$. If $\mathcal{A}_0^{s\varepsilon} = \{0\}$, then $s = s_j$ for some j and the quintuples with $M = \widetilde{M}$ are of the form (1) and (2) below. If $s\varepsilon(M_j) = M_j$, then $s \in \Omega(M_j)s_3$ and

$\mathcal{A}_M^{s\varepsilon} = \{0\}$, and this gives (3) below. The quintuples for $\widetilde{G}$, besides those corresponding to discrete automorphic representations, are:

(1) $(\widetilde{M}, \widetilde{G}, \{0\}, \mu, s_3) : \mu = (\mu_1, \mu_2, \mu_3)$ is a character of $\widetilde{M}\backslash\widetilde{\mathbf{M}}$ such that $\mu_j\overline{\mu}_j = 1$ for $j = 1, 2, 3$.

(2) $(\widetilde{M}, \widetilde{G}, \{0\}, \mu, s_j) : j = 1, 2$ and $\mu = (\xi, \xi, \xi)$, where ξ is a character of $\widetilde{M}\backslash\widetilde{\mathbf{M}}$ such that $\xi\overline{\xi} = 1$.

(3) $(M_j, \widetilde{G}, \{0\}, \sigma, s) : s \in \Omega(M_j)s_3$ and σ is a discrete automorphic representation of M_j such that $s\varepsilon(\sigma) \sim \sigma$.

Consider quintuples of type (3) for G. The group M_j is isomorphic to $\mathrm{GL}_2 \times \mathrm{GL}_1$ over E. Suppose that $\sigma = \sigma_2 \otimes \chi_1$, where σ_2 is discrete on $\mathrm{GL}_2(E)$ and χ_1 is a Hecke character of $\mathrm{GL}_1(E)$. The condition $s_3\varepsilon(\sigma) \sim \sigma$ implies that χ_1 is trivial on $N_{E/F}(I_E)$ and σ_2 is ε-invariant on $\mathrm{GL}_2(E)$. By Theorem 11.5.2, the restriction of the central character of σ_2 to I_F is trivial and σ_2 is the base change of a discrete L-packet ρ_2 on $U(2)$ which is not of the form $\rho(\theta)$, by Theorem 11.5.1. Since $\sigma_2 \otimes \chi_1$ is trivial on I_F, χ_1 is also trivial on I_F. Hence σ is the base change with respect to either ψ_H or $\psi_{H'}$ of a discrete L-packet ρ on H which is not of the form $\rho(\theta)$. We identify M_3 with $\widetilde{H}$ and let P denote a parabolic subgroup containing $\widetilde{H}$. By the multiplicity one theorem for GL(2), the space I_σ^P is isomorphic to the representation induced by σ, which we denote by I_σ. Let $I_\sigma(\varepsilon)$ denote the operator $M_{P|\varepsilon(P)}(1) \circ I_\sigma^P(\varepsilon)$. We have $|\Omega_0|^{-1} \cdot |\Omega(M_j)| = \frac{1}{3}$, and $(s\varepsilon - 1)$ acts on $\mathcal{A}_{M_j}/\mathcal{A}_G$ by -1, so that $\Delta(s) = 2$. Since the M_j are conjugate for $1 \leq j \leq 3$, the remark of §13.4 shows that the total contribution of quintuples of type (3) is equal to:

$$\frac{1}{2}\sum \mathrm{Tr}(I_\sigma(\phi)I_\sigma(\varepsilon))$$

where the sum is over the set $\{\sigma\}$ of automorphic representations of $\widetilde{H}$ such that σ is the base change by either ψ_H or $\psi_{H'}$ of a discrete L-packet ρ on H which is not of the form $\rho(\theta)$.

In the case of quintuples of type (1) and (2), if $\mu = (\mu_1, \mu_2, \mu_3)$, then for each j, either $\mu_j|C_F = 1$ or $\mu_j|C_F = \omega_{E/F}$. Since $\mu_1\mu_2\mu_3$ is trivial on C_F, either $\mu_j|C_F = 1$ for $j = 1, 2, 3$ or $\mu_j|C_F = \omega_{E/F}$ for precisely two values of j. These two cases will be denoted by $\mu|C_F = 1$ and $\mu|C_F = \omega_{E/F}$, respectively. We call μ regular if the μ_j are distinct and semi-regular if precisely two of the μ_j coincide. If all of the μ_j coincide, we call μ singular.

By §13.4, the contribution of a quintuple depends only on its Ω_0-orbit. If μ is regular or semi-regular, then $s = s_3$ and if μ is singular, $s = s_1, s_2$, or s_3. Furthermore, $\Delta(s_3) = 4$ and $\Delta(s_1) = \Delta(s_2) = 1$. The total contribution of quintuples of type (1) and (2) is the sum of the terms

$$\frac{1}{4} \sum_{\mu \text{ regular}} \mathrm{Tr}(I_\mu^{\widetilde{B}}(\phi) M_{\widetilde{B}|\widetilde{B}}(s_3) \circ I_\mu^{\widetilde{B}})$$

$$\frac{1}{8} \sum_{\mu \text{ semi-regular}} \mathrm{Tr}(I_\mu^{\widetilde{B}}(\phi) M_{\widetilde{B}|\widetilde{B}}(s_3) \circ I_\mu^{\widetilde{B}})$$

$$\frac{1}{24} \sum_{\mu \text{ singular}} \mathrm{Tr}(I_\mu^{\widetilde{B}}(\phi) M_{\widetilde{B}|\widetilde{B}}(s_3) \circ I_\mu^{\widetilde{B}}) + \frac{1}{6} \sum_{j=1,2} \sum_{\mu \text{ singular}} \mathrm{Tr}(I_\mu(\phi) M_{\widetilde{B}|\widetilde{B}}(s_j) \circ I_\mu^{\widetilde{B}})$$

where the sums are over a set of representatives for the Ω_0-orbits.

By the results of Keys-Shahidi ([KyS]), the intertwining operator $M_{\widetilde{B}|\widetilde{B}}(s)$ can be written as a product of a normalized intertwining operator $R(s)$ and the scalar factor:

$$\text{(13.5.1)} \qquad \lim_{z \to 1} \prod_{\substack{\alpha > 0 \\ s\alpha < 0}} L(z, \mu_\alpha) L(-z, \mu_{\overline{\alpha}}^{-1})^{-1}$$

where $L(z, \mu_\alpha) = L(z, \mu_i \mu_j^{-1})$ if α is the root $\alpha(d(a_1, a_2, a_3)) = a_i/a_j$. If χ is a Hecke character of C_E such that $\chi\overline{\chi} = 1$, then $L(z, \chi) = L(z, \chi^{-1})$ and the terms of the product for which μ_α is non-trivial have limit equal to 1. This scalar is thus equal to 1 if μ is regular. In the regular case, let $I_\mu(\varepsilon)$ denote the operator $R(s_3) \circ I_\mu^{\widetilde{B}}(\varepsilon)$.

Suppose that μ is not regular. If $s\mu = \mu$, then $R(s)$ acts by the identity by [KyS], Theorem 4.2. If μ is semi-regular, the product in (13.5.1) reduces to $\zeta(z)\zeta(-z)^{-1}$. If μ is singular, the product is equal to $\zeta(z)\zeta(-z)^{-1}$ if $s = s_1$ or s_2 and to $(\zeta(z)\zeta(-z)^{-1})^3$ if $s = s_3$. The limit is -1 in these cases. We can assume μ chosen within its Ω_0-orbit so that $\mu_1 = \mu_3$. Then $\varepsilon\mu = \mu$ and $s_3\mu = \mu$. In this case, let $I_\mu(\varepsilon)$ denote the operator $\varphi(g) \to \varphi(\varepsilon^{-1}(g))$. Then $M_{\widetilde{B}|\widetilde{B}}(s) \circ I_\mu^{\widetilde{B}}(\varepsilon)$ is equal to $-I_\mu(\varepsilon)$. The contribution of the non-regular characters is equal to the sum of terms (5),(6), and (3′) below.

Finally, observe that a discrete automorphic representation π contributes $\mathrm{Tr}(\pi(\phi)\pi(\varepsilon))$ to $\theta_{\widetilde{G}}$, by the multiplicity one theorem for GL(3). The next proposition summarizes the above discussion. For $\rho \in \Pi(\mathbf{H})$, we denote $\psi_H(\rho)$ and $\psi'_H(\rho)$ by $\tilde{\rho}$ and $\tilde{\rho}'$, respectively.

PROPOSITION 13.5.1: *$\theta_{\widetilde{G}}(\phi)$ is equal to the sum of the following terms:*

(1) $\sum_{\dim(\pi)=1} \mathrm{Tr}(\pi)\pi(\varepsilon))$

(2) $\sum \mathrm{Tr}(\pi(\phi)\pi(\varepsilon))$

(3) $\frac{1}{2}\sum' \mathrm{Tr}(I_{\widetilde{\rho'}}(\phi)I_{\widetilde{\rho'}}(\phi)I_{\widetilde{\rho'}}(\varepsilon))$

(4) $\frac{1}{4} \sum_{\substack{\mu \text{ regular}, \\ \mu|C_F=1}} \mathrm{Tr}(I_\mu(\phi)I_\mu(\varepsilon))$

(5) $-\frac{1}{8} \sum_{\substack{\mu \text{ semi-regular}, \\ \mu|C_F=\omega_{E/F}}} \mathrm{Tr}(I_\mu(\phi)I_\mu(\varepsilon))$

(6) $-\frac{3}{8} \sum_{\mu \text{ singular}} \mathrm{Tr}(I_\mu(\phi)I_\mu(\varepsilon))$

(1′) $\frac{1}{2}\sum' \mathrm{Tr}(I_{\tilde{\rho}}(\phi)I_{\tilde{\rho}}(\varepsilon))$

(2′) $\frac{1}{4} \sum_{\substack{\mu \text{ regular}, \\ \mu|C_F=\omega_{E/F}}} \mathrm{Tr}(I_\mu(\phi)I_\mu(\varepsilon))$

(3′) $-\frac{1}{8} \sum_{\substack{\mu \text{ semi-regular}, \\ \mu|C_F=1}} \mathrm{Tr}(I_\mu(\phi)I_\mu(\varepsilon)$

The sum (1) is over the ε-invariant characters of $\widetilde{G}\backslash\widetilde{\mathbf{G}}$ and sum (2) is over the set of infinite-dimensional discrete ε-invariant representations of $\widetilde{G}$. Sums (3) and (1′) are over discrete L-packets ρ on H which are not of the form $\rho(\theta)$. The sums over μ are over sets of representatives for $\Omega(M_0)$-orbits, chosen so that $\mu_1 = \mu_3$ if μ is not regular.

13.6 The quintuples corresponding to discrete terms in the trace formula for G are as follows:

(1) $(M, G, \{0\}, \sigma, w)$: σ is a character of $M\backslash\mathbf{M}$ such that $w\sigma = \sigma$.

(2) $(G, G, \{0\}, \sigma, 1)$: σ is discrete on G.

Replacing G by H, we obtain the quintuples for H. The distributions θ_G and θ_H are equal to a sum of two terms.

$$\sum m(\pi)\mathrm{Tr}(\pi(f))$$

$$\frac{1}{4}\sum_{\chi=w\chi} \mathrm{Tr}(M(\chi)I_\chi(f)) \tag{13.6.1}$$

where π ranges over the discrete automorphic representations, χ ranges over the characters of $M\backslash\mathbf{M}$ such that $\chi = w\chi$, and $M(\chi)$ is the intertwining operator associated to w acting on I_χ. We write $M(\chi)$ as $M_G(\chi)$ or $M_H(\chi)$ to make the group explicit. A character $\chi = (\chi_1, \chi_2)$ satisfies $\chi = w\chi$

if and only if $\chi_1\overline{\chi}_1 = 1$. We will write $\chi|C_F = 1$ or $\omega_{E/F}$ according as $\chi_1|C_F = 1$ or $\omega_{E/F}$. The next proposition follows immediately from Proposition 11.2.1(b) and Proposition 11.5.1(c).

PROPOSITION 13.6.1: *$S\theta_H(f)$ is equal to the sum of the following terms:*

(1′) $\sum' \mathrm{Tr}(\rho(f))$

(2′) $\frac{1}{2}\sum \mathrm{Tr}(\rho(\theta)(f))$

(3′) $-\frac{1}{4} \sum_{\substack{\chi = w\chi \\ \chi|C_F = 1}} \mathrm{Tr}(I_\chi(f))$

The first sum is over the discrete L-packets on H which are not of the form $\rho(\theta)$ and the second is over a set of representatives for the H-equivalence classes of H-regular characters θ of $C\backslash\mathbf{C}$.

Let S be a finite set of places of F, containing the infinite places, such that v is unramified in E for all $v \notin S$. By an eigenvalue package (e.v.p), we shall mean a collection

$$t = t_S = \{t_v : v \notin S\}$$

where t_v is a homomorphism of the Hecke algebra $\mathcal{H}_v$ in $\mathbf{C}$. If $f \in \otimes_{v\notin S}\mathcal{H}_v$ (restricted direct product), let $f^\wedge(t) = \Pi t_v(f_v)$. This definition holds for any group. We regard t_v as an orbit of unramified characters in $\Pi^u(T_v)$ modulo the action of $\Omega(T,G)$ (cf. §4.5). If $\pi = \otimes\pi_v$ is an automorphic representation which is unramified for $v \notin S$, then π defines an e.v.p. $t(\pi) = t_S(\pi) = \{t_v(\pi)\}_{v\notin S}$ as follows: $t_v(\pi)$ is the orbit in $\Pi^u(T_v)$ such that π_v is isomorphic to the unique unramified constituent of $i_G(\chi)$ for any χ in the orbit t_v. A map ψ of L-groups defines a transfer $t \to \psi(t)$ of e.v.p.'s on the source group to e.v.p.'s on the target group. We will often use the strong multiplicity one theorem for GL_n [JS], which states that two cuspidal representations π and π' of GL_n coincide if $t_v(\pi) = t_v(\pi')$ for almost all v. We identify two e.v.p.'s if they are equal everywhere.

PROPOSITION 13.6.2: *$S\theta_G(f)$ is equal to the sum of the following terms:*

(1) $\sum_{\dim(\pi)=1} \mathrm{Tr}(\pi(f))$

(2) $\sum_2 m(\pi)\mathrm{Tr}(\pi(f))$

(3) $\sum_3 m(\pi)\mathrm{Tr}(\pi(f)) - \frac{1}{2}\sum' \mathrm{Tr}(\rho(f^H))$

(4) $\sum_4 m(\pi)\mathrm{Tr}(\pi(f)) - \frac{1}{4} \sum_{\theta \text{ regular}} \mathrm{Tr}(\rho(\theta)(f^H))$

(5) $-\frac{1}{8} \sum_{\substack{\chi = w\chi \\ \chi|C_F = \omega_{E/F}}} \mathrm{Tr}(I_\chi(f))$

(6) $-\frac{1}{4} \sum_{\chi_1 = 1} \mathrm{Tr}(I_\chi(f))$

(7) $\sum_7 m(\pi)\mathrm{Tr}(\pi(f)) + \frac{1}{4} \sum_{\theta \text{ semi-regular}} (\mathrm{Tr}(M(\tilde{\theta})I_{\tilde{\theta}}(f)) - \mathrm{Tr}(\rho(\theta)(f^H)))$

The sum in (1) *is over the set of one-dimensional representations of* $G\backslash\mathbf{G}$. *The sums* $\sum_j$ *for* $j = 2, 3, 4, 7$ *are over certain sets of cuspidal representations of* G *defined below.*

The terms of $-\dfrac{1}{2}S\theta_H(f^H)$ are spread out over lines (3), (4), (5), and (7). In line (3), the second sum is over $\rho \in \Pi(\mathbf{H})$ which are not of the form $\rho(\theta)$ and corresponds to (1′) in Proposition 13.6.1. We divide (2′) of Proposition 13.6.1 into two parts, placing the terms corresponding to regular θ in (4) and those corresponding to semi-regular θ in (7). The sum in (4) is over a set of representatives for the H-equivalence classes of regular θ.

Term (13.6.1) for G is spread out over (5), (6) and (7). If $\chi|C_F = 1$ and $\chi_1 \neq 1$, then χ is of the form $\tilde{\theta}$ for a semi-regular character θ of $C\backslash\mathbf{C}$ and the corresponding part of (13.6.1) occurs in the second sum of (7). The sum is over a set of representatives for the semi-regular characters modulo H-equivalence (cf. §12.1). If $\chi_1 = 1$, then I_χ is irreducible and $M_G(\chi)$ acts by -1 by [KyS], Theorem 5.1. The part corresponding to such χ occurs in (6). Finally, (3′) for H is combined with the part of (13.6.1) for G associated to χ such that $\chi|C_F = \omega_{E/F}$ in line (5). If $\chi|C_F = 1$, then $\chi\mu|C_F = \omega_{E/F}$, and $i_G(\chi\mu)$ is irreducible. In this case, $M_G(\chi\mu)$ acts by -1 by [KyS], and we obtain the sum over such χ of

$$\frac{1}{4}\mathrm{Tr}(M_G(\chi\mu)I_{G,\chi\mu}(f)) + \frac{1}{8}\mathrm{Tr}(I_{H,\chi}(f^H)) = -\frac{1}{8}\mathrm{Tr}(I_{G,\chi\mu}(f))$$

This shows up in (5).

In lines (3), (4), and (7), each term in the second sum defines an e.v.p. on H and the sum $\sum_j$ in line (j) is over the set of discrete π on G such that $t(\pi)$ is equal to the transfer via ξ_H of some e.v.p. which occurs in the second sum. Finally, the sum in (2) is over representations in the discrete representations of G which do not occur in any other line.

To prove the proposition, we show that a discrete representation of G occurs in at most one line. It is clear that an e.v.p. which occurs in lines (1) or (2) does not occur in any other line. A direct check shows that if t_1 and t_2 are e.v.p's occurring in lines (j_1) and (j_2), respectively, where

$4 \le j_i \le 7$ and $j_1 \neq j_2$, then $\psi_G \circ \xi_H(t_1) \neq \psi_G \circ \xi_H(t_2)$. Hence t_1 does not occur in line (j_2). An e.v.p. occurring in line (3) in no other line by (b) and (c) of the following lemma.

LEMMA 13.6.3: *Let ρ_1, ρ_2 be distinct discrete L-packets on H.*

(a) *The representations $\tilde{\rho}_1, \tilde{\rho}_1', \tilde{\rho}_2, \tilde{\rho}_2'$ define distinct e.v.p.'s on $\widetilde{H}$. Furthermore, $\xi_{\widetilde{H}}(t(\tilde{\rho}_1))$, $\xi_{\widetilde{H}}(t(\tilde{\rho}_1'))$, $\xi_{\widetilde{H}}(t(\tilde{\rho}_2))$, and $\xi_{\widetilde{H}}(t(\tilde{\rho}_2'))$ are distinct unless $\rho_j = \rho(\theta_j)$, where θ_1 and θ_2 are G-conjugate. In the latter case, $\xi_{\widetilde{H}}(t(\tilde{\rho}_1))$ and $\xi_{\widetilde{H}}(t(\tilde{\rho}_1'))$ are distinct. If χ is a unitary character of $\widetilde{B}$, then $t(i_{\widetilde{G}}(\chi)) \neq \xi_{\widetilde{H}}(t(\tilde{\rho}_1)), \xi_{\widetilde{H}}(t(\tilde{\rho}_1'))$ if ρ_1 is not of the form $\rho(\theta)$.*

(b) *The representations $\xi_H(\rho_1), \xi_H(\rho_2)$ define distinct e.v.p.'s on G unless $\rho_1 = \rho(\theta_1)$ and $\rho_2 = \rho(\theta_2)$, where θ_1 and θ_2 are G-conjugate.*

(c) *Let χ be a unitary character of $M\backslash\mathbf{M}$. If ρ_1 is not of the form $\rho(\theta)$, then $\xi_H(\rho_1)$ and I_χ define distinct e.v.p.'s on G.*

Proof: Observe first that (b) follows from (a) by the commutative diagram of §4.8. To prove (a), we recall the classification theorem of Jacquet-Shalika for GL_n ([JS], Theorem 4.4). Let P_1 and P_2 be parabolic subgroups of GL_n with Levi factors M_1 and M_2. Let σ_j be a cuspidal representation of M_j and let π_j be an automorphic constituent of $\mathrm{Ind}(\mathrm{GL}_n(\mathbf{A}), P_j(\mathbf{A}), \sigma_j)$. The theorem states that if π_1 and π_2 define the same e.v.p. in GL_n, then (M_1, σ_1) and (M_2, σ_2) are conjugate.

The representations $\tilde{\rho}_1, \tilde{\rho}_1', \tilde{\rho}_2, \tilde{\rho}_2'$ are distinct by Theorem 11.5.1(a). As remarked in §4.8, $\xi_{\widetilde{H}}$ is equivalent to an L-map corresponding to an embedding of a Levi factor of type (2,1) in GL_3. Suppose that $\psi_{\widetilde{H}}(t(\tilde{\rho}_1)) = \psi_{\widetilde{H}}(t(\tilde{\rho}_1'))$. This cannot happen if $\tilde{\rho}_1$ and $\tilde{\rho}_1'$ are cuspidal, by the Jacquet-Shalika theorem, or if $\dim(\rho_1) = 1$, by a direct check. The remaining case, $\rho_1 = \rho(\theta)$, is also clear by a direct check. The remaining assertions in (a) follow similarly. Part (c) follows from (a) since $\xi_{\widetilde{H}}(\psi_H'(t(\rho))) = \psi_G(\xi_H(t(\rho)))$.

13.7 By Theorem 10.3.1, we have the main equality

$$\theta_{\widetilde{G}}(\phi) - S\theta_G(f) - \frac{1}{2}S\theta_H(\phi^H) = 0 .$$

By Propositions 13.5.1, 13.6.1, and 13.6.2, the right hand side of this main equality is equal to the sum of the following terms:

(1) $\displaystyle\sum_{\dim(\tilde{\pi})=1} \mathrm{Tr}(\tilde{\pi}(\phi)\tilde{\pi}(\varepsilon)) - \sum_{\dim(\pi)=1} \mathrm{Tr}(\pi(f))$

(2) $\sum \mathrm{Tr}(\tilde{\pi}(\phi)\tilde{\pi}(\varepsilon)) - \sum_2 m(\pi)\mathrm{Tr}(\pi(f))$

$$\text{(3)} \quad \tfrac{1}{2}{\textstyle\sum}' \mathrm{Tr}(I_{\widetilde{\rho'}}(\phi)I_{\widetilde{\rho'}}(\varepsilon)) - {\textstyle\sum}_3 m(\pi)\mathrm{Tr}(\pi(f)) + \tfrac{1}{2}{\textstyle\sum}' \mathrm{Tr}(\rho(f^H))$$

$$\text{(4)} \quad \tfrac{1}{4} \sum_{\substack{\mu \text{ regular}, \\ \mu|C_F=1}} \mathrm{Tr}(I_\mu(\phi)I_\mu(\varepsilon)) - {\textstyle\sum}_4 m(\pi)\mathrm{Tr}(\pi(f)) + \tfrac{1}{4} \sum_{\theta \text{ regular}} \mathrm{Tr}(\rho(\theta)(f^H))$$

$$\text{(5)} \quad -\tfrac{1}{8} \sum_{\substack{\mu \text{ semi-regular}, \\ \mu|C_F=\omega_{E/F}}} \mathrm{Tr}(I_\mu(\phi)I_\mu(\varepsilon)) + \tfrac{1}{8} \sum_{\substack{\chi={}^w\chi, \\ \chi|C_F=\omega_{E/F}}} \mathrm{Tr}(I_\chi(f))$$

$$\text{(6)} \quad -\tfrac{3}{8} \sum_{\mu \text{ singular}} \mathrm{Tr}(I_\mu(\phi)I_\mu(\varepsilon)) + \tfrac{1}{4} \sum_{\chi_1=1} \mathrm{Tr}(I_\chi(f)) + \tfrac{1}{8} \sum_{\chi_1=1} \mathrm{Tr}(I_\chi(\phi^H))$$

$$\text{(7)} \quad -{\textstyle\sum}_7 m(\pi)\ \mathrm{Tr}(\pi(f)) - \tfrac{1}{4} \sum_{\theta \text{ semi-regular}} (\mathrm{Tr}(M(\tilde{\theta})I_{\tilde{\theta}}(f)) - \mathrm{Tr}(\rho(\theta)(f^H))$$

$$\text{(1')} \quad \tfrac{1}{2}{\textstyle\sum}' \mathrm{Tr}(I_{\tilde{\rho}}(\phi)I_{\tilde{\rho}}(\varepsilon)) - \tfrac{1}{2}{\textstyle\sum}' \mathrm{Tr}(\rho(\phi^H))$$

$$\text{(2')} \quad \tfrac{1}{4} \sum_{\substack{\mu \text{ regular}, \\ \mu|C_F=\omega_{E/F}}} \mathrm{Tr}(I_\mu(\phi)I_\mu(\varepsilon)) - \tfrac{1}{4} \sum_{\theta\ H \text{ regular}} \mathrm{Tr}(\rho(\theta)(\phi^H))$$

$$\text{(3')} \quad -\tfrac{1}{8} \sum_{\substack{\mu \text{ semi-regular}, \\ \mu|C_F=1}} \mathrm{Tr}(I_\mu(\phi)I_\mu(\varepsilon)) + \tfrac{1}{8} \sum_{\substack{\chi={}^w\chi, \\ \chi_1|C_F=1,\chi_1\neq 1}} \mathrm{Tr}(I_\chi(\phi^H))$$

Each term in each sum above defines an e.v.p. on $\widetilde{G}$ via functoriality with respect to the L-maps defined in §4. We now recall an argument of Langlands ([L1], p. 208–211) which will be used repeatedly in this chapter. Let S be a finite set of places of F including the infinite places. Let $t_{S,j}$ be a sequence of e.v.p.'s and let $\alpha_j \in \mathbf{C}$ be such that the sum

$$\sum \alpha_j f^\wedge(t_{S,j})$$

is absolutely convergent and equal to zero for all $f \in \otimes_{v \notin S} \widetilde{\mathcal{H}}_v$. Then $\alpha_j = 0$ for all j. We refer to this argument as "separating by Hecke eigenvalues". Observe that the map ψ_G is injective on e.v.p.'s.

PROPOSITION 13.7.1: *Each of the above terms* (1)-(8) *and* (1′) – (3′) *is equal to zero.*

The proof of this proposition will be given after the proof of Proposition 13.8.2 below.

13.8 Proofs of local results. Throughout this section, let E'/F' be a fixed quadratic extension of local fields and denote by E/F and w a global quadratic extension and a place w of F such that E_w/F_w is isomorphic to E'/F'. Below, we will embed various local situations in a global situation in order to deduce local results from the trace formula. When this is done, we implicitly assume that a choice of global central character ω such that ω_w is the given central character of the local situation has been made. We shall repeatedly use the following consequence of [JL], Lemma 16.1.1 (cf. [LL], page 768).

PROPOSITION 13.8.1: *Let X be a countable set of irreducible unitary representations of a reductive group G with central character ω and for $\pi \in X$, let $a(\pi) \in \mathbf{C}^*$. Suppose that the following sum is absolutely convergent and is equal to zero:*

$$\sum_{\pi \in X} a(\pi) Tr(\pi(f)) = 0$$

for all $f \in C(G, \omega)$. Then X is empty.

PROPOSITION 13.8.2: *In the local situation, suppose that $\rho_0 \in \Pi(H)$ is unitary. Then there is a choice of $I_{\tilde{\rho}_0}(\varepsilon)$ such that* $\mathrm{Tr}(I_{\tilde{\rho}_0}(\phi) I_{\tilde{\rho}_0}(\varepsilon)) = \mathrm{Tr}(\rho_0(\phi^H))$.

Proof: We first prove the result for ρ_0 square-integrable.

Let ρ be a cuspidal L-packet on H which is not of the form $\rho(\theta)$. Suppose that ρ_v is unramified for v outside a finite set S of places of F and let $t_S = t_S(I_{\tilde{\rho}})$. By Lemma 13.6.3, t_S occurs in line (1′) and possibly in the second sum in line (2), but in no other line of §13.7. Note that t_S cannot occur in the first sum in line (2) by the Jacquet-Shalika theorem. Separating by Hecke eigenvalues, we conclude that the sum of all terms in the lines (1) - (3′) which correspond to t is equal to zero, and we obtain an equality

$$\frac{1}{2}\mathrm{Tr}(I_{\tilde{\rho}}(\phi) I_{\tilde{\rho}}(\varepsilon)) = \frac{1}{2}\mathrm{Tr}(\rho(\phi^H)) + \sum m(\pi)\mathrm{Tr}(\pi(f)) \tag{13.8.1}$$

where the sum is over the set of cuspidal π on G such that $\psi_G(t(\pi_v)) = t_v(I_{\tilde{\rho}})$ for all $v \notin S$.

Choose E/F so that there are two places w_1 and w_2 such that E_{w_j}/F_{w_j} is isomorphic to E'/F' for $j = 1, 2$, and such that all infinite places (other than w_1 and w_2 if F' is archimedean) split in E. Let ρ be a cuspidal L-packet on H which is not of the form $\rho(\theta)$ such that $\rho_{w_j} = \rho_0$ for $j = 1, 2$ and ρ_v is a p.s. L-packet for all places of F which do not split in E other than w_1 and w_2. The existence of such a ρ is an easy consequence of the trace formula (cf. §13.11, where a similar result is proved for G). The condition that ρ not be of the form $\rho(\theta)$ is insured by choosing ρ_v to be supercuspidal at a finite place v which splits in E.

For $v \notin S$, we have

$$\mathrm{Tr}(I_{\tilde{\rho}_v}(\phi_v) I_{\tilde{\rho}_v}(\varepsilon)) = \mathrm{Tr}(\rho_v(\phi_v^H)) = \mathrm{Tr}(\pi_v(f_v))$$

for all π in the sum in (13.8.1), for a suitable choice of $I_{\tilde{\rho}_v}(\varepsilon)$, and hence

$$\frac{1}{2}\prod_{v\in S}\mathrm{Tr}(I_{\tilde{\rho}_v}(\phi_v)I_{\tilde{\rho}_v}(\varepsilon))=\prod_{v\in S}\mathrm{Tr}(\rho_v(\phi_v^H))+\sum m(\pi)\prod_{v\in S}\mathrm{Tr}(\pi_v(f_v)).$$

However, since ρ_v is a p.s. L-packet for $v\neq w_1,w_2$ which split in E, $\mathrm{Tr}(I_{\tilde{\rho}_v}(\phi_v)I_{\tilde{\rho}_v}(\varepsilon))=\mathrm{Tr}(\rho_v(\phi_v^H))$ for all $v\neq w_1,w_2$ by Propositions 4.10.2 and 4.13.2. Similarly, for all $v\neq w_1,w_2$, there exists a representation π'_v of G_v such that $\mathrm{Tr}(\rho_v(\phi_v^H))=\mathrm{Tr}(\pi'_v(f_v))$. If v remains prime and $\rho_v=i_{H_v}(\chi)$, then $\pi'_v=i_{G_v}(\chi)$, and if v splits, then π'_v is the unique representation such that $\psi_G(\pi'_v)=I_{\tilde{\rho}}$ (cf. §4.13). Proposition 13.8.1 implies that $\pi_v=\pi'_v$ for all π occurring in the sum in (13.8.1), and we obtain an equality of the form

(13.8.2)
$$\frac{1}{2}\prod_{j=1,2}\mathrm{Tr}(I_{\tilde{\rho}_0}(\phi_{w_j})I_{\tilde{\rho}_0}(\varepsilon))-\frac{1}{2}\prod_{j=1,2}\mathrm{Tr}(\rho_0(\phi_{w_j}^H))=\sum m(\pi)\prod_{j=1,2}\mathrm{Tr}(\pi_j(f_{w_j})),$$

again for a suitable choice of $I_{\tilde{\rho}_0}(\varepsilon)$. Since ρ_0 is square integrable, its character is non-zero on the elliptic set and we can choose ϕ_{w_2} with support on the elliptic regular set such that $\mathrm{Tr}(\phi_{w_2}^H)\neq 0$ and all stable ε-orbital integrals of ϕ_{w_2} vanish. Then we can take $f_{w_2}\equiv 0$ and for all ϕ_{w_1}, we have:

$$\mathrm{Tr}(I_{\tilde{\rho}_0}(\phi_{w_1})I_{\tilde{\rho}_0}(\varepsilon))=c\,\mathrm{Tr}(\rho_0(\phi_{w_1}^H))$$

for some non-zero constant c. Plugging back into (13.8.2), we obtain

$$(c^2-1)\prod_{j=1,2}\mathrm{Tr}(\rho(\phi_{w_j}^H))=2\sum m(\pi)\prod_{j=1,2}\mathrm{Tr}(\pi_{w_j}(f_{w_j}))\,.$$

Choose $\phi_{w_1}=\phi_{w_2}$ with support on the elliptic set so that $\mathrm{Tr}(\rho(\phi_{w_j}^H))\neq 0$ and $f_j=0$ as above. This gives a contradiction unless $c=\pm 1$. Since $I_{\tilde{\rho}_0}(\varepsilon)$ is only determined up to a sign, this proves the proposition for square-integrable ρ_0. If ρ_0 is not square-integrable, it is either a p.s. L-packet or one-dimensional. In the former case, the character identity follows from Proposition 4.10.2. In the latter case, it follows easily from the type of argument used to prove Corollary 12.3.5.

We now prove Proposition 13.7.1. Observe that (1) is equal to zero by Proposition 12.4.1. Consider (3′). If μ is semi-regular and $\mu|C_F=1$, then (up to an element of the Weyl group) $\mu=\chi\circ N$ for a unique character χ of $M\backslash\mathbf{M}$ such that $\chi_1|C_F=1$ and $\chi_1\neq 1$. Hence (3′) vanishes by

Proposition 4.10.2. Similarly, if μ is semi-regular and $\mu|C_F = \omega_{E/F}$, then, up to an element of the Weyl group, $\mu = \chi \circ N$ for a unique character χ of $M\backslash\mathbf{M}$ such that $\chi|C_F = \omega_{E/F}$, and the sums in line (5) cancel by proposition 4.10.2. If μ is singular, then $\mu = \chi \circ N$ for a unique character χ of $M\backslash\mathbf{M}$ such that $\chi_1 = 1$ and the sums in line (6) add to zero by Proposition 4.10.2.

To complete the proof, we show that an e.v.p. on $\widetilde{G}$ occurs in at most one of the remaining lines. It will then follow that each line is separately equal to zero by separation by Hecke eigenvalues.

If $\tilde{\pi}$ occurs in (2), then $t(\tilde{\pi})$ does not coincide with an e.v.p. induced from any Levi factor by the Jacquet-Shalika theorem and hence does not appear in any of the remaining lines. Suppose that π_0 occurs in $\sum_2$ of line (2). Then $\psi_G(t(\pi_0))$ occurs in no other lines except possibly (1′) or (2′). To show that this does not happen, suppose that $\psi_G(t(\pi_0)) = t(I_{\tilde{\rho}})$ for some $\tilde{\rho}$ in (1′). Separating by Hecke eigenvalues, we obtain an equality

$$\frac{1}{2}\mathrm{Tr}(I_{\tilde{\rho}}(\phi)I_{\tilde{\rho}}(\varepsilon)) - \frac{1}{2}\mathrm{Tr}(\rho(\phi^H)) = \sum m(\pi)\mathrm{Tr}(\pi(f))$$

where the sum is over the set of cuspidal π such that $t(\pi) = t(\pi_0)$. By Proposition 13.8.2, $\mathrm{Tr}(I_{\tilde{\rho}}(\phi)I_{\tilde{\rho}}(\varepsilon)) = \pm\mathrm{Tr}(\rho(\phi^H))$ and hence either the left-hand side is equal to zero, in which case the sum is empty, or we have an equality:

$$-\mathrm{Tr}(\rho(\phi^H)) = \sum m(\pi)\mathrm{Tr}(\pi(f)) \ .$$

If ρ_v is a p.s. L-packet for all v which remain prime in E, then, as in the previous proof, there exists a representation π' of $\mathbf{G}$ such that $\mathrm{Tr}(\rho(\phi^H)) = \mathrm{Tr}(\pi'(f))$ and this contradicts the above equality since the $m(\pi)$ are non-negative. If ρ_v is square-integrable for some v which remains prime in E, then, as in the previous proof, there is a choice of ϕ_v for which $\mathrm{Tr}(\rho_v(\phi_v^H)) \neq 0$ and $f_v = 0$. This is again a contradiction to the above equality. This shows that $\psi_G(t(\pi_0))$ does not occur in (1′). A similar argument shows that it does not occur in (2′). This shows that (2) is equal to zero.

We are left with (3), (4), (7) (1′), and (2′). By Proposition 13.6.3(a), an e.v.p. which occurs in (3) or (1′) does not occur in any other line, hence (3) and (1′) vanish. No e.v.p. occurs in both (4) and (2′). If θ is semi-regular, then $t(I_\mu)$ occurs in (7), where $\mu = \mu'(\theta)$ (cf. §12.1). Since μ is semi-regular, $t(I_\mu)$ does not occur in (4) or (2′), and hence (4), (2′), and (7) are equal to zero. This completes the proof of Proposition 13.7.1.

We now begin the proof of Theorem 13.1.1. Let $\theta = \psi \otimes \varphi\psi \otimes \varphi\psi$ be a semi-regular character of $C\backslash\mathbf{C}$. By [KyS], Theorem 5.1,

$$\mathrm{Tr}(M(\tilde{\theta})I_{\tilde{\theta}}(f)) = \prod(\chi_{\pi^1(\theta v)}(f_v) - \chi_{\pi^2(\theta v)}(f_v))$$

for all semi-regular characters θ of $C\backslash\mathbf{C}$. If v is finite and unramified in E and θ_v is unramified, then φ_v is trivial. In this case, $\rho(\theta_v)$ is a p.s. representation, $I_{\tilde{\theta}_v}$ is irreducible, and the relation

$$(*) \qquad \chi_{\pi^1(\theta v)}(f_v) - \chi_{\pi^2(\theta v)}(f_v) = \chi_{\rho(\theta v)}(f_v^H)$$

holds ($\pi^2(\theta_v)$ does not exist in this case). Similarly, if v splits in E, or if θ_v is singular, then $(*)$ holds and $\pi^2(\theta_v)$ does not exist. It also holds for v such that $E_v/F_v = \mathbf{C}/\mathbf{R}$ by Proposition 12.3.2.

Let S be a finite set of places and suppose that θ_v is unramified and $f_v \in \mathcal{H}_v$ for all $v \notin S$. Line 13.7(7) is equal to zero and this equality breaks up into a separate equality for each e.v.p. occurring in the sum. Distinct semi-regular characters give rise to distinct e.v.p.'s on G. By Proposition 13.7.1, we obtain the equality:

$$\prod_{v \in S}(\chi_{\pi^1(\theta_v)}(f_v) - \chi_{\pi^2(\theta_v)}(f_v)) + 4\sum m(\pi)\prod_{v\in S}\chi_{\pi_v}(f_v) = \prod_{v\in S}\chi_{\rho(\theta_v)}(f_v^H)$$

where the sum is over π such that $t_v(\pi) = t_v(\pi^1(\theta_v))$ for all $v \notin S$. We may further replace S by any finite set such that for all $v \notin S$, at least one of the four conditions is satisfied: θ_v is either singular or unramified, v is infinite, or v splits in E.

Assume F is p-adic and fix a semi-regular character $\theta' = \psi' \otimes \psi'\varphi \otimes \psi'\varphi$ of $C(F')$. We can choose E/F and w such that E_w/F_w is isomorphic to E'/F' and E/F is a CM field. Let $\mu(E)$ be the group of roots of unity of E. If v is a place of F, let $\mathcal{O}_v^1$ be the set of $x \in E_v^*$ such that $N_{E_v/F_v}(x) = 1$ and $|x|_w = 1$ for all places w of E dividing v. Then $E^1 \cap \Pi_v\mathcal{O}_v^1 = \mu(E)$ and $\mu(E)\backslash\Pi_v\mathcal{O}_v^1$ embeds as a subgroup of finite index in $E^1\backslash\mathbf{E}^1$, where $\mathbf{E}^1$ is the group of adelic points of E^1. The group $\Pi'\mathcal{O}_v^1$ (product over finite v) embeds as a closed subgroup of $\mu(E)\backslash\Pi_v\mathcal{O}_v^1$ and hence every character of $\Pi'\mathcal{O}_v^1$ extends to a character of $E^1\backslash\mathbf{E}^1$. Choose a character φ of $E^1\backslash\mathbf{E}^1$ such that $\varphi_w = \varphi'$ and φ_v is trivial for all finite $v \neq w$. Let ψ be a character of $E^1\backslash\mathbf{E}^1$ such that $\psi_w = \psi'$, and set $\theta = \psi \otimes \psi\varphi \otimes \psi\varphi$. Then $\theta_w = \theta'$ and $(*)$ holds for all $v \neq w$. The set S may be taken to consist of w alone. The hypothesis of Lemma 12.7.1 is therefore satisfied and $(*)$ holds for $v = w$.

For $\rho = \pi^1(\theta')$, we define $\Pi(\rho) = \{\pi^1(\theta'), \pi^2(\theta')\}$. Then the character identity of Theorem 13.1.1(2) holds with $\langle\rho,\pi\rangle$ defined as in Proposition 13.1.3(c).

PROPOSITION 13.8.3: *Assume E'/F' is a p-adic. Let $\rho_0 \in \Pi^2(H)$ and assume that ρ_0 is not of the form $\rho(\theta)$ for θ semi-regular. Then there exists a unique set $\Pi(\rho)$ of square-integrable representations of cardinality* 2Card(ρ_0) *and for each $\pi \in \Pi(\rho_0)$, a sign $\langle\pi,\rho_0\rangle = \pm 1$ such that the character identity of Theorem* 13.1.1(2) *holds.*

Proof: Let ρ be a cuspidal L-packet on H which is not of the form $\rho(\theta)$ for any character θ of $C\backslash\mathbf{C}$. Consider the terms in 13.7(3) which correspond to the e.v.p. t defined by $I_{\tilde{\rho}'}$ on $\widetilde{G}$. By lemma 13.6.3, of the terms in the first and third sums, only those corresponding to $\widetilde{\rho'}$ and ρ yield t, and we obtain the equality:

$$\text{(13.8.3)} \qquad \mathrm{Tr}(I_{\widetilde{\rho'}}(\phi)I_{\widetilde{\rho'}}(\varepsilon)) = 2\sum m(\pi)\mathrm{Tr}(\pi(f)) - \mathrm{Tr}(\rho(f^H))$$

where the sum is over cuspidal π on G such that $\psi_G(t(\pi)) = t$.

We may assume that E/F is a CM extension and that ω_v is unramified for all finite $v \neq w$. Assume first that ρ_0 is not of the form $\rho(\theta)$. A standard application of the trace formula for H (cf. [L$_1$], page 227) shows that there exists a cuspidal L-packet ρ on H with the following properties: (i) for all infinite v, ρ_v is a discrete series L-packet on H_v which is not of the form $\rho(\theta)$ with θ semi-regular, (ii) for finite $v \neq w, \rho_v$ is unramified, (iii) $\rho_w = \rho_0$.

Let v be an infinite place. By (i), $\Pi(\rho_v)$ is a discrete series L-packet, and if $\Pi(\rho_v) = \{\pi_{1v}, \pi_{2v}, \pi_{3v}\}$, then

$$\mathrm{Tr}(\rho_v(f_v^H)) = \sum_{1\leq j\leq 3} \langle\rho_v, \pi_{jv}\rangle \; \mathrm{Tr}(\pi_{jv}(f_v))$$

by Proposition 12.3.2. For fixed v, the signs $\langle\rho_v, \pi_{jv}\rangle$ for $j = 1,2,3$ are not all equal. We can assume that

$$\langle\rho_v, \pi_{1v}\rangle = 1 \;, \; \langle\rho_v, \pi_{2v}\rangle = -1.$$

and also that ρ is chosen so that $\Pi(\rho_v)$ is integrable. Let f_{jv} be a pseudo-coefficient for π_{jv}. Fix one infinite place u and let $f_u = f_{1v} - f_{2v}$. The stable orbital integrals of f_u vanish and we can take $\phi_u = 0$. Furthermore,

$$\mathrm{Tr}(\rho_u(f_u^H)) = 2, \; \mathrm{Tr}(\pi_{ju}(f_u)) = (-1)^{j+1} \quad \text{for} \quad j = 1,2$$

and $\mathrm{Tr}(\pi(f_u)) = 0$ for irreducible $\pi \neq \pi_{1u}, \pi_{2u}$. For all infinite places $v \neq u$, let $f_v = f_{1v}$. Then

$$\mathrm{Tr}(\rho_v(f_v^H)) = \mathrm{Tr}(\pi_{1v}(f_v)) = 1$$

and $\mathrm{Tr}(\pi(f_v)) = 0$ for $\pi \neq \pi_{1v}$. The equality (13.8.3) becomes

$$2\sum m(\pi)\varepsilon_\pi \mathrm{Tr}(\pi^\infty(f^\infty)) = 2\mathrm{Tr}(\rho^\infty((f^\infty)^H))$$

where $\varepsilon_\pi = \pm 1$, $f^\infty = \Pi f_v$ (product over the finite places) and π^∞ and ρ^∞ are the finite parts of π and ρ_1 respectively. Since ρ_v is unramified for finite $v \neq w$, $\pi_v = \xi_H(\rho_v)$ for all $v \neq w$ and all π occurring in the sum and we obtain our equality of the form

$$\sum a(\pi_w)\ \mathrm{Tr}(\pi_w(f_w)) = \ \mathrm{Tr}(\rho_0(f_w^H))$$

for some $a(\pi_w) \in \mathbf{Z}$. The hypothesis of Lemma 12.7.2 is thus satisfied for ρ_0. The existence of $\Pi(\rho_0)$ and the character identity if ρ_0 is not of the form $\rho(\theta)$ follow.

Now let θ_0 be a regular character of $C(F')$ and let $\rho_0 = \rho(\theta_0)$. We construct, again by the trace formula, a cuspidal L-packet ρ on H with properties (i) and (iii) as above, such that at a finite number of places, ρ_v belongs to $\Pi^2(H_v)$ but is not of the form $\rho(\theta)$, and such that (ii) holds for the remaining finite places $v \neq w$. Then ρ is not globally of the form $\rho(\theta)$ and we obtain (13.8.3). Using the character identities for $\Pi(\rho_v)$ for $\rho_v \in \Pi^2(H_v)$ not of the form $\rho(\theta)$ already obtained, the argument above goes through. Lemma 12.7.2 implies the existence of $\Pi(\rho_0)$ satisfying the character identity such that $\mathrm{Card}(\Pi(\rho_0)) = 2\mathrm{Card}(\rho_0)$. This completes the proof of Proposition 13.8.3. Proposition 13.1.2(a) also follows.

We have now shown the existence of $\xi_H(\rho)$ satisfying the character identity for all $\rho \in \Pi^2(H)$. The uniqueness of $\xi_H(\rho)$ is immediate from the linear independence of characters. Suppose that $\rho \in \Pi(H)$ is neither square-integrable nor one-dimensional. Then $\rho = JH(i_H(\chi\mu^{-1}))$ for some character χ and the character identity is satisfied with $\xi_H(\rho) = JH(i_G(\chi))$, provided that $\chi_1(\alpha) \neq \|\alpha\|^{\pm 1}$. This completes the proof of Theorem 13.1.1. All of the statements of Proposition 13.1.3 follow from Lemma 12.7.2 and 12.7.3, except the assertion that the elements of $\pi(\rho)$ have the same formal degree (this will be shown in the course of the proof of Lemma 13.8.7 below) and part (c), which has already been proved above. Proposition 13.1.2(a) has also been shown, and Proposition 13.1.4 follows from Corollary 12.7.4 and Proposition 13.1.3(d).

PROPOSITION 13.8.4: *If $\tilde{\pi} \in E_\varepsilon(\widetilde{G})$ is supercuspidal, then there exists a unique element $\pi \in E(G)$ such that $\psi_G(\pi) = \tilde{\pi}$. The representation π is supercuspidal and does not belong to $\Pi(\rho)$ for any $\rho \in \Pi(H)$.*

Proof: We need the following lemma, whose proof is postponed to the end of the chapter.

LEMMA 13.8.5: *Let π' be a supercuspidal representation in $E_\varepsilon(\widetilde{G}_{/E'})$. Then there exists a global extension E/F and a cuspidal ε-invariant representation $\tilde{\pi}$ of G over E such that E_w/F_w is isomorphic to E'/F', $\tilde{\pi}_w = \pi'$, and such that π_v is unramified at all places of F which do not split in E.*

Let $\tilde{\pi}$ be as in the lemma. Separating the equality 13.7(2) = 0 according to Hecke eigenvalues gives

$$\text{Tr}(\tilde{\pi}(\phi)\tilde{\pi}(\varepsilon)) = \sum m(\pi)\ \text{Tr}(\pi(f)) \tag{13.8.4}$$

where the sum is over cuspidal π on G such that $\psi_G(t(\pi)) = t(\tilde{\pi})$. By the choice of $\tilde{\pi}$, for all $v \neq w$ there is an L-packet Π_v on G_v such that $\text{Tr}(\tilde{\pi}_v(\phi_v)\tilde{\pi}_v(\varepsilon)) = \text{Tr}(\Pi_v(f_v))$ and, by Proposition 13.8.1, (13.8.4) yields an equality $\text{Tr}(\pi'(\phi)\pi'(\varepsilon)) = \sum a(\pi_w)\text{Tr}(\pi_w(f))$ where $a(\pi_w) \in \mathbf{Z}$. The hypothesis of Lemma 12.7.5 is satisfied, and $\text{Tr}(\pi'(\phi)\pi'(\varepsilon)) = \text{Tr}(\pi_w(f_w))$ for some supercuspidal representation π_w. It follows that $\text{Tr}(\pi_w(f_w))$ is a stable distribution. If π_w belongs to $\Pi(\rho)$ for some $\rho \in \Pi(H)$, then ρ is square-integrable. But in this case, χ_{π_w} is not orthogonal to χ_ρ^G and this contradicts Corollary 12.5.4.

PROPOSITION 13.8.6: *For all $\rho \in \Pi^2(H_w)$, there is an integer $\alpha(\rho)$ such that*

$$\alpha(\rho)\chi_{\Pi(\rho)}(f_w) = Tr(I_{\tilde{\rho}'}(\phi_w)I_{\tilde{\rho}'}(\varepsilon))$$

for some choice of $I_{\tilde{\rho}'}(\varepsilon)$

Proof: If $\rho = \rho(\theta)$, where θ is semi-regular, then $\chi_{\Pi(\rho)}$ is a trace on a principal series representation and the equality holds with $\alpha(\rho) = 1$ for a suitable choice of $I_{\tilde{\rho}'}(\varepsilon)$ by Proposition 4.10.2

Suppose that $\rho_0 \in \Pi^2(H_w)$ is not of the form $\rho(\theta)$ with θ semi-regular. We put ourselves in the situation of the proof of Proposition 13.8.3. Let $f_u = f_{1u} + f_{2u}$ and $f_v = f_{1v}$ for v infinite, $v \neq u$. Then $\text{Tr}(\rho_u(f_u^H)) = 0$

and $\mathrm{Tr}(I_{\tilde\rho'_u}(\phi_u)I_{\tilde\rho'_u}(\varepsilon)) = \pm 2$, by Proposition 12.3.4, and (13.8.3) yields an equality of the form:

$$\mathrm{Tr}(I_{\tilde{\rho}'_0}(\phi_w)I_{\tilde{\rho}'_0}(\varepsilon)) = \sum_{\pi_w \in X} b(\pi_w)\ \mathrm{Tr}(\pi_w(f_w)) \tag{13.8.5}$$

where X is a countable set of representations and the $b(\pi_w)$ are integers that all have the same sign. We can assume $I_{\tilde\rho'_0}(\varepsilon)$ chosen so that the $b(\pi_w)$ are positive. On the other hand, if we choose $f_v = f_{1v}$ for all infinite v, then $\mathrm{Tr}(I_{\tilde\rho'_v}(\phi_v)I_{\tilde\rho'_v}(\varepsilon)) = \mathrm{Tr}(\rho_v(f_v^H)) = 1$ and (13.8.3) and (13.8.5) together yield an equality:

$$\begin{aligned}\mathrm{Tr}(I_{\tilde{\rho}'_0}(\phi_w)I_{\tilde{\rho}'_0}(\varepsilon)) &= \sum_{\pi_w \in X} b(\pi_w)\mathrm{Tr}(\pi_w(f_w) \\ &= 2\sum_{\pi_w \in X'} c(\pi_w)\mathrm{Tr}(\pi_w(f_w)) - \mathrm{Tr}(\rho_0(f_w^H))\end{aligned}$$

where X' is a countable set of representations and the $c(\pi_w) \in \mathbf{Z}$. Since $\Pi(\rho')$ has been shown to exist and $\mathrm{Tr}(\rho'(f_w^H))$ is a sum over $\pi \in \Pi(\rho')$ of $\pm\mathrm{Tr}(\pi(f_w))$, this equality implies that X contains each member of $\Pi(\rho')$ with coefficient equal to an odd integer.

Suppose that ρ_0 is not of the form $\rho(\theta)$. Then $\Pi(\rho_0) = \{\pi_1, \pi_2\}$. Let f_j be a pseudo-coefficient for π_j and let $f_w = f_1 - f_2$. Then $\Phi(\gamma, f_w) = 0$ if γ is non-elliptic and

$$\Phi(\gamma, f_w) = \chi_{\pi_1}(\gamma) - \chi_{\pi_2}(\gamma) = \chi_\rho^G(\gamma)$$

if γ is elliptic for a suitable numbering of π_1, π_2. Proposition 12.5.3 implies that the integral of f_w against a stable class function is zero. The right-hand side of (13.8.5) defines a stable distribution since it depends only on ϕ_w and hence $b(\pi_1) = b(\pi_2)$.

The equality (13.8.5) therefore has the form

$$\begin{aligned}&\mathrm{Tr}(I_{\tilde{\rho}'_0}(\phi_w)I_{\tilde{\rho}'_0}(\varepsilon)) \\ &\quad = \alpha(\rho_0)\sum_{\pi \in \Pi(\rho_0)} \mathrm{Tr}(\pi(f_w)) + \sum_{\pi \in X(\rho_0)} a(\pi)\ \mathrm{Tr}(\pi(f_w))\end{aligned} \tag{13.8.6}$$

where $X(\rho_0)$ is a set of irreducible representations such that $X(\rho_0) \cap \Pi(\rho_0)$ is empty and $\alpha(\rho_0)$, $a(\pi)$ are integers.

LEMMA 13.8.7: *An equality of the form* (13.8.6) *holds for all* $\rho_0 \in \Pi^2(H)$.

Proof: We need to verify the lemma for $\rho_0 = \rho(\theta)$, where θ regular. The above argument gives an equality

(13.8.7)
$$\begin{aligned}&\mathrm{Tr}(I_{\mu'(\theta)}(\phi_w)I_{\mu'(\theta)}(\varepsilon))\\ &\quad= \sum_{\pi\in\Pi(\rho(\theta))} \alpha(\pi)\mathrm{Tr}(\pi(f_w)) + \sum_{\pi\in X(\rho(\theta))} a(\pi)\mathrm{Tr}(\pi(f_w))\end{aligned}$$

where $\alpha(\pi)$, $a(\pi)$ are positive integers. Furthermore, $X(\rho(\theta)) \cap \Pi(\rho)$ is empty. Here we use that $I_{\mu'(\theta)} = I_{\tilde{\rho}'_0}$.

We digress to prove the following lemma. Suppose that $\theta = \theta_1 \otimes \theta_2 \otimes \theta_3$ is a regular character of C over F'. Let $\theta' = \theta_1 \otimes \theta_3 \otimes \theta_2$, $\theta'' = \theta_2 \otimes \theta_1 \otimes \theta_3$ and let $\rho_1 = \rho(\theta)$, $\rho_2 = \rho(\theta')$, $\rho_3 = \rho(\theta'')$. The ρ_j are distinct, but $I_{\mu'(\theta)} = I_{\mu'(\theta')} = I_{\mu'(\theta'')}$.

LEMMA 13.8.8: $\Pi(\rho_1) = \Pi(\rho_2) = \Pi(\rho_3)$.

Proof: We first show that there exists a (formal) linear combination

$$L(\rho_1) = \sum_{\pi\in\Pi(\rho_1)} a_\pi \cdot \pi$$

such that the restriction to the elliptic set of the associated character defines a stable distribution. By Corollary 12.5.4, this is the case if it is orthogonal to χ_ρ^G for all $\rho \in \Pi^2(H)$. If $\Pi(\rho) \cap \Pi(\rho_1)$ is empty, then $L(\rho_1)$ and χ_ρ^G are clearly orthogonal. This holds, in particular if $\rho = \rho(\theta)$ where θ is semi-regular.

Suppose that ρ is not of the form $\rho(\theta)$ with θ semi-regular. Then the twisted characters of $I_{\tilde{\rho}'}$ and $I_{\tilde{\rho}'_1}$ under the ε-elliptic norm are orthogonal by Proposition 12.6.2 unless $I_{\tilde{\rho}'}$ and $I_{\tilde{\rho}'_1}$ are equivalent. This is the case if and only if $\rho \in \{\rho_1, \rho_2, \rho_3\}$. By (13.8.6) and (13.8.7), we have relations of the type

$$\mathrm{Tr}(I_{\tilde{\rho}'}(\phi)I_{\tilde{\rho}'}(\varepsilon)) = \sum_{\pi\in\Pi(\rho)} a(\pi)\mathrm{Tr}(\pi(f)) + \sum_{\pi\in X(\rho)} a(\pi)\mathrm{Tr}(\pi(f))$$

$$\mathrm{Tr}(I_{\tilde{\rho}'_1}(\phi)I_{\tilde{\rho}'_1}(\varepsilon)) = \sum_{\pi\in\Pi(\rho_1)} b(\pi)\mathrm{Tr}(\pi(f)) + \sum_{\pi\in X(\rho_1)} b(\pi)\mathrm{Tr}(\pi(f))$$

where the $a(\pi)$ and $b(\pi)$ are positive. By Proposition 12.5.5, at most finitely many elliptic representations occur on the right hand sides of these equations and no elliptic representation occurs in both equations if $\rho \notin$

$\{\rho_1, \rho_2, \rho_3\}$. Hence there are at most 3 elements $\rho \in \Pi^2(H)$ such that $\Pi(\rho) \cap \Pi(\rho_1)$ is non-empty. Since $\text{Card}(\Pi(\rho_1)) = 4$, we can choose the a_π so that $L(\rho_1)$ is stable. Similarly, we choose a stable combination $L(\rho_2)$.

We now show that $\Pi(\rho_1)$ and $\Pi(\rho_2)$ either coincide or are disjoint. Suppose that $\Pi(\rho_1)$ and $\Pi(\rho_2)$ are not disjoint. By Lemma 12.7.2, the functions $\chi^G_{\rho_j}$ are linear combinations of characters of square-integrable representations with coefficients equal to ± 1 and they are orthogonal by Proposition 12.5.2. It follows from the orthogonality relations that $\Pi(\rho_1) \cap \Pi(\rho_2)$ must have an even number of elements, and hence, if $\Pi(\rho_2)$ do not coincide, then $\Pi(\rho_1) \cap \Pi(\rho_2) = \{\pi_1, \pi_2\}$. In this case, the orthogonality relations imply that there exist distinct square-integrable representations π, π', π'', π''' different from π_1 and π_2 such that for a suitable numbering of ρ_1 and ρ_2,

$$\pm\chi^G_{\rho_1}(f^H) = \text{Tr}(\pi_1(f)) + \text{Tr}(\pi_2(f)) + \text{Tr}(\pi(f)) \pm \text{Tr}(\pi'(f))$$
$$\pm\chi^G_{\rho_2}(f^H) = \text{Tr}(\pi_1(f)) - \text{Tr}(\pi_2(f)) \pm \text{Tr}(\pi''(f)) \pm \text{Tr}(\pi'''(f)) \ .$$

Since $L(\rho_i)$ is orthogonal to $\chi^G_{\rho_j}$ for $i,j = 1,2$, the only possibilities for the $L(\rho_j)$ are

$$L(\rho_1) = a_1(\pi_1 + \pi_2) + b_1\pi + c_1\pi'$$
$$L(\rho_2) = a_2(\pi_1 - \pi_2) + b_2\pi'' + c_2\pi'''$$

for some a_i, b_i, c_i, where $b_i c_i \neq 0$ for $i = 1,2$. Let f_w be a pseudo-coefficient corresponding to the linear combination $A_1 L(\rho_1) + A_2 L(\rho_2)$, where $A_j \in \mathbf{C}$. For some choice of A_1 and A_2 not both zero, the application of (13.8.7) to f_w gives zero and thus

$$(*) \qquad \text{Tr}(\tilde{\pi}(\phi_w)\tilde{\pi}(\varepsilon)) = 0 \ .$$

for $\tilde{\pi} = I_{\mu'(\theta)}$. Note that if the A_j are not both zero, then $f_w \neq 0$. The orthogonality relations imply that the representations $\{\pi_1, \pi_2, \pi, \pi', \pi'', \pi''''\}$ do not occur in any equality of type (13.8.6) or (13.8.7) other than the one for $I_{\mu'(\theta)}$. It follows that (*) also holds for $\tilde{\pi} = I_{\tilde{\rho}'}$ for all $\rho \in \Pi^2(H_w)$. Proposition 13.8.4 implies that (*) holds for all supercuspidal representations in $\tilde{\pi} \in E_\varepsilon(\widetilde{G}_w)$. It also holds for all principal series representations and Steinberg representations.

The function $\Phi^M(\gamma, f_w)$ is identically zero since the orbital integrals of f_w vanish off of the elliptic set. We now show that this is a necessary and sufficient condition that there exist a choice of ϕ_w corresponding to f_w such that $\phi^H_w = 0$. The vanishing of $\Phi^M(\gamma, f_w)$ implies that of $\Phi^M_\varepsilon(\delta, \phi_w)$

and $\Phi^M(\gamma, \phi_w^H)$. Let $\gamma_0 \in H_w$ be an elliptic element and suppose that $N(\delta_0) = \gamma_0$, where $\delta_0 \in \widetilde{H}_w$.

Suppose first that γ_0 is singular in H_w. Then $\Phi^{\rm st}(\gamma, \phi_w^H) = c\phi_w^H(\gamma_0)$ for γ near γ_0 by the germ expansion, since $\Phi^M(\gamma_0, \phi_w^H) = 0$. If γ_0 is non-central in G, then $\phi_w^H(\gamma_0)$ is a multiple of $\Phi_\varepsilon^\kappa(\delta_0, \phi_w)$ by Proposition 8.4.1(b). By the germ expansion, $\Phi_\varepsilon^{\rm st}(\delta, \phi_w)$ is a multiple of $\Phi^{\rm st}(\delta_0, \phi_w)$ for δ near δ_0. The condition $\Phi_\varepsilon^\kappa(\delta_0, \phi_w) = 0$ imposes no constraint on $\Phi_\varepsilon^{\rm st}(\delta, \phi_w)$ and we can assume that $\phi_w^H(\gamma_0) = 0$. If γ_0 is central in G, then $\phi_w^H(\gamma_0)$ is a multiple of the distribution of Lemma 8.4.3(b). We can assume that $\phi_w^H(\gamma_0) = 0$ without imposing a constraint on the stable distribution of 8.4.3(a) or on $\Phi_\varepsilon^{\rm st}(\delta, \phi_w)$ for δ_0, near δ_0, by the germ expansion. Similarly, if γ_0 is H-regular but singular in G, then $\Phi^{\rm st}(\delta_0, \phi_w)$ is a multiple of the local integral in Propositon 7.4.1(c) by Proposition 8.4.2(a). In this case, the stable distributions supported in the orbits whose closure contains δ_0 are $\Phi^{\rm st}(\delta_0, \phi_w)$ and

$$\int\limits_{\widetilde{Z}_w M_w \backslash \widetilde{M}_w} \left(\int\limits_{F_w^*} \int\limits_{N\gamma_w \backslash \widetilde{N}_w} \phi^K(m^{-1}n^{-1}\delta n(t)\varepsilon(nm))|t|d^*t d \right) \|\alpha_3(m)\|^{-2} dm \, .$$

These, and hence also $\Phi_\varepsilon^{\rm st}(\delta, \phi_w)$ for δ near δ_0, can be prescribed arbitrarily while setting $\phi_w^H(\gamma_0) = 0$. Finally, if γ_0 is (G, H)-regular, the orbital integrals $\Phi_\varepsilon^{\rm st}(\delta, \phi_w)$ and $\Phi_\varepsilon^\kappa(\delta, \phi_w)$ can be prescribed independently of one another for δ such that $N(\delta)$ is near γ_0. We can therefore assume that $\phi_w^H = 0$.

Then (*) holds for $\tilde{\pi} = I_{\tilde{\rho}}$ for all $\rho \in \Pi^2(H_w)$, by Proposition 13.8.2. In all, we have shown that (*) holds for all $\tilde{\pi} \in E_\varepsilon(\widetilde{G}_w)$. By the density theorem in the twisted case [KR], all twisted orbital integrals of ϕ_w vanish. This implies that all stable orbital integrals of f_w vanish. However, for regular elliptic γ, $\Phi^{\rm st}(\gamma, f_w)$ is a multiple of the character of $A_1L(\rho_1) + A_2L(\rho_2)$ and its vanishing violates the orthogonality relations between the characters for the representations $\pi_1, \pi_2, \pi, \pi', \pi'', \pi'''$. It follows that $\Pi(\rho_0)$ and $\Pi(\rho_2)$ are either disjoint or coincide.

If $\Pi(\rho_1)$ and $\Pi(\rho_2)$ are disjoint, then $L(\rho_1)$ and $L(\rho_2)$ are obviously linearly independent and the argument of the preceding paragraph can be repeated. This completes the proof of Lemma 13.8.8.

Let $\Pi = \Pi(\rho_1)$ and suppose that $\Pi = \{\pi_1, \pi_2, \pi_3, \pi_4\}$. Then

$$\chi^G_{\rho_j} = \sum_{1 \leq k \leq 4} \langle \rho_j, \pi_k \rangle \chi_{\pi_k}$$

where $\langle \rho_j, \pi_k \rangle = \pm 1$ for $1 \leq j \leq 4$. By the orthogonality relations and Proposition 12.5.2,

$$\langle \chi^G_{\rho_i}, \chi^G_{\rho_j} \rangle = \sum \langle \rho_i, \pi_k \rangle \langle \rho_j, \pi_k \rangle = 0$$

for $i \neq j$. Furthermore,

$$(**) \qquad \sum_{k=1}^{4} d(\pi_k) \langle \rho_j, \pi_k \rangle = 0$$

for $1 \leq j \leq 3$, where $d(\pi_k)$ is the formal degree of π_k by Proposition 12.7.2. For fixed j, the signs $\langle \rho_j, \pi_k \rangle$ are not all equal and hence the vector $\varepsilon_j = (\langle \rho_j, \pi_k \rangle)$ must contain two $+1$'s and two -1's. For if not, say $\varepsilon_1 = \pm(1,1,1,-1)$, then the orthogonality relations would imply that, up to ordering, $\varepsilon_2 = \pm(1,1,-1,1)$ and $\varepsilon_3 = \pm(1,-1,1,1)$, and

$$\pm\varepsilon_1 \pm \varepsilon_2 \pm \varepsilon_3 = (3,1,1,1)$$

for some choice of signs. But this contradicts (**) since the $d(\pi_k)$ are positive. It follows that the π_j can be numbered so that the ε_j can be numbered so that the ε_j are as follows:

$$\varepsilon_1 = (1,1,-1,-1), \quad \varepsilon_2 = (1,-1,1,-1), \quad \varepsilon_3 = \pm(1,-1,-1,1) .$$

It will be shown below that $\varepsilon_3 = (1,-1,-1,1)$.

Proposition 12.5.3 implies that the elliptic inner product of

$$\sum_{1 \leq k \leq 4} \langle \rho_j, \pi_k \rangle \chi_{\pi_k}$$

against a stable class function is zero for $1 \leq j \leq 3$. The right-hand side of (13.8.7) defines, by restriction of its character, a stable class function on the elliptic set. Taking its inner product with the above sums, we get:

$$\sum_{1 \leq k \leq 4} \alpha(\pi_k) \langle \rho_j, \pi_k \rangle = 0 ,$$

and the $\alpha(\pi_k)$ are all equal. This proves Lemma 13.8.7. It also follows from (**) that the $d(\pi_k)$ all coincide.

We now complete the proof of Proposition 13.8.6. We have to show that $X(\rho_0)$ is empty. If θ is a semi-regular character and f_w is a pseudo-coefficient for the L-packet $\Pi(\rho(\theta)) = \{\pi_1, \pi_2\}$, then (13.8.6) gives

$$\operatorname{Tr}(I_{\tilde{\rho}_0'}(\phi_w) I_{\tilde{\rho}_0'}(\varepsilon)) = \pm(a(\pi_1) - a(\pi_2)) \operatorname{Tr}(\pi_1(f_w)) .$$

The left-hand side vanishes by the orthogonality relations, since $\rho_0 \neq \rho(\theta)$ by assumption. Hence $a(\pi_1) = a(\pi_2)$. As in §12.7, assume that if π is square-integrable, then at most one of π, π^{nt} occurs in $X(\rho_0)$. By the orthogonality relations, at most finitely many $\pi \in X(\rho_0)$ are square-integrable or of the form π^{nt} for π square-integrable. We therefore assume that $X(\rho_0)$ consists of p.s. representations and finitely-many square-integrable representations. The orthogonality of the twisted characters of $I_{\tilde{\rho}_1'}, I_{\tilde{\rho}_2'}$ imply that a square-integrable representation occurs in at most one set $X(\rho_1)$ and that $X(\rho_1) \cap \Pi(\rho_2) = \varnothing$ if $\rho_1, \rho_2 \in \Pi^2(H)$. Similarly, by Proposition 12.4.1, $X(\rho_1)$ does not contain any representations of the form $\operatorname{St}_G(\psi)$.

We now show that $X(\rho_0)$ consists entirely of p.s. representations. If $\pi' \in X(\rho_0)$ is square-integrable, then it is supercuspidal and does not belong to $\Pi(\rho)$ for any $\rho \in \Pi^2(H_w)$, by the above remarks. Hence $\chi_{\pi'}$ is stably invariant by the orthogonality relations and Corollary 12.5.4. Set

$$f_w = \alpha(\rho_0)|\Pi(\rho_0)| f_{\pi'} - a(\pi') \sum_{\pi \in \Pi(\rho_0)} f_\pi$$

where $f_{\pi'}, f_\pi$ are pseudo-coefficients. Then (13.8.6) gives

$$(***) \qquad \operatorname{Tr}(\tilde{\pi}(\phi_w)\tilde{\pi}(\varepsilon)) = 0$$

for $\tilde{\pi} = I_{\tilde{\rho}_0'}$ and also for $\tilde{\pi} = I_{\tilde{\rho}_1'}$ for all $\rho_1 \in \Pi^2(H_w)$, since $\Pi(\rho_0) \cup \{\pi'\}$ is disjoint from $X(\rho_1)$. If $\tilde{\pi} \in E_\varepsilon(\widetilde{G}_w)$ is square-integrable, there exists a representation π'' of G_w such that $\psi_G(\pi'') = \pi$. By the orthogonality relations for the twisted characters of $\tilde{\pi}$ and $I_{\tilde{\rho}_0'}$, π'' does not belong to $X(\rho_0)$, and (***) holds for $\tilde{\pi}$. It follows that (***) holds for all ε-invariant representations of $\widetilde{G}_w$ which are either supercuspidal, principal series, or of the form $I_{\tilde{\rho}'}$ with $\rho \in \Pi^2(H)$. The argument in the proof of Lemma 13.8.8 shows that all stable orbital integrals of f vanish. However, this violates the orthogonality of $\chi_{\pi'}$ and χ_Π. Hence $X(\rho_0)$ consists entirely of p.s. By Lemma 12.7.7, the sum over $X(\rho_0)$ defines a distribution given by integration of F_f against an integrable function on M. To conclude that $X(\rho_0)$ is empty, argue as in the proof of Lemma 12.7.2. Proposition 13.8.6 follows.

PROPOSITION 13.8.9: *For all* $\rho \in \Pi^2(H_w), \alpha(\rho) = 1$.

Proof: Let ρ be a cuspidal L-packet on H. We have already shown that

$$\mathrm{Tr}(\rho(f^H)) = \sum_{\pi \in \Pi(\rho)} \langle \rho, \pi \rangle \, \mathrm{Tr}(\pi(f))$$

If ρ is not of the form $\rho(\theta)$, (13.8.3) can be written as

$$(13.8.8) \qquad 2\sum m(\pi)\,\mathrm{Tr}(\pi(f)) = \alpha(\rho)\chi_{\Pi(\rho)}(f) + \sum_{\pi \in \Pi(\rho)} \langle \rho, \pi \rangle \, \mathrm{Tr}(\pi(f))$$

by Proposition 13.8.6, where $\alpha(\rho) = \Pi\alpha(\rho_v)$. Here $\alpha(\rho_v) = 1$ if ρ_v is unramified. The sum on the left is over the set of discrete automorphic representations $\pi = \otimes\pi_v$ such that $\pi_v \in \Pi(\rho_v)$ for almost all v. It follows that for such π, $\pi_v \in \Pi(\rho_v)$ for all v. Since $m(\pi) \geq 0$ for all $\pi \in \Pi(\rho)$, and $\langle \rho, \pi \rangle = -1$ for some $\pi \in \Pi(\rho)$, $\alpha(\rho)$ must be positive.

A transfer of automorphic representations from H to G was constructed in [GP] using the Weil representation. It is compatible with the L-map ξ_H for some choice of μ in the following sense: if $\sigma = \otimes\sigma_v$ is cuspidal on H, then σ transfers to a cuspidal representation $\pi = \otimes\pi_v$ such that $\xi_H(\sigma_v) = \pi_v$ for almost all v such that σ_v is unramified. Furthermore, by [GP], Propositions 2.4 and 8.5, if π is related to σ in this way, then π has a global Whittaker model and $m(\pi) = 1$. In particular, for every cuspidal ρ which is not of the form $\rho(\theta)$, there exists a $\pi \in \Pi(\rho)$ such that $m(\pi) = 1$.

Now fix a square-integrable L-packet ρ' on H over F'. For any positive integer n, there exists a global E/F and a set $X = \{w\}$ of n places of F such that E_w/F_w is isomorphic to E'/F' for all $w \in X$. There exists a cuspidal L-packet ρ_0 on H which is not of the form $\rho(\theta)$ such that $\rho_{0w} = \rho'$ for all $w \in X$. If $\alpha(\rho') \neq 1$, then $\alpha(\rho_0) > 3$ for $n \geq 2$ and (13.8.8) implies that $m(\pi) > 1$ for all $\pi \in \Pi(\rho)$. This contradicts the remark of the previous paragraph. Therefore $\alpha(\rho') = 1$ and Proposition 13.8.9 is proved.

We now show that $\varepsilon_3 = (1, -1, -1, 1)$, in the notation of the proof of Lemma 13.8.7. Let $\varphi = \varphi_1 \otimes \varphi_2 \otimes \varphi_3$ be a regular character of $C\backslash\mathbf{C}$ and let $\varphi' = \varphi_1 \otimes \varphi_3 \otimes \varphi_2$, $\varphi'' = \varphi_2 \otimes \varphi_1 \otimes \varphi_3$. Let E/F and w be as above and assume that $\varphi_w = \theta$, $\varphi'_w = \theta'$, $\varphi''_w = \theta''$. Set $\rho_1 = \rho(\varphi)$, $\rho_2 = \rho(\varphi')$, $\rho_3 = \rho(\varphi'')$, and let $\mu = \mu(\varphi)$. The L-packets $\Pi(\rho_j)$ coincide by Lemma 13.8.8. Denote this L-packet by Π. By the character identities for I_μ and ρ_j, we see that the piece of the equality 13.7(4) $= 0$ corresponding to the e.v.p. $t(I_\mu)$ is

$$\sum_{\pi\in\Pi} m(\pi)\ \mathrm{Tr}(\pi(f)) = \frac{1}{4}\ \mathrm{Tr}(I_\mu(\phi)I_\mu(\varepsilon)) + \frac{1}{4}\sum_{1\leq j\leq 3} \mathrm{Tr}(\rho_j(f^H))$$

By Propositions 13.8.3 and 13.8.9, this gives

$$4\sum_{\pi\in\Pi} m(\pi)\ \mathrm{Tr}(\pi(f)) = \sum_{\pi\in\Pi}\left(1+\sum_{j=1}^{3}\langle\rho_j,\pi\rangle\right)\ \mathrm{Tr}(\pi(f)) \tag{13.8.9}$$

and hence

$$m(\pi) = \frac{1}{4}\sum_{s\in\widehat{\Pi}}\langle s,\pi\rangle \tag{13.8.10}$$

in the notation of §13.2.

We may assume that the φ_j are unramified at all finite places $v \neq w$ (cf. the discussion prior to Proposition 13.8.3). For each finite $v \neq w$, let π_v^0 be the unique unramified element in Π_v. Then $\langle\rho_{jv},\pi_v^0\rangle = 1$ for $1 \leq j \leq 3$, as follows immediately from the character identities applied to the units in the Hecke algebras. For archimedean v, choose $\pi_v^0 \in \Pi_v$ such that $\langle\rho_{jv},\pi_v^0\rangle = 1$ for $1 \leq j \leq 3$. Such a π_v^0 exists by Proposition 12.3.2 if $E_v/F_v = \mathbf{C}/\mathbf{R}$ and it exists trivially in the remaining archimedean cases.

Let $\Pi_w = \{\pi_1,\pi_2,\pi_3,\pi_4\}$. If $\varepsilon_3 = (-1,1,1,-1)$, then

$$\sum_{s\in\widehat{\Pi}_w}\langle s,\pi_k\rangle = \begin{cases} 2 & \text{if } 1\leq k\leq 3 \\ -2 & \text{if } k=4. \end{cases}$$

Set $\pi = \otimes\pi_v \in \Pi$, where $\pi_v = \pi_v^0$ if $v \neq w$ and $\pi_w = \pi_4$. By (13.8.10), we obtain $m(\pi) = -2$. This is impossible, and hence $\varepsilon_3 = (1,-1,-1,1)$.

PROPOSITION 13.8.10: *Let π' be a supercuspidal representation of G which does not belong to $\Pi(\rho)$ for any $\rho \in \Pi^2(H)$. Then there exists a unique supercuspidal representation $\tilde{\pi} \in E_\varepsilon(\widetilde{G})$ such that $\psi_G(\pi') = \tilde{\pi}$.*

Proof: With E/F and w as before, an application of the trace formula shows that there exists a cuspidal representation π_0 of G such that $\pi_{0w} = \pi'$ and π_{0v} is unramified at all finite places $v \neq w$ which remain prime in E (cf. §13.11). We can assume that no infinite place ramifies in E. The representation π_0 occurs in line (2) of 13.7 and we obtain an equality

$$\mathrm{Tr}(\tilde{\pi}(\phi)\tilde{\pi}(\varepsilon)) = \sum m(\pi)\mathrm{Tr}(\pi(f)) \tag{13.8.11}$$

where the sum is over π which define the same e.v.p. as π_0. By the choice of π_0, for all $v \neq w$, there exists a representation π_v of G_v such that $\mathrm{Tr}(\tilde{\pi}_v(\phi_v)\tilde{\pi}_v(\varepsilon)) = \mathrm{Tr}(\pi_v(f_v))$ and, by Proposition 13.8.1, we reduce to an equality of the form

$$\mathrm{Tr}(\tilde{\pi}_w(\phi_w)\tilde{\pi}_w(\varepsilon)) = \sum m(\pi)\mathrm{Tr}(\pi_w(f_w)) .$$

where π_{0w} appears in the right. If $\tilde{\pi}_w$ is supercuspidal, then $\tilde{\pi}_w = \psi_G(\pi'')$ for some supercuspidal π'' by Proposition 13.8.4. In this case, the sum in (13.8.11) must reduce to a single term and $\pi'' = \pi_{0w}$. Equality (13.8.11) shows, in any case, that $\tilde{\pi}_w$ is ε-elliptic. If it is not supercuspidal, then it must be of the form $I_{\tilde{\rho}}$ or $I_{\tilde{\rho}'}$ for some $\rho \in \Pi(H_w)$. The former possibility is precluded by the character identity of Proposition 13.8.2, while the latter is precluded by the character identity of Proposition 13.8.6. This proves Proposition 13.8.10.

We have also shown the existence of lifts. For $\rho \in \Pi^2(H)$, we have $\psi_G(\pi(\rho)) = I_{\tilde{\rho}'}$ by Proposition 13.8.9 and this is Proposition 13.2.2(c). Parts (b) and (d) of Proposition 13.2.2 follow from Proposition 12.4.1 and Lemma 12.7.6, respectively. Part (a) follows from Propositions 13.8.4, 13.8.10, and 13.2.2(c). The uniqueness of lifts also follows from linear independence of characters.

13.9 Non-cuspidal spectrum. According to the theory of Eisenstein series, the discrete non-cuspidal spectrum is spanned by the residues of Eisenstein series. Since the rank of G is equal to 1, the possible residues are well-known. Let $\chi = (\varphi, \psi)$ be a unitary character of $M\backslash\mathbf{M}$ and set:

$$M(s) = \frac{L(s,\varphi)\ L(2s,\varphi'\omega_{E/F})}{L(s+1,\varphi)\ L(2s+1,\varphi'\omega_{E/F})}$$

where φ' is the restriction of φ to C_F. Let $\chi(s) = \chi\|\ \|^s$. The poles of Eisenstein series, as a function of $\chi(s)$, coincide with the poles of its constant term, and these are given by the poles of $M(s)$ ([Lai], §3). A discrete non-cuspidal automorphic representation π of G is isomorphic to the unique irreducible quotient of $i_G(\chi(s))$ for some $\chi(s)$ with $\mathrm{Re}(s) \geq 0$ such that $M(s)$ has a pole. This occurs only in the following two cases:

(i) φ is trivial and $s = 1$.

(ii) $\varphi' = \omega_{E/F}$, $s = \frac{1}{2}$, and $L(\frac{1}{2}, \varphi) \neq 0$.

In the first case, π is the one-dimensional representation defined by the character $\psi \circ \det_G$ of $G\backslash\mathbf{G}$. In the second case, define a one-dimensional representation ξ of $H\backslash\mathbf{H}$ by $\xi(h) = \eta(\det_0(h))\psi(\det(h))$ where $\eta(a/\bar{a}) = \varphi\mu^{-1}(a)$. Then $\pi_v = \pi^n(\xi_v)$ for all v. This proves Theorem 13.3.6(a) and shows that $\pi^n(\xi)$ occurs in the discrete non-cuspidal spectrum if $L(\frac{1}{2},\varphi) \neq 0$.

13.10 Proofs of global results. Let $\tilde{\pi}$ be an ε-invariant cuspidal representation of $\widetilde{G}$. Applying separation of eigenvalues to 13.7(2), we obtain an equality:

$$\mathrm{Tr}(\tilde{\pi}(\phi)\tilde{\pi}(\varepsilon)) = \sum m(\pi)\mathrm{Tr}(\pi(f)) \tag{13.10.1}$$

where π ranges over the set of discrete π on G such that $\psi_G(t(\pi)) = t(\tilde{\pi})$.

We now show that for all v, there is an L-packet Π_v on G_v such that $\chi_{\tilde{\pi}_v\varepsilon}(\phi_v) = \chi_\Pi(f_v)$. This is immediate if $\tilde{\pi}_v$ is unramified or if v splits in E. Suppose that v is finite. The classification of ε-invariant representations of §12.4 together with the results on local base change in §13.2 shows that Π_v exists unless $\tilde{\pi}_v = I_{\tilde{\rho}}$ for some $\rho \in \Pi(H)$ such that ρ is either one-dimensional or square-integrable. But in this case, $\chi_{\tilde{\pi}_v\varepsilon}(\phi_v) = \chi_\rho(\phi_v^H)$ and this is incompatible with (13.10.1) (cf. the proof of Proposition 13.8.2). The remaining case to check is when $E_v/F_v = \mathbf{C}/\mathbf{R}$. Again we can rule out $\tilde{\pi}_v = I_{\tilde{\rho}}$ for $\rho \in \Pi^2(H_v)$ and then Π_v exists by the remarks of §12.3.

Let $\Pi = \otimes\Pi_v$. Then $\pi \in \Pi$ for all π in the sum in (13.10.1) and we obtain the equality

$$\sum_{\pi\in\Pi}\langle 1,\pi\rangle\mathrm{Tr}(\pi(f)) = \sum m(\pi)\mathrm{Tr}(\pi(f)) \ .$$

particular, $m(\pi) = \langle 1,\pi\rangle$ for all $\pi \in \Pi$. This rules out the possibility that $\pi^n(\xi_v)$ occurs as a local component of some $\pi \in \Pi$, where v is a place of F that remains prime in E and ξ_v is a one-dimensional representation of H_v. For if it did, then $\Pi_v = \{\pi^n(\xi_v), \pi^s(\xi_v)\}$ and we could choose $\pi = \otimes\pi_w \in \Pi$ such that $\langle 1,\pi_w\rangle = 1$ for all $w \neq v$ and $\pi_w = \pi^s(\xi_v)$. We would then have $\langle 1,\pi_w\rangle = \langle 1,\pi\rangle = -1$, which is impossible since $m(\pi) \geq 0$. Therefore $\langle 1,\pi\rangle = 1$ for all $\pi \in \Pi$ and Theorem 13.3.3 follows. This also shows that if $\pi^n(\xi_v)$ occurs as a local component of a discrete π, then π occurs in line (3) of 13.7.

Suppose that $\theta_1 = \theta' \otimes \theta'' \otimes \theta'''$ is a regular character of $C\backslash\mathbf{C}$ and let $\theta_2 = \theta' \otimes \theta''' \otimes \theta''$, $\theta_2 = \theta'' \otimes \theta' \otimes \theta'''$. Let $\boldsymbol{\mu} = \boldsymbol{\mu}'(\theta_1)$ and set $\Pi = \Pi(\theta_1)$.

Applying separation of eigenvalues to 1.3.7(4), we obtain

$$\sum m(\pi)\mathrm{Tr}(\pi(f)) = \frac{1}{4}\mathrm{Tr}(I_\mu(\phi)I_\mu(\varepsilon)) + \frac{1}{2}\sum_{j=1}^{3}\mathrm{Tr}(\rho(\theta_j)(f^H)) \tag{13.10.2}$$

The sum is over discrete π such that $\psi_G(t(\pi)) = t(I_\mu)$. By the local character identities, the right-hand side is equal to

$$\pm\sum_{\pi\in\Pi}\langle 1,\pi\rangle\mathrm{Tr}(\pi(f)) + \sum_{\pi\in\Pi}\sum_{j=1}^{3}\langle\rho_j,\pi\rangle\mathrm{Tr}(\pi(f))$$

where $\rho_j = \rho(\theta_j)$. In fact the minus sign cannot occur, for if it did, (13.10.2) would yield $m(\pi) < 0$ for some $\pi \in \Pi$. This proves the multiplicity formula of Theorem 13.3.7 for Π. A similar argument applied to 13.7(3) proves Theorem 13.3.7 for $\rho \in \Pi(\mathbf{H})$ which are not of the form $\rho(\theta)$. Theorems 13.3.1 and 13.3.2 are also proved. Theorem 13.3.6 follows from §13.9 and the above discussion. Theorems 13.3.5 and 13.3.4 are a consequence of the strong multiplicity one theorem for GL(2) and GL(3) and Lemma 13.6.3. This completes the proofs of the results of §13.3.

13.11 Proof of Lemma 13.8.5: Let E/F be a global extension such that E_w/F_w is isomorphic to E'/F' for some finite place w. Let u be a finite place which splits in E and set $\phi = \phi_w \times \phi_u \times \phi'$, where ϕ' is a product of functions ϕ_v for $v \neq w, u$, such that ϕ_v is in the Hecke algebra if v remains prime in E. Let ϕ_w be a matrix coefficient of π'. Choose ϕ_u with support in the ε-elliptic regular subset of $\widetilde{G}_u$ and such that the unstable orbital integrals $\Phi_\varepsilon^\kappa(\delta, \phi_u)$ vanish for all ε-elliptic regular δ in G_u. We can then take $\phi^H = 0$. We further specify ϕ_u and ϕ' below.

Since π' is supercuspidal, it can occur as the local component only of representations occuring in line (2) of Proposition 13.5.1, and therefore:

$$\theta_{\widetilde{G}}(\phi) = \sum \mathrm{Tr}(\tilde{\pi}(\phi)\tilde{\pi}(\varepsilon)), \tag{13.11.1}$$

where the sum is over the ε-invariant cuspidal representations $\tilde{\pi}$ of $\widetilde{G}$ such that $\tilde{\pi}_w = \pi'$ and $\tilde{\pi}_v$ is unramified for all $v \neq w$ which do not split in E.

The functions $\Phi_\varepsilon^M(\delta, \phi_v)$ are identically zero for $v = w, u$, and this gives $J_{\widetilde{M}}(\phi) = 0$. Since ϕ_w is matrix coefficient of a supercuspidal representation the local operator $I_{\sigma,\lambda}(\phi_w)$ (§2.5) vanishes (cf. [C₃], §3), hence $\theta_{\widetilde{M}}(\phi) = 0$ and we obtain $\theta_{\widetilde{G}}(\phi) = J_{\widetilde{G}}(\phi)$. By the condition on ϕ_u, $J_{\widetilde{G}}(\mathcal{O}_{st}, \phi)$ vanishes unless $\mathcal{O}_{st}$ is ε-elliptic regular, and $J_{\widetilde{G}}(\phi)$ is a sum of stable ε-elliptic regular orbital integrals of ϕ.

To finish the proof, we must show that ϕ can be chosen as above so that $J_{\widetilde{G}}(\phi) \neq 0$. We first show that there exist $\delta_0 \in \widetilde{G}$ such that $\Phi_\varepsilon^{st}(\delta_0, \phi_w) \neq 0$ and δ_0 is ε-elliptic regular at u. As above, $\mathrm{Tr}(I_{\tilde{\rho}}(\phi_w) I_{\tilde{\rho}}(\varepsilon)) = 0$, hence $\mathrm{Tr}(\rho(\phi_w^H)) = 0$ for all $\rho \in \Pi^2(H_w)$, by Proposition 13.8.2. By the completeness of the stable characters $\{\chi_\rho : \rho \in \Pi^2(H_w)\}$ (cf. Corollary 12.5.4), all stable orbital integrals of ϕ_w^H vanish, and hence $\Phi_\varepsilon^\kappa(\delta, \phi_w) = 0$ for ε-regular elliptic δ and non-trivial κ. This implies that $\Phi_\varepsilon^{st}(\delta, \phi_w) \neq 0$ if $\Phi_\varepsilon(\delta, \phi_w) \neq 0$, and since $\Phi_\varepsilon(\delta, \phi_w)$ is not identically zero on the ε-elliptic regular set, there exists an ε-elliptic regular element δ_0 such that $\Phi_\varepsilon^{st}(\delta_0, \phi_w) \neq 0$. By the remarks of §3.11, we can assume, up to stable ε-conjugacy, that $\delta_0 \in \widetilde{T}'_w$, where $\widetilde{T}'_w$ is the centralizer in $\widetilde{G}_w$ of a Cartan subgroup T'_w of G_w. Let T'_u be an elliptic Cartan subgroup in G_u. There exists a Cartan subgroup T of G which is conjugate to T'_w and T'_u in G_w and G_u, respectively. It suffices to let T be the centralizer of a regular element $X \in \mathrm{Lie}(G)$ which is close to regular elements in $\mathrm{Lie}(T'_w)$ and $\mathrm{Lie}(T'_u)$, respectively. Then δ_0 is ε-conjugate to an element of $\widetilde{T}_w$.

It is clear that $\widetilde{T}$ is dense in $\widetilde{T}_w$ ($\widetilde{T}$ is isomorphic to a product of multiplicative groups of field extensions of E) and hence we can assume that $\delta_0 \in \widetilde{T}$ and that $\Phi^{st}(\delta_0, \phi_w) \neq 0$. Choose ϕ_u so that $\Phi_\varepsilon^{st}(\delta_0, \phi_u) \neq 0$.

Next, we show it is possible to choose δ_0 so that it lies in a maximal compact subgroup at all places $v \neq w$ which do not split in E. Assume, for simplicity, that T is of type (3), so that $\widetilde{T}$ is isomorphic to K^*, where K/E is a cubic extension. The other cases are similar. Since δ_0 is ε-elliptic, w remains prime in K. Let Y be the set of places of K which lie over a prime of F which splits in E and let Y' be the remaining places of K. Let

$$X = \{v' \in Y' : |\delta_0|_{v'} \neq 1\} .$$

Since Y is a set of primes of density 1, it generates the generalized ideal class group of K with respect to any fixed modulus. Hence, given $v_0 \in X$, there exists $\alpha \in K^*$ such that α is a prime element at v_0, $|\alpha - 1|_{v'}$ is small if v' lies above w or u, and $|\alpha|_{v'} = 1$ if $v' \in Y'$ and $v' \neq v_0$. Replacing δ_0 by $\delta_0 \alpha^n$ for suitable n, we can assume that δ_0 is a unit at v_0 and hence lies in a maximal compact subgroup of $\widetilde{G}_{v_0}$. Since X is finite, the assertion follows by induction.

We now argue as in [BDKV],§A.2 and [K], pg. 35. For each place v of F, let C_v be a compact subset of $\widetilde{G}_v$ and let $C = \Pi C_v$. Assume that $\mathrm{supp}(\phi_w) \subset \widetilde{Z}_w C_w$ and $\mathrm{supp}(\phi_u) \subset \widetilde{Z}_u C_u$. Assume further that $C_v = \widetilde{K}_v$ for almost all v and for all v which remain prime in E. There are only finitely many stable ε-semisimple conjugacy classes modulo $\widetilde{Z}$ of elements $\delta \in \widetilde{G}$ such that $\Omega_\delta = \{g^{-1}\delta\varepsilon(g) : g \in \widetilde{\mathbf{G}}\}$ intersects $\widetilde{\mathbf{Z}}C$. Choose C so that $\Omega_{\delta_0} \cap \widetilde{\mathbf{Z}}C$ is non-empty. By choosing the support of ϕ_u and C_u sufficiently small, we can assume that $\Omega_\delta \cap \widetilde{\mathbf{Z}}C$ is empty unless δ is $\widetilde{G}_u$-conjugate to δ_0 modulo $\widetilde{Z}_u$. In particular, if $\Omega_\delta \cap \widetilde{\mathbf{G}}$ is non-empty, then δ is stably ε-conjugate to δ_0 modulo $\widetilde{Z}$. We can choose ϕ' so that ϕ has support in $\widetilde{\mathbf{Z}}C$, $\Phi_\varepsilon^{\mathrm{st}}(\delta_0, \phi') \neq 0$, and ϕ'_v is the unit of the Hecke algebra for all v which remain prime in F. In this case, $J_{\widetilde{G}}(\phi)$ reduces to a single term which is a non-zero multiple of $\Phi_\varepsilon^{\mathrm{st}}(\delta_0, \phi)$ and all representations π occuring in (13.11.1) above satisfy the conditions of the lemma.

CHAPTER 14

Comparison of inner forms

Let $G = U(3)$ and let G' be an inner form of G defined by a pair (D, α). We assume that G' is not isomorphic to G over F. The purpose of this chapter is to establish a correspondence between global L-packets on G' and G.

14.1 Conjugacy classes. We fix an inner isomorphism $\psi : \underline{G}' \to \underline{G}$. The map ψ induces an F-isomorphism between the centers of $\underline{G}'$ and $\underline{G}$ and we identify the center of G' with Z. If $\mathcal{O}_{\mathrm{st}}(\gamma')$ is a stable conjugacy class of semisimple elements in G', then the conjugacy class of $\psi(\gamma')$ is defined over F and intersects G in a stable conjugacy class $\mathcal{O}_{\mathrm{st}}(\gamma)$ by Steinberg's theorem. We obtain an injective map from stable semisimple classes in G' to a subset of the set of stable semisimple classes in G. We write $\gamma' \leftrightarrow \gamma$ if $\mathcal{O}_{\mathrm{st}}(\gamma')$ corresponds to $\mathcal{O}_{\mathrm{st}}(\gamma)$. If $\gamma \in G$, we will say that γ occurs in G' if $\gamma' \leftrightarrow \gamma$ for some $\gamma' \in G'$.

14.2 Stable local transfer. Let v be a place of F. Then α extends to an involution of the second kind on $D_v = D \otimes_F F_v$. If v remains prime in E, then α defines an isomorphism of D_v with its opposite algebra. This implies that the class of D_v in the Brauer group has order two and since $\dim_E(D) = 9$, D_v is the split algebra. If v splits in E, then $D_v = D_w \oplus D_{w'}$, where w and w' are the places of E lying above v, and α interchanges the two factors. It follows from §1.9 that G'_v is isomorphic to one of the following:

(i) G_v, if v is finite and does not split in E

(ii) G_v or $U_3(\mathbf{R})$, if E_v/F_v is $\mathbf{C}/\mathbf{R}$

(iii) D^*_w, if v splits in E, where w is a place of E dividing v.

Let S_0 be the set of infinite places v of F such that G'_v is isomorphic to $U_3(\mathbf{R})$ and let S be the set of places v such that D_w is ramified for $w|v$.

If $v \in S \cup S_0$, then stable conjugacy coincides with conjugacy in G'_v and we define $\Phi^{\mathrm{st}}(\gamma', f'_v) = \Phi(\gamma', f_v)$. For $v \notin S \cup S_0$, $\Phi^{\mathrm{st}}(\gamma', f'_v)$ is the usual stable orbital integral. We define a local transfer $f'_v \to f_v$ by the requirement

$$(14.2.1) \qquad \Phi^{\mathrm{st}}(\gamma, f_v) = \begin{cases} \Phi^{\mathrm{st}}(\gamma', f'_v) & \text{if } \gamma' \leftrightarrow \gamma \\ 0 & \text{if } \gamma \text{ does not occur in } G' \end{cases}$$

for all regular semisimple $\gamma \in G_v$.

If $v \notin S \cup S_0$, then the cocycle $\{\psi \circ \sigma(\psi)^{-1} : \sigma \in \Gamma\}$ splits in $H^1(F_v, G_{\mathrm{ad}})$ and there exists $g \in G_{\mathrm{ad}}(\overline{F}_v)$ such that $\psi_v = \mathrm{ad}(g) \circ \psi$ is an F_v-isomorphism of G'_v with G_v. It is well-defined up to conjugacy by an element of $G_{\mathrm{ad}}(F_v)$. By Lemma 3.5.3(a), G_v maps onto $G_{\mathrm{ad}}(F_v)$ and hence ψ_v is well-defined up to G_v-conjugacy. In particular, the equivalence classes of representations of G_v and G'_v are canonically identified. We fix a choice of ψ_v and hence forth identify G'_v with G_v for all $v \notin S \cup S_0$. Under this identification, we take f_v equal to f'_v. Then (14.2.1) is obviously satisfied, since $\psi_v(\gamma')$ is stably conjugate to γ if $\gamma' \leftrightarrow \gamma$. For almost all $v, G'(\mathcal{O}_v)$ and $G(\mathcal{O}_v)$ are hyperspecial maximal compact subgroups and ψ_v may be chosen so that it maps $G'(\mathcal{O}_v)$ isomorphically onto $G(\mathcal{O}_v)$. We assume this is the case for almost all v. The resulting isomorphism between the Hecke algebras with respect to $G'(\mathcal{O}_v)$ and $G(\mathcal{O}_v)$ is independent of the choice of ψ_v.

If $v \in S$, the existence of f_v is well-known ([R$_2$], §2) and is a special case of the transfer of functions between multiplicative groups of division algebras and GL_n. If $v \in S_0$, the existence of f_v follows from results of Shelstad ([S$_1$]).

From now on, f_v will denote a function that corresponds to f'_v.

14.3. Endoscopic transfer

The local transfer factors of [LS] are, in general, only defined up to a non-zero scalar factor. In our treatment of the transfer $f \to f^H$ in §4.9, we made use of a fixed embedding of H into G. This enabled us to work explicitly and choose a canonical transfer factor locally at every place. This transfer factor is characterized by the property

$$\mathrm{Tr}(i_G(\chi_\mu)(f)) = \mathrm{Tr}(i_H(\chi)(f^H))$$

If μ and ω are unramified, this is equivalent to the property that $\hat{\xi}_H(f)$ is a transfer of f for $f \in \mathcal{H}(G, \omega)$.

In general, if G is connected reductive over a global field and H is an endoscopic group given with an embedding $\xi : {}^LH \to {}^LG$, then there exists a global collection $\{\Delta^v_{G/H}(\gamma_H, \gamma)\}$ of transfer factors by [LS]. The individual factors $\Delta^v_{G/H}$ are only defined up to a non-zero scalar which depends on certain choices of auxiliary data. However if $\gamma_H \in H$ and $\gamma = (\gamma_v) \in \mathbf{G}$, then the product $\Pi_v \Delta^v_{G/H}(\gamma_H, \gamma_v)$ is defined ($\Delta^v_{G/H}(\gamma_H, \gamma_v) = 1$ for *a.a.v.*) and its value is independent of the choices. For almost all places v, G and H are unramified and a map $f \to \hat{\xi}(f)$ on suitable Hecke algebras is defined. We can assume that the Hecke algebras are defined relative to $G(\mathcal{O}_v)$ and $H(\mathcal{O}_v)$, which are hyperspecial maximal compact subgroups for *a.a.v.* In this case, there is at most one transfer factor such that $\hat{\xi}(f)$ is a transfer of f. This choice, which we call the canonical transfer factor, can be explicitly determined and it has been verified by Hales ([Hl]) that for any global collection $\{\Delta^v_{G/H}\}$ as above, $\Delta^v_{G/H}$ is canonical for almost all v.

We now continue with $G = U(3)$, $H = U(2) \times U(1)$, and G' as before. We fix a global collection of transfer factors which we denote by $\{\Delta'_v(\gamma, \gamma')\}$, where $\gamma \in H_v$ and $\gamma' \in G'_v$. The transfer $f'_v \to f'^H_v$ from functions on G'_v to functions on H_v is defined by the requirement

$$\sum_{\{\gamma'\}} \Delta'_v(\gamma, \gamma') \Phi(\gamma', f'_v) = \Phi^{\mathrm{st}}(\gamma, f'^H_v)$$

where $\gamma \in H_v$ is regular semisimple and $\{\gamma'\}$ is a set of representatives for the conjugacy classes in G'_v such that $\gamma' \leftrightarrow \gamma$. We denote the left-hand side of this equality by $\Phi^\kappa(\gamma', f'_v)$. If $v \notin S \cup S_0$, then G'_v is isomorphic to G_v and the existence of f'^H_v has already been established in §4. If $v \in S$, then we take $f'^H_v = 0$, because $\mathcal{O}_{\mathrm{st}}(\gamma)$ does not occur in H_v. Indeed, the non-central elements in G'_v then belong to Cartan subgroups of type (3), i.e., which split over a cubic extension. If $v \in S_0$, the existence of f'^H_v follows from the results of Shelstad ([S$_1$]), where again the technique of Clozel-Delorme is used to obtain f'^H of compact support.

14.4 Local correspondence. Suppose that $v \in S \cup S_0$. An L-packet on G'_v is defined to be a set consisting of a single irreducible, admissible representation of G'_v (which is necessarily finite-dimensional since G'_v is compact modulo Z_v). Let $\Pi^2(G_v)$ be the set of square-integrable L-packets on G_v. There is a bijection $\psi'_v : \Pi(G'_v) \to \Pi^2(G_v)$ characterized by the relation $\mathrm{Tr}(\Pi(f'_v)) = \mathrm{Tr}(\Pi(f_v))$.

If $v \in S$ and w is a place of E dividing v, then G'_v and G_v are canonically isomorphic to D^*_w and $\mathrm{GL}_3(E_w)$, respectively. The existence of the bijection ψ'_v is a special case of the results of [BDKV], [R$_2$].

Suppose that $v \in S_0$. The existence of ψ_v follows from results of Shelstad ([S$_1$]), but is also easy to check directly in this special case. Suppose that

$\varphi = \varphi(a,b,c)$ where $a \geq b \geq c$, in the notation of §12.3. Let $m = a - b$ and $n = b - c$. If $mn \neq 0$, then a finite-dimensional representation F_φ of G_v is defined. The inner twisting ψ induces a bijection between the finite-dimensional representations of G_v and G'_v, since it determines an isomorphism between the associated complex groups. We donote also by F_φ the corresponding representation of G'_v. The L-packet $\Pi(\varphi)$ is square-integrable if and only if $mn \neq 0$ and if $mn = 0$, then $\psi'_v(F_\varphi) = \Pi(\varphi)$.

If $v \notin S \cup S_0$, then we identify $\Pi(G'_v)$ with $\Pi(G_v)$ by means of ψ_v. For all v, we identify the sets of one-dimensional representations of G'_v and G_v with $\mathrm{Hom}(E^1_v, \mathbf{C}^*)$.

PROPOSITION 14.4.1: (a) *Let $v \in S_0$. Suppose that $m = 1$ (resp., $n = 1$) and let $\xi = \xi(a,c,b)$ (resp., $\xi = \xi(b,a,c)$). Then*

$$\chi_{\Pi(\xi)}(f) = \begin{cases} -\mathrm{Tr}(F_\varphi(f')) & \text{if } mn \neq 0 \\ 0 & \text{if } mn = 0 . \end{cases}$$

(b) *If $v \in S$, then $\mathrm{Tr}(\Pi(f)) = 0$ if Π is an infinite-dimensional L-packet on G_v which does not belong to $\Pi^2(G_v)$.*

(c) *If $\psi \in \mathrm{Hom}(E^1, \mathbf{C}^*)$, then $\mathrm{Tr}(\psi(f)) = \mathrm{Tr}(\psi(f'))$.*

Proof: The orbital integrals of f are zero on M_v and hence f has zero trace on all principal series representations. By the proof of Proposition 12.3.3,

$$\chi_{\Pi(\xi)}(f) = \mathrm{Tr}(i_G(\chi)(f)) - \mathrm{Tr}(\Pi(\varphi)(f))$$

(where χ is as in the proof) and hence $\chi_{\Pi(\xi)}(f) = -\mathrm{Tr}(\Pi(\varphi)(f))$. Part (a) follows. For (b), observe that the orbital integrals of f vanish off the elliptic set and the character of an infinite dimensional unitary representation of $G_v = \mathrm{GL}_3(F_v)$ vanishes on the elliptic set unless it is square-integrable. To prove (c), assume first that $v \in S_0$. Choose φ so that $\psi = F_\varphi$. Then

$$\mathrm{Tr}(F_\varphi(f)) - \mathrm{Tr}(\Pi(\varphi)(f)) = \mathrm{Tr}(\pi_1(f)) - \mathrm{Tr}(\pi_2(f)) - \mathrm{Tr}(\pi_3(f))$$

where $\pi_1 = i_G(\chi_\varphi)$, $\pi_2 = i_G(\chi^+_\varphi)$, and $\pi_3 = i_G(\chi^-_\varphi)$. The right-hand side vanishes, as above, and hence $\mathrm{Tr}(\Pi(\varphi)(f)) = \mathrm{Tr}(F_\varphi(f'))$, as required. If $v \in S$, then $\mathrm{Tr}(\psi(f')) = \mathrm{Tr}(\mathrm{St}_G(\psi)(f))$ since ψ maps to $\mathrm{St}_G(\psi)$ under the local correspondence, and the assertion follows since the characters of ψ and $\mathrm{St}_G(\psi)$ agree on the elliptic set.

Our next goal is to calculate the transfer of L-packets $\Pi(H_v) \to \Pi(G'_v)$ associated to $f'_v \to f'^H_v$ for $v \in S_0$. It depends on the choice of transfer factor Δ'_v, which we fix for the purposes of explicit calculation as follows. Let T_v be a compact Cartan subgroup of H_v. We fix an embedding $i: T_v \to G'_v$ and also view T_v as a subgroup of G_v. Set

$$\Delta''_v(\gamma, i(\gamma)) = \tau(\gamma)D_G(\gamma)D_H(\gamma)^{-1} = \Delta_v(\gamma,\gamma) .$$

If γ' is stably conjugate to $i(\gamma)$ in G'_v and γ_H is stably conjugate to γ in H_v, we set

$$\Delta''_v(\gamma_H, \gamma') = \kappa(\text{inv}(i(\gamma), \gamma'))\Delta''_v(\gamma, \gamma') .$$

This specifies Δ''_v completely and it is easy to check that Δ''_v arises from the construction of [LS] for a suitable choice of data.

PROPOSITION 14.4.2: *Let $v \in S_0$. Relative to the transfer factor Δ''_v, the following character identities hold:*

(a) *Let ρ be an infinite-dimensional L-packet on H_v. Then* $\text{Tr}(\rho(f'^H)) = 0$ *unless* $\Pi(\rho) \in \Pi^2(G_v)$.

(b) *Let $a > b > c$ be integers and let $\varphi = \varphi(a,b,c)$. Then*

$$\text{Tr}(\rho(f'^H)) = \begin{cases} -\text{Tr}(F_\varphi(f')) & \textit{if} \quad \rho = \rho(a,b,c) \\ \text{Tr}(F_\varphi(f')) & \textit{if} \quad \rho = \rho(b,a,c) \quad \textit{or} \quad \rho(a,c,b). \end{cases}$$

(c) *Let ξ be a one-dimensional representation of H_v. Assume that $\xi = \xi(a,c,b)$ or $\xi(b,a,c)$ where $a \geq b \geq c$. Let $m = a - b$, $n = b - c$. Then*

$$\text{Tr}(\xi(f'^H)) = \begin{cases} -\text{Tr}(F_\varphi(f)) & \textit{if} \quad mn \neq 0 \\ 0 & \textit{if} \quad mn = 0 \ , \end{cases}$$

where $\varphi = \varphi(a,b,c)$.

Proof: We drop the subscript v in the proof. It is clear that $\text{Tr}(\rho(f'^H)) = 0$ if ρ is infinite-dimensional and not square-integrable. In this case, χ_ρ vanishes on the elliptic regular set and $\Phi(\gamma, f'^H)$ vanishes off of the elliptic regular set.

Let a, b, c be integers such that $a > c$ let $\rho = \rho(a,b,c)$. Set:

$$\begin{aligned} d_G(\gamma) &= (\gamma_1\gamma_2\gamma_3)^{-1}(\gamma_1 - \gamma_2)(\gamma_2 - \gamma_3)(\gamma_1 - \gamma_3) \\ &= (\gamma_1/\gamma_3)(1 - \gamma_2\gamma_1^{-1})(1 - \gamma_3\gamma_2^{-1})(1 - \gamma_3\gamma_1^{-1}) \end{aligned}$$

and $d_H(\gamma) = (\gamma_1 - \gamma_3)$. By definition,

$$\tau(\gamma) = \gamma_2(\gamma_1\gamma_3)^t[(\gamma_2\gamma_1^{-1} - 1)(1 - \gamma_2\gamma_3^{-1})]^{-1}|(\gamma_2\gamma_1^{-1} - 1)(1 - \gamma_2\gamma_3^{-1})|$$

and hence

$$\tau(\gamma)D_H(\gamma)D_G(\gamma)^{-1} = (\gamma_1\gamma_2)^t d_H(\gamma)d_G(\gamma)^{-1} .$$

The Langlands parametrization is so defined that ρ is the unique L-packet on H such that:

$$\chi_\rho(\gamma) = -d_H(\gamma)^{-1}[\gamma_1^{a-t}\gamma_2^b\gamma_3^{c-t} - \gamma_1^{c-t}\gamma_2^b\gamma_3^{a-t}]$$

It follows that:

$$\tau(\gamma)D_H(\gamma)D_G(\gamma)^{-1}\chi_\rho(\gamma) = -d_G(\gamma)^{-1}\left[\gamma_1^a\gamma_2^b\gamma_3^c - \gamma_1^c\gamma_2^b\gamma_3^a\right]$$

Let $\beta(\gamma) = \gamma_1^a\gamma_2^b\gamma_3^c$. The sum

$$d_G(\gamma)^{-1} \sum_{w\in\Omega(T,G)} sgn(w)\beta(w\gamma w^{-1})$$

is zero if a, b, c are not distinct. If a, b, c are distinct, then it is equal to $\varepsilon\mathrm{Tr}(F_\varphi(\gamma'))$, where $\varepsilon = 1$ if $a > b > c$ and $\varepsilon = -1$ if $b > a$ or $c > b$, by the Weyl character formula. We now apply Lemma 12.5.1 (it is stated for a p-adic field, but it is valid in the archimedean case). Since κ is trivial in this case, the pull-back $\chi_\rho^{G'}(\gamma')$ of χ_ρ to G' is equal to

$$\begin{aligned} &D_G(\gamma)^{-1}\sum \tau(w\gamma w^{-1})D_H(w\gamma w^{-1})\chi_\rho(w\gamma w^{-1}) \\ &\quad = -\sum d_G(w\gamma w^{-1})[\beta(w\gamma w^{-1}) - \beta(sw\gamma w^{-1}s^{-1})] \end{aligned}$$

where the sums are over $w \in \Omega_F(T,H)\backslash\Omega_F(T,G)$, and s is the non-trivial element in $\Omega_F(T,H)$. This is equal to

$$-d_G(\gamma) \sum_{w\in\Omega_F(T,G)} sgn(w)\beta(w\gamma w^{-1}) = \begin{cases} -\varepsilon\mathrm{Tr}(F_\varphi(\gamma')) & \text{if } a,b,c \text{ are disti[illegible]} \\ 0 & \text{otherwise.} \end{cases}$$

It follows that $\mathrm{Tr}(\rho(f'^H))$ vanishes if a, b, c are not distinct and otherwise is given by statement (b). To prove (c), let $\rho = \rho(b, a, c)$ (resp., $\rho = \rho(a, c, b)$) if $\xi = \xi(b, a, c)$ (resp., $\xi = \xi(a, c, b)$). Then $\{\rho_1, \xi\}$ is the set of constituents of a principal series representation of H. The trace of f'^H is zero on the principal series and hence $\mathrm{Tr}(\rho(f'^H)) = -\mathrm{Tr}(\xi(f'^H))$.

14.5 The trace formula. Since G' is anisotropic, $T_{G'}(f')$ is the trace of $\rho(f')$ on $L(G')$ (defined with respect to the central character ω). The χ-expansion is given by

$$\theta_{G'}(f') = \sum m(\pi)\mathrm{Tr}(\pi(f')$$

where the sum is over the spectrum of $L(G')$. The $\mathcal{O}$-expansion is equal to

$$J_{G'}(f') = \sum \varepsilon(\gamma)^{-1} m(\mathbf{Z}G_\gamma \backslash \mathbf{G}_\gamma) \Phi(\gamma, f')$$

where the sum is over on set of representatives for the conjugacy classes in G' modulo Z.

We assume, from now on, that $f = \Pi f_v$ and $f'^H = \Pi_v f_v'^H$, where $f_v' \to f_v$ and $f_v' \to f_v'^H$ relative to the transfer factor Δ_v', for all v. For almost all finite v, of course, $f_v', f_v'^H$, and f_v are the units in their respective Hecke algebras.

Let $\mathcal{O}_{\mathrm{st}} = \mathcal{O}_{\mathrm{st}}(\gamma_0)$ be a stable semisimple conjugacy class in G and define

$$J(\mathcal{O}_{\mathrm{st}}, f') = \begin{cases} \varepsilon_{\mathrm{st}}(\gamma_0)^{-1} m(\mathbf{Z}G_{\gamma_0} \backslash \mathbf{G}_{\gamma_0}) \sum_{\gamma \in \mathcal{C}'} \Phi(\gamma, f) & \text{if } \mathcal{O}_{\mathrm{st}} \text{ occurs in } G' \\ 0 & \text{otherwise} \end{cases}$$

where $\mathcal{C}'$ is a set of representatives for the conjugacy classes in G' within the stable class defined by γ_0, and $\varepsilon_{\mathrm{st}}(\gamma_0)$ is the number of $z \in Z$ such that γ_0 is stably conjugate to $z\gamma_0$.

THEOREM 14.5.1: (a) *Let $\mathcal{O}_{\mathrm{st}}$ be an elliptic stable conjugacy class in G. Then*

$$J(\mathcal{O}_{\mathrm{st}}, f') = \mathrm{SJ}(\mathcal{O}_{\mathrm{st}}, f) + \frac{1}{2} \sum \mathrm{SJ}(\mathcal{O}'_{\mathrm{st}}, f'^H)$$

where the sum is over the stable classes $\mathcal{O}'_{\mathrm{st}}$ in H which transfer to $\mathcal{O}_{\mathrm{st}}$.

(b) $\theta_{G'}(f') = S\theta_G(f) + \frac{1}{2} S\theta_H(f'^H)$

Proof: Let $\gamma_0 \in \mathcal{O}_{\mathrm{st}}$. If γ_0 is regular, we refer to [L$_2$] where the stabilization is carried out for the elliptic regular terms in general.

LEMMA 14.5.2: *Assume that E/F is global and that $D = M_3(E)$. Let $\gamma_0 \in H$ and suppose that γ_0 is singular in G. Let $\gamma' \in G$ correspond to γ_0.*

(a) *If γ_0 is regular in H, then $\Phi^{\mathrm{st}}(\gamma_0, f'^H) = 0$.*

(b) *If γ_0 is central in H but not central in G, then*

$$\Phi^\kappa(\gamma', f') = f'^H(\gamma_0), \ \Phi^{\mathrm{st}}(\gamma', f') = \Phi^{\mathrm{st}}(\gamma_0, f),$$

where κ denotes the non-trivial element $\mathfrak{R}(\gamma_0/F)$.

(c) *If γ_0 is central in G, then $f'^H(\gamma_0) = 0$.*

Proof: Let T be a Cartan subgroup of type (1) in H. If $v \in S_0$, then

$$\lim_{\gamma' \to \gamma_0} D_G(\gamma') \Phi(\gamma', f_v') = 0 \ .$$

since G'_v is compact and $D_G(\gamma_0) = 0$. Let T'' be the set of $\gamma \in T_v$ which are singular in G but regular in H. Then $D_H(\gamma) \neq 0$ for $\gamma \in T''$, hence $\Phi^{st}(\gamma, f_v'^H) = 0$ for all $\gamma \in T''$. This proves (a). If γ_0 is central in G', let γ' approaches γ within T''. The limit formula for H implies that $f_v'^H(\gamma_0) = 0$, and (c) follows.

If γ_0 is central in H but not central in G, then the method used in the proof of Proposition 8.2.1 gives the relation $\Delta_{G/H}(\gamma_0)\Phi(\gamma_0, f'_v) = f_v'^H(\gamma_0)$. Proposition 8.2.1(a) and the product formula imply that $\Phi^\kappa(\gamma', f') = f'^H(\gamma_0)$. The equality $\Phi^{st}(\gamma', f') = \Phi^{st}(\gamma_0, f)$ is a special case of [Kt$_6$], Proposition 2. It can be proved directly by the method used in §8.2. Part (b) follows.

We return to the proof of Theorem 14.5.1. If $v \in S_0 \cup S$, then $\Phi^M(\gamma, f_v) \equiv 0$ and hence $\Phi^M(\gamma, f) \equiv 0$. Furthermore, $\int_{F^*} f_v^K(\gamma n(t))\omega_{E/F}(t)|t|^2 d^*t = 0$. If $v \in S$, this follows from [R$_2$], Lemma 2.6. If $v \in S_0$, then this integral can be expressed as an integral of traces of principal series representations by the remark of §8.3, and it vanishes since the traces are expressible in terms of $\Phi^M(\gamma, f_v)$. By Proposition 10.1.2(b),

$$\mathrm{St}(\mathcal{O}_{st}, f) = \begin{cases} m(\mathbf{Z}G\backslash\mathbf{G})f(\gamma_0) & \text{if } \gamma_0 \text{ is central} \\ \dfrac{1}{2}m(\mathbf{Z}H\backslash\mathbf{H})\Phi^{st}(\gamma_0, f) & \text{if } \gamma_0 \text{ is singular but not central.} \end{cases}$$

If γ_0 is central in G, then $f'_v(\gamma_0) = f_v(\gamma_0)$. This follows from [R$_2$], Lemma 3.3 if $v \in S$ and from the limit formulas (cf. §8.4) if $v \in S_0$. Furthermore, $f'^H(\gamma_0) = 0$ by Lemma 14.5.2(c). Therefore $SJ(\mathcal{O}'_{st}, f'^H) = 0$ by Proposition 10.1.2(a) and (a) follows for $\mathcal{O}_{st} = \mathcal{O}_{st}(\gamma_0)$.

Suppose that γ_0 is singular but not central. If $D \neq M_3(E)$, then $\mathcal{O}_{st}$ does not occur in G'. For $v \in S$, $G_v = \mathrm{GL}_3(F_v)$ and γ_0 is not elliptic in G_v, in which case $\Phi^{st}(\gamma_0, f_v) = 0$ by [R$_2$], Lemma 2.6. If $D = M_3(E)$, then $\mathcal{O}_{st}$ occurs in G'. The conjugacy classes in G' corresponding to γ_0 are parametrized by the set $\mathfrak{D}_{G'}(G'_{\gamma'_0}/F)$ of $\xi \in F^*/NE^*$ such that ξ is negative at places $v \in S_0$ by Proposition 3.8.1(d) and the contribution of $\mathcal{O}_{st}$ to the trace formula for G' is:

$$m(\mathbf{Z}H\backslash\mathbf{H}) \sum_{\delta \in \mathfrak{D}_{G'}(G'_{\gamma'_0}/F)} \Phi(\gamma^\delta, f')$$
$$= \frac{1}{2}m(\mathbf{Z}H\backslash\mathbf{H})[\Phi^{st}(\gamma', f') + \Phi^\kappa(\gamma', f')]$$

Here, $\gamma' \in G'$ corresponds to γ_0 and γ^δ is an element in the conjugacy class corresponding to $\delta \in \mathfrak{D}_{G'}(G'_{\gamma'_0}/F)$. We may assume that γ_0 is central in H.

By Lemma 14.5.2(b) and the above, we see that

$$\frac{1}{2}m(\mathbf{Z}H\backslash\mathbf{H})\Phi^{st}(\gamma',f') = SJ(\mathcal{O}_{st},f)$$
$$\frac{1}{2}m(\mathbf{Z}H\backslash\mathbf{H})\Phi^{\kappa}(\gamma',f') = \frac{1}{2}SJ(\mathcal{O}'_{st},f'^{H})$$

where $\mathcal{O}'_{st}$ is the central class $\{\gamma_0\}$ in H. There is an additional stable class $\mathcal{O}'_{st}(\gamma_1)$ in H which transfers to $\mathcal{O}_{st}$, but $SJ(\mathcal{O}'_{st}(\gamma_1),f'^{H}) = 0$ by Lemma 14.5.2(a). This completes the proof of (a).

From (a) and the equality $\theta_{G'}(f') = J_{G'}(f')$, we obtain

$$\theta_{G'}(f') = SJ_G(f) + \frac{1}{2}SJ_H(f'^H)\,.$$

On the other hand

$$\mathrm{SJ}_G(f) = S\theta_G(f) + S\theta_M(f) - \mathrm{SJ}_M(f)$$
$$\mathrm{SJ}_H(f'^H) = S\theta_H(f'^H) + S\theta_{M_H}(f'^H) - \mathrm{SJ}_{M_H}(f'^H)$$

and hence:

$$\theta_{G'}(f') - [S\theta_G(f) + \frac{1}{2}S\theta_H(f'^H)]$$

(14.5.1)

$$= \left[S\theta_M(f) + \frac{1}{2}S\theta_{M_H}(f'^H)\right] - \left[SJ_M(f) + \frac{1}{2}SJ_{M_H}(f'^H)\right]$$

Let $S\mathcal{I}_M(\gamma,f,v) = S\mathcal{I}(\gamma,f_v)\Pi_{w\neq v}\Phi^M(\gamma,f_w)$. Observe $S\mathcal{I}_M(\gamma,f,v)$ and $\mathcal{I}_{M_H}(\gamma,f'^H,v)$ vanish if $G'_v = G_v$ because there exists a place $w \neq v$ such that $G'_w \neq G_w$ and $\Phi^M(\gamma,f_w) = \Phi^M_H(\gamma,f'^H_w) = 0$. It follows that

$$\mathrm{SJ}_M(f) + \frac{1}{2}SJ_{M_H}(f'^H) = \sum_v \sum_{\gamma\in Z\backslash M}\left[S\mathcal{I}_M(\gamma,f,v) + \frac{1}{2}\mathcal{I}_{M_H}(\gamma,f'^H,v)\right]$$

where the first sum is over v such that $G'_v \neq G_v$. If $v \in S_0$, then

$$S\mathcal{I}_M(\gamma,f,v) + \frac{1}{2}\mathcal{I}_{M_H}(\gamma,f'^H,v)$$

extends to a smooth function on $\mathbf{M}$, by Corollary 9.3.4. If v is finite and $G'_v \neq G_v$, then $f'^H = 0$ and $S\mathcal{I}_M(\gamma,f,v)$ extends to a smooth function on $\mathbf{M}$, by Proposition 9.4.2. The argument of comparison of measures used in the proof of Theorem 10.3.1 implies that the right-hand side of (14.5.1) is zero.

14.6 The global correspondence

THEOREM 14.6.1: *The following equalilty holds:*

$$\text{Tr}(\rho_d(f')) = \sum_n(\Pi)\text{Tr}(\Pi(f)) + \frac{1}{2}\sum n(\rho)\text{Tr}(\rho(f'^H)) \tag{14.6.1}$$

where Π *(resp.,* ρ*) ranges over the subset of* $\Pi(\mathbf{G})$ *(resp.,* $\Pi(\mathbf{H})$*) of packets with central character* ω*.*

Proof: By Theorem 14.5.1, $\text{Tr}(\rho_d(f')) = S\theta(f) + \frac{1}{2}S\theta_H(f'^H)$. The traces of f and f'^H are zero on all principal series representations. It follows from Proposition 13.6.2 and the multiplicity formulas of Theorem 13.3.7 that $S\theta(f)$ is equal to $\sum n(\Pi)\text{Tr}(\Pi(f))$. Similarly, $S\theta_H(f'^H)$ is equal to $\sum n(\rho)\text{Tr}(\rho(f^H))$ by Propositions 13.6.1 and 11.2.1.

By a global L-packet on G' we will mean a tensor product $\Pi' = \otimes\Pi'_v$, where $\Pi'_v \in \Pi(G'_v)$ for $v \in S \cup S_0$ and $\Pi'_v \in \Pi'(G_v)$ for $v \notin S \cup S_0$, such that Π'_v is unramified for almost all v. Let $\Pi(\mathbf{G}')$ denote the set of Π' such that π occurs discretely for some $\pi \in \Pi'$. If Π' is infinite-dimensional, let $\psi'(\Pi') = \otimes\psi'_v(\Pi'_v)$. If Π' is one-dimensional, corresponding to a character ψ of $E^1\backslash\mathbf{E}^1$, we let $\psi'(\Pi')$ denote the one-dimensional representation of $\mathbf{G}$ defined by ψ. By the definition of ψ'_v and Proposition 14.4.1(c), we have $\text{Tr}(\Pi'(f')) = \text{Tr}(\psi'(\Pi')(f)$.

Let S' be a finite set of places of F containing $S \cup S_0$. For $v \notin S$ we have identified G'_v and G_v and a collection $t_{S'} = \{t_v\}_{v\notin S'}$, where $t_v \in \text{Hom}(\mathcal{H}_v, \mathbf{C})$ can be regarded as an e.v.p. on either G' or G. It is immediate from (14.6.1) (and the argument of separation of eigenvalues of §13.7) that $t_{S'}$ is of the form $t_{S'}(\pi)$ for a discrete representation π of G' if and only if $t_{S'} = t_{S'}(\Pi)$ for some $\Pi \in \Pi(\mathbf{G})$. Furthermore, Π is unique by Theorem 13.3.5. Let $\Pi \in \Pi(\mathbf{G})$ and let $t_{S'} = t_{S'}(\Pi')$, where S' is chosen large enough so that π_v is unramified for $v \notin S'$. From (14.6.1), we obtain

$$\sum m(\pi)\text{Tr}(\pi(f)) = n(\Pi)\text{Tr}(\Pi(f)) + \frac{1}{2}\sum_{\substack{\rho\in\Pi(\mathbf{H})\\ \Pi(\rho)=\Pi}} n(\rho)\text{Tr}(\rho(f'^H)). \tag{14.6.2}$$

Let $\Pi_s(\mathbf{G}')$, $\Pi_e(\mathbf{G}')$, and $\Pi_a(\mathbf{G}')$ denote the sets of L-packets Π' such that the e.v.p. $t(\Pi')$ coincides with $t(\Pi)$ for some Π belonging to $\Pi_s(\mathbf{G})$, $\Pi_e(\mathbf{G})$, and $\Pi_a(\mathbf{G}')$, respectively. These sets are disjoint (Theorem 13.3.5) and every discrete representation belongs to an L-packet in one of them i.e., their union is $\Pi(\mathbf{G}')$.

PROPOSITION 14.6.2: *The map ψ' defines a bijection between $\Pi_s(\mathbf{G}')$ and*

$$\{\Pi \in \Pi_s(\mathbf{G}) : \dim(\Pi) = 1 \text{ or } \Pi_v \in \Pi^2(G_v) \text{ for all } v \in S \cup S_0\} .$$

Proof: It is clear that ψ' defines a bijection between the sets of one-dimensional automorphic representations of G' and G. Suppose that $\Pi \in \Pi_s(\mathbf{G})$ is infinite-dimensional. Then the right-hand side of (14.6.2) reduces to a single term $\mathrm{Tr}(\Pi(f))$. If $\mathrm{Tr}(\Pi(f)) \neq 0$ for some f in the image of $f' \to f$, then χ_{Π_v} does not vanish identically on the elliptic regular set in G_v. If $v \in S$, then $\Pi_v = \{\pi_v\}$ where π_v is square-integrable, since the only elliptic representations of $\mathrm{GL}_3(F_v)$ that are not square-integrable are one-dimensional. If $v \in S_0$, then Π_v is either square-integrable or of the form $\Pi(\xi_v)$ for some one-dimensional representation ξ_v of H_v. However, the latter possibility is ruled out by Theorem 13.3.6(c). It follows from the local correspondence that $\Pi = \psi'(\Pi')$ for some $\Pi' \in \Pi_s(\mathbf{G}')$ and the left-hand side of (14.6.2) is equal to $\mathrm{Tr}(\Pi'(f'))$. The theorem follows.

THEOREM 14.6.3: *Suppose that $D \neq M_3(E)$. Then $\Pi(\mathbf{G}) = \Pi_s(\mathbf{G})$. In particular, no representation of the form $\pi^n(\xi)$ occurs as the local component of an automorphic representartion of G'.*

Proof: In this case, $f'^H = 0$. If $v \in S$, then $\mathrm{Tr}(\Pi(\rho_v)(f)) = 0$ for all ρ_v since $\Pi(\rho_v)$ consists of a representation induced from a proper parabolic subgroup and the orbital integrals of f vanish off of the elliptic set. The sum on the right-hand side of (14.6.1) reduces to a sum over $\Pi \in \Pi_s(\mathbf{G})$ which, by the argument in the previous proof, lie in the image of ψ'.

For the rest of this chapter, assume that $D = M_3(E)$. The transfer $f' \to f'^H$ has been defined relative to a fixed global set of transfer factors $\{\Delta'_v\}$. We now determine the global transfer on automorphic representations associated to $f' \to f'^H$. Let $\{\Delta_v\}$ denote the set of transfer factors for the pair (G, H) defined in §4.9 and for $v \notin S_0$, let Δ''_v be the transfer factor for (G'_v, H_v) defined by $\Delta''_v(\gamma_H, \gamma') = \Delta_v(\gamma_H, \psi_v(\gamma'))$. Let T be a Cartan subgroup of H such that T_v is compact for all $v \in S_0$. Then T embeds in G' and we fix one embedding $i : T \to G'$. Let $\gamma \in T_v$ be a regular element, viewed as an element of H_v and G_v. Define $\Delta''_v(\gamma, i(\gamma)) = \Delta_v(\gamma, \gamma)$ for $v \in S_0$. For all v, there is a constant c_v such that $\Delta'_v = c_v \Delta''_v$. For almost all v, Δ''_v and Δ'_v are canonical and $c_v = 1$. Let $c = \Pi c_v$. We now show that $c = \pm 1$. If $v \notin S_0$, then

$$\Delta''_v(\gamma, i(\gamma)) = \Delta_v(\gamma, \psi_v(i(\gamma))) = \kappa(\gamma, \psi_v(i(\gamma)))\Delta_v(\gamma, \gamma)$$

where γ is viewed as an element of G via $H \subset G$. The character $\kappa(\gamma, \psi_v(i(\gamma)))$ is equal to ± 1 and it is $+1$ for almost all v. By a global property of transfer factors (cf. §4.9), $\Pi_v \Delta'_v(\gamma, i(\gamma)) = 1$. On the other hand

$$\prod_v \Delta'_v(\gamma, i(\gamma)) = (\prod_v c_v)(\prod_v \Delta''_v(\gamma, i(\gamma))) = \pm \prod_v c_v$$

since $\Pi_v \Delta_v(\gamma, \gamma) = 1$. Hence $c = \pm 1$.

Let $\xi \in \Pi(\mathbf{H})$ be a one-dimensional representation. For $v \in S_0$, there exist integers $a_v \geq b_v \geq c_v$ such that either $a_v - b_v = 1$ and $\xi_v = \xi(a_v, c_v, b_v)$ or $b_v - c_v = 1$ and $\xi_v = \xi(b_v, a_v, c_v)$. Let $m_v = a_v - b_v$, $n_v = b_v - c_v$. If $m_v \cdot n_v \neq 0$, then the finite-dimensional representation F_φ, for $\varphi = \varphi(a_v, b_v, c_v)$, exists. In this case, set $\Pi'(\xi_v) = \{F_\varphi\}$. If $v \notin S_0$, set $\Pi'(\xi_v) = \Pi(\xi_v)$. Globally, set $\Pi'(\xi) = \otimes \Pi'(\xi_v)$. Let $N = \mathrm{Card}(S_0)$.

THEOREM 14.6.4: *Let $\Pi' \in \Pi_a(\mathbf{G}')$. Then there exist a one-dimensional representation $\xi \in \Pi(\mathbf{H})$ such that $m_v \cdot n_v \neq 0$ for all $v \in S_0$ and $\Pi' = \Pi'(\xi)$. The multiplicity $m(\pi)$ of an element $\pi = \otimes \pi_v \in \Pi'(\xi)$ is equal to one if the cardinality of the the finite set $\{v \notin S_0 : \pi_v = \pi^s(\xi_v)\}$ is congruent to N* mod 2 *and is equal to zero otherwise.*

Proof: If $\Pi' \in \Pi_a(\mathbf{G})$, then there exists a unique ξ such that the e.v.p.'s $t(\Pi')$ and $t(\Pi(\xi))$ coincide. The equality (14.6.2) yields

$$\sum m(\pi)\mathrm{Tr}(\pi(f')) = \frac{1}{2}\mathrm{Tr}(\Pi(\xi)(f)) + \frac{1}{2}\mathrm{Tr}(\xi(f'^H)), \tag{14.6.3}$$

where π ranges over the discrete representations such that $t(\pi) = t(\Pi')$. If $m_v n_v = 0$ for some $v \in S_0$, then $\mathrm{Tr}(\Pi(\xi_v)(f_v) = 0$ and $\mathrm{Tr}(\xi_v(f_v'^H)) = 0$ by Propositions 14.4.1(a) and 14.4.2(c). This is impossible if the sum on the left-hand side is non-empty, and hence $m_v n_v \neq 0$ for $v \in S_0$. Denote by F_v the unique element in $\Pi'(\xi_v)$, for $v \in S_0$. By the same propositions, the above remarks on transfer factors, and Proposition 13.1.4, the right-hand side of (14.6.3) is equal to

$$\begin{aligned} &\frac{1}{2}(-1)^N \prod_{v \in S_0} \mathrm{Tr}(F_v(f'_v)) \prod_{v \notin S_0} (\mathrm{Tr}(\pi^n(\xi_v)(f_v)) - \mathrm{Tr}(\pi^2(\xi_v)(f_v))) \\ &+ \frac{1}{2}(-1)^N c \prod_{v \in S_0} \mathrm{Tr}(F_v(f'_v)) \prod_{v \notin S_0} (\mathrm{Tr}(\pi^n(\xi_v)(f_v)) + \mathrm{Tr}(\pi^2(\xi_v)(f_v))). \end{aligned}$$

Recall that if $f_v \in \mathcal{H}_v$, where $v \notin S_0$, then $\mathrm{Tr}(\pi^2(\xi_v)(f_v)) = 0$, so each of the above expressions is a finite sum of traces for functions f such that $f_v \in \mathcal{H}_v$ for v outside a given set of places. By (14.6.1), the coefficients of the traces are non-negative integers. This implies that $\frac{1}{2}(-1)^N(c+1)$ and $\frac{1}{2}(-1)^N(c-1)$ are both non-negative integers and hence $(-1)^N c \geq 0$. It was shown above that $c = \pm 1$ and this shows that $(-1)^N c = 1$. The theorem follows from this and (14.6.3).

Let $\rho \in \Pi(\mathbf{H})$ and suppose that $\Pi(\rho_v) \in \Pi^2(G_v)$ for all $v \in S_0$. There exists an L-packet $\Pi'(\rho_n) = \{\pi'_v\}$ on G'_v such that $\psi'(\Pi'(\rho_v)) = \Pi(\rho_v)$, by the local correspondence. By Proposition 14.4.2(b), there is a unique sign $\langle \rho_v, \pi'_v \rangle = \pm 1$ such that

$$\mathrm{Tr}(\rho_v(f'^H_v)) = -c_v \langle \rho_v, \pi'_v \rangle \mathrm{Tr}(\pi'_v(f'_v)) \ .$$

In fact $\langle \rho_v, \pi'_v \rangle = 1$ if $\rho_v = \rho(a,b,c)$ and $\langle \rho_v, \pi'_v \rangle = -1$ if $\rho_v = \rho(b,a,c)$ or $\rho(a,c,b)$, in the notation of Proposition 14.4.2(b). Define $\Pi'(\rho) = \otimes \Pi'(\rho_v)$, where $\Pi'(\rho_v) = \Pi(\rho_v)$ for $v \notin S_0$, and define the global sign $\langle \rho, \pi' \rangle$ as a product $\Pi \langle \rho_v, \pi'_v \rangle$ for $\pi' \in \Pi'(\rho_v)$. For $v \notin S_0$, we identify π'_v with a representation of G_v as in §14.2 and define $\langle \rho_v, \pi'_v \rangle$ as in §13.1. Set $\langle 1, \pi'_v \rangle = \langle 1, \pi' \rangle = 1$. Let

$$\widehat{\Pi}' = \{\rho \in \Pi(\mathbf{H}) : \Pi' = \Pi'(\rho)\} \cup \{1\} \ .$$

For $s \in \widehat{\Pi}'$ and $\pi' \in \Pi'$, $\langle s, \pi' \rangle$ is defined. By Theorem 13.3.4, $|\widehat{\Pi}'| = 2$ or 4 for Π' of the form $\Pi'(\rho)$.

THEOREM 14.6.5: *Let $\Pi' \in \Pi_e(\mathbf{G}')$. Then $\Pi' = \Pi'(\rho)$ for some $\rho \in \Pi(\mathbf{H})$ such that $\Pi(\rho_v) \in \Pi^2(G_v)$ for all $v \in S_0$. For $\pi' \in \Pi'$,*

$$m(\pi') = |\widehat{\Pi}'|^{-1} \sum_{s \in \Pi'} \langle s, \pi \rangle.$$

Proof: There exists $\Pi \in \Pi_e(\mathbf{G})$ such that $t(\Pi') = t(\Pi)$ and (14.6.2) yields

$$(14.6.4) \qquad \sum m(\pi)\mathrm{Tr}(\pi(f')) = |\widehat{\Pi}|^{-1} \left[\mathrm{Tr}(\Pi(f)) + \sum \mathrm{Tr}(\rho(f'^H))\right]$$

where π ranges over the discrete representations of G' such that $t(\pi) = t(\Pi')$ and ρ ranges over the set X of elements of $\Pi(\mathbf{H})$ such that $\pi(\rho) = \Pi$. As in the proof of Proposition 14.6.2, we see that $\Pi_v \in \Pi^2(G_v)$ for all $v \in S_0$. Hence $\Pi'(\rho)$ exists for $\rho \in X$ and

$$\mathrm{Tr}(\rho'(f'^H)) = (-1)^N c \sum_{\pi' \in \Pi'(\rho)} \langle \rho, \pi' \rangle \mathrm{Tr}(\pi'(f')) \,.$$

As shown in the proof of Theorem 14.6.4 $(-1)^N c = 1$. Hence the right-hand side of 14.6.4, is equal to

$$\left|\widehat{\Pi}\right|^{-1} \sum_{s \in \widehat{\Pi}} \sum_{\pi' \in \Pi''} \langle s, \pi' \rangle \ \mathrm{Tr}(\pi'(f'))$$

where $\Pi'' = \Pi'(\rho)$ for any $\rho \in X$. We must have $\Pi'' = \Pi'$ and the theorem follows.

CHAPTER 15

Additional results

In this chapter we discuss the local Langlands correspondence and a vanishing theorem for the cohomology of certain arithmetic subgroups of $U(2,1)$.

15.1 Local Langlands correspondence. Let G be a group and let H be a subgroup of G of index 2. Fix an element $s \in G - H$. If ρ is a representation of H, let ρ^s denote the representation $\rho(shs^{-1})$. The isomorphism class of ρ^s is independent of the choice of s. Let ρ^* be the contragredient of ρ ($\rho^*(g) = \rho(g^{-1})^*$, where $\rho(g)^*$ denotes the adjoint of $\rho(g)$).

LEMMA 15.1.1: *Let ρ be an irreducible finite-dimensional representation of H such that ρ^* and ρ^s are equivalent. Let $A \in \mathrm{Hom}_H(\rho^*, \rho^s)$ be an isomorphism. Then $\rho(s^2) = c(\rho)A \cdot A^{*-1}$, where $c(\rho) = \pm 1$. Furthermore, $c(\rho)$ depends only on the isomorphism class of ρ and not on the choice of s or A.*

Proof: By definition, $A\rho^*(g)A^{-1} = \rho(sgs^{-1})$. Hence

$$A^{*-1}\rho(g)A^* = \rho^*(sgs^{-1}) = A^{-1}\rho(s^2gs^{-2})A$$

and $AA^{*-1} = c\rho(s^2)$ for some $c \in \mathbf{C}^*$ by Schur's lemma. We have $A^{*-1}A = c^{-1}\rho^*(s^2) = c^{-1}A\rho(s^2)A^{-1}$ and this shows that $c^2 = 1$. The operator A is unique up to a non-zero scalar, hence c is independent of the choice of A. For all $h \in H$,

$$\rho(h)A(\rho(h)A)^{*-1} = \rho(h)A\rho^*(h)A^{-1}(AA^{*-1}) = c\rho(hs)^2 .$$

This shows that c is independent of the choice of s, and the lemma is proved.

The invariant $c(\rho)$ can be defined more directly as follows. There is a canonical extension of $\varphi \otimes \varphi^s$ to a representation of G obtained by letting s act via $v \otimes w \to w \otimes \rho(s^2)v$. If ρ^* is isomorphic to ρ^s, then $\rho \otimes \rho^s$ contains a unique H-invariant line on which G acts. The action of G on this line is trivial if and only if $c(\rho) = 1$.

Let E/F be a quadratic extension of p-adic fields and let $G = U(n)$. Fix an element $s \in W_F$ whose projection to Γ is σ. Let

$$\tau : W_E \to \mathrm{GL}_n(\mathbf{C}) \times W_E$$

be an L-map, and let τ_0 be the projection of τ onto the first factor. Let $\varepsilon(\tau)$ denote the map $w \to \Phi_n\ {}^t\tau_0(s^{-1}ws)^{-1}\Phi_n^{-1} \times w$. The equivalence class of $\varepsilon(\tau)$ is independent of the choice of s and $\varepsilon(\tau)$ is equivalent to τ if and only if τ_0^* is equivalent to τ_0^s. If $\varepsilon(\tau)$ is equivalent to τ, let $c(\tau) = c(\tau_0)$. Let $\det(\tau)$ denote the determinant of τ_0. Recall that μ is a character of E^* whose restriction to F^* is $\omega_{E/F}$.

LEMMA 15.1.2: *Assume that τ is irreducible and that it is equivalent to $\varepsilon(\tau)$.*

(a) *τ extends to an L-map of W_F into ${}^LG$ if and only if $c(\tau) = 1$. The extension, if it exists, is unique up to equivalence.*

(b) *$c(\tau \otimes \mu) = -c(\tau)$. Precisely one of the two representations $\{\tau, \tau \otimes \mu\}$ extends to an L-map of W_F into ${}^LG$.*

(c) *The restriction of* $\det(\tau)$ *to F^* is trivial or $\omega_{E/F}$. If n is even, then the restriction of* $\det(\tau)$ *to F^* is trivial. If n is odd, then $c(\tau) = 1$ if and only if the restriction of* $\det(\tau)$ *to F^* is trivial.*

Proof: The map τ extends to an L-map of W_F into ${}^LG$ if and only if there exists $A \in \mathrm{GL}_n(\mathbf{C})$ such that $A\Phi_n\,{}^t\tau_0(w)^{-1}\ \Phi_n^{-1}A^{-1} = \tau_0(sws^{-1})$ and $(A \times s)^2 = \tau(s^2)$. If A satisfies the first condition, then $(A \times s)^2 = c(\tau)\tau_0(s^2) \times s^2$ by Lemma 15.5.1. Hence τ extends if and only if $c(\tau) = 1$. To check the uniqueness of the extension, note that different choices of A lead to $\widehat{G}$-conjugate extensions since $\lambda\tau(s)\lambda^{-1} = \lambda^2 A \times s$ for any scalar λ. This proves (a). Since $s^2 \in W_F$ and projects to an element of C_F which is not a norm from $E, \mu(s^2) = -1$ and (b) follows. For (c), observe that $\det(\tau_0(s^2)) = c(\tau)^n = \pm 1$. If n is even, $\det(\tau(s^2)) = 1$ and $\det(\tau)$ is trivial on F^*. If n is odd, then $\det(\tau)$ is trivial on F^* if and only if $c(\tau) = 1$.

For the remainder of this section, let $n = 2$ or 3 and let $G = U(n)$. Let Π_n^s be the set of supercuspidal L-packets on G and let $\sum_n^s$ be the set of $\widehat{G}$-conjugacy classes of L-maps $\tau : W_F \to {}^LG$ such that $im(\tau)$ is not contained in the L-group ${}^LB$ of a Borel subgroup.

The local Langlands conjecture asserts that there exists a bijection Φ_E between the set of irreducible supercuspidal representations of $\mathrm{GL}_n(E)$ and the set of equivalence classes of irreducible (continuous) representations of

W_E into $\mathrm{GL}_n(\mathbf{C})$. It is known for $n = 2, 3$ ([Hn]). Furthermore, $\Phi_E(\varepsilon(\pi)) = \varepsilon(\Phi_E(\pi))$ for all π. The Langlands conjecture also predicts the existence of a bijection $\Phi_n : \Pi_n^s \to \sum_n^s$ which is compatible with base change, in the sense that $\Phi_n(\Pi)_E = \Phi_E(\Pi_E)$, where Π_E denotes the base change lifting of Π to $\mathrm{GL}_n(E)$.

Let $\sum_n^{ss}$ be the set of $\tau \in \sum_n^s$ such that $(\tau_E)_0$ is irreducible, where τ_E is the restriction of τ to W_E. Let Π_n^{ss} be the set of supercuspidal L-packets Π on $U(n)$ such that Π_E is supercuspidal and let $\Pi_n^{\varepsilon s}$ be the set of ε-invariant supercuspidal representations π with central character ω_π such that $\omega_\pi | F^*$ is trivial. The base change lift defines a bijection between Π_3^{ss} and $\Pi_3^{\varepsilon s}$ by Theorem 13.2.1.

PROPOSITION 15.1.2: *Let $G = U(3)$. There exists a unique bijection $\Phi_3 : \Pi_3^{ss} \to \sum_3^{ss}$ which is compatible with base change.*

Proof: If $\Pi \in \Pi_3^{ss}$ and $\tau' = \Phi_E(\Pi_E)$, then τ' is equivalent to $\varepsilon(\tau')$ and $\det(\tau')$ is trivial on F^* (since $\det(\tau')$ corresponds to ω_π if $\Phi_E(\pi) = \tau'$). By Lemma 15.1.2, τ' has an extension to a map $\tau : W_F \to {}^LG$ which is unique up to $\widehat{G}$-conjugacy. The proposition follows with $\Phi_3(\Pi) = \tau$.

In the case $n = 2$, we can also define a bijection $\Phi_2 : \Pi_2^{ss} \to \sum_2^{ss}$. Let $\tau' = \Phi_E(\Pi_E)$ and let τ'' be the unique element of $\{\tau', \tau' \otimes \mu\}$ which extends to a map τ of W_F into ${}^LG$ (Lemma 15.1.2). Set $\Phi_2(\Pi) = \tau$. The map Φ_2 is a bijection by Proposition 11.4.1 and it is compatible with base change if and only if $\tau' = \tau''$ for all Π. This is expected, but we have not tried to check it.

It is possible to extend Φ_3 to a bijection between Π_3^s and $\sum_3^s$. Let $\rho = \rho_0 \otimes \rho_1 \in \Pi^2(H)$, where $\rho_0 \in \Pi_2^s$ and ρ_1 is a character of $U(1)$, and let $\Pi = \Pi(\rho)$. Let $\tau_1 : W_F \to {}^LU(1)$ be an L-map corresponding to ρ_1 and assume that $\tau_1(w) = \alpha(w) \times w$, $\alpha(w) \in \mathrm{GL}_1(\mathbf{C})$. If $\rho_0 \in \Pi_2^{ss}$, let $\tau_0 = \Phi_2(\rho_0)$ and set $\tau(w) = \beta(w) \times \alpha(w) \times w$, where $\tau_0(w) = \beta(w) \times w$, $\beta(w) \in \mathrm{GL}_2(\mathbf{C})$. We then define $\Phi_3(\Pi) = \xi_H \circ \tau$. If ρ_0 does not belong to Π_2^{ss}, then $\Pi = \Pi(\theta)$ for some regular character θ of C. Let $\tau_C : W_F \to {}^LC$ be L-map associated to θ and set $\Phi_3(\Pi) = \xi_H \circ \xi_C \circ \tau_C$. This defines an injective map from Π_3^s to $\sum_3^s$ (cf. Proposition 13.1.2(c)).

To show that Φ_3 is a bijection, let $\tau : W_F \to {}^LG$ be a map such that $(\tau_E)_0$ is reducible, where $(\tau_E)_0$ is the projection of τ_E onto $\mathrm{GL}_3(\mathbf{C})$. Up to conjugacy, we can assume that $im(\tau_E) \subset \widehat{H} \times W_E$. If $(\tau_E)_0$ stabilizes a unique line, then $\tau(s)$ normalizes $\widehat{H} \times W_E$ and $im(\tau) \subset \widehat{H} \rtimes W_F$. It follows that $\tau = \xi_H \circ \tau_H$ for some $\tau : W_F \to {}^L H$. Suppose that $\tau(w) = \beta(w) \times \alpha(w) \times w$ where $\beta(w) \in \mathrm{GL}_2(\mathbf{C})$ and $\alpha(w) \in \mathrm{GL}_1(\mathbf{C})$, and let $\tau_0(w) = \beta(w) \times w$, $\tau_1(w) = \alpha(w) \times w$. Then $\tau_0 \in \sum_2^{ss}$ and $\tau_0 = \Phi_2(\rho_0)$ for some $\rho_0 \in \Pi^2(H)$. Similarly, τ_1 corresponds to a character ρ_1 of $U(1)$. In this case, $\tau = \Phi_3(\Pi(\rho))$ where $\rho = \rho_0 \otimes \rho_1$. If $(\tau_E)_0$ can be diagnalized, it is straightforward to check that $\tau = \xi_H \circ \xi_C \circ \tau_C$ for some L-map $\tau_C : W_F \to {}^L C$ corresponding to a regular character θ of C. In this case, $\tau = \Phi_3(\Pi(\theta))$.

15.2 Representations with cohomology. In this section, let $G = U_{2,1}(\mathbf{R})$. We use the notation of §12.3. Let $\varphi = \varphi(a,b,c)$ where $a > b > c$ and let F_φ be the corresponding finite-dimensional representation of G. Let F_φ^* be the contragredient of F_φ. Let π be an irreducible unitary representation of G and let $\mathcal{G} = \mathrm{Lie}(G)$. We denote the relative $(\mathcal{G}, K)$-cohomology of $\pi_0 \otimes F_\varphi^*$, where π_0 is the subspace of K-finite vectors in π, by $H^*(\mathcal{G}, K, \pi \otimes F_\varphi^*)$. It is well-known ([BW]) that $H^*(\mathcal{G}, K, \pi \otimes F_\varphi^*)$ is zero unless the center of $\mathfrak{U}(\mathcal{G})$ acts by the same character on π and F_φ.

PROPOSITION 15.2.1: *Suppose that $H^*(\mathcal{G}, K, \pi \otimes F_\varphi^*) \neq 0$. Then one of the following occurs*

(a) $\pi \in \Pi_\varphi$. *In this case*

$$H^j(\mathcal{G}, K, \pi \otimes F_\varphi^*) = \begin{cases} 0 & \text{if } j \neq 2 \\ \mathbf{C} & \text{if } j = 2. \end{cases}$$

(b) $a - b = 1$ *and* $\pi = J_\varphi^-$ *or* $b - c = 1$ *and* $\pi = J_\varphi^+$. *In this case*

$$H^j(\mathcal{G}, K, \pi \otimes F_\varphi^*) = \begin{cases} 0 & \text{if } j \neq 1,3 \\ \mathbf{C} & \text{if } j = 1,3. \end{cases}$$

(c) $\dim(F_\varphi) = 1$ *and* $\pi = F_\varphi$. *In this case,*

$$H^j(\mathcal{G}, K, \pi \otimes F_\varphi^*) = \begin{cases} 0 & \text{if } j \neq 0,2,4 \\ \mathbf{C} & \text{if } j = 0,2,4. \end{cases}$$

Proof: [BW], Theorem 4.11.

15.3 Automorphic representations with cohomology.

Let E/F be a CM extension of number fields and let G' be an inner form of G defined by a pair (D, α). Let S'_∞ be the set of infinite places v of F such that $G'_v = U_{2,1}(\mathbf{R})$. Let π be a discrete automorphic representation. By Theorem 13.3.6(c) and the results of §14.4, if $\pi_v = J_\varphi^+$ or J_φ^- for some

φ and some $v \in S'_\infty$, then π belongs to an L-packet $\Pi(\xi)$, where ξ is a one-dimensional automorphic representation of H. In particular, if $D \neq M_3(E)$, then no such π occur discretely.

The group G' gives rise to a family of congruence subgroups in the projective group $PU_{2,1}(\mathbf{R})$ in the standard way. For each open compact subgroup K of $\Pi_{v<\infty} G'(F_v)$, let Γ_K denote the image in $PU_{2,1}(\mathbf{R})$ of $\{\gamma \in G'(\mathbf{Q}) : \gamma \in K\}$. The next theorem follows from the remark of the previous paragraph and the relation between cohomology of discrete subgroups and automorphic forms ([BW]).

THEOREM 15.3.1: *Suppose that* $F = \mathbf{Q}$. *If* $D \neq M_3(E)$, *then* $H^1(\Gamma_K, \mathbf{C}) = 0$ *for all* Γ_K *as above.*

This generalizes a result of Rapoport and Zink ([RZ]). It was proved by them for certain Γ_K using geometric methods. The approach taken here was suggested by Langlands ([L$_2$], p. 65). Theorem 13.3.6(c) also applies if E/F is a CM extension. Let K_∞ be a maximal compact subgroup of $G'_\infty = \Pi G'_v$ (product over the infinite places of F) and let $m = \frac{1}{2}\dim(G'_\infty | K_\infty)$. Let Γ_K be a congruence subgroup of the projective group associated to G'_∞ and an open compact subgroup K of $\Pi_{v<\infty} G'_v(F_v)$ as above. If $D \neq M_3(E)$, then the cohomology groups $H^j(\Gamma_K, \mathbf{C})$ coincide with the $(\mathfrak{G}, K)$-cohomology of the trivial representation of G'_∞ for $j \neq m$.

[A] Afternoon seminar on the trace formula (Kottwitz, Langlands, Rogawski, Shahidi, Shelstad). IAS notes, 1984.

[A_1] Arthur, J.: Some tempered distributions on semisimple groups of real rank one. Annals of Math. 100 (1974), 553-584.

[A_2] _______, On a family of distributions obtained from orbits. Can. J. Math. 38, no. 1 (1986), 179-214.

[A_3] _______, A trace formula for reductive groups, I. Duke Math J. 45, no. 4 (1978), 911-952.

[A_4] _______, Unipotent automorphic representation: conjectures. In: Orbites unipotentes et répresentations, II. Astérisque 171-172 (1989), 13-71.

[A_5] _______, The trace formula in invariant form. Annals of Math. 114 (1981), 1-74.

[AC] Arthur, J. and Clozel, L.: *Simple algebras, base change, and the advanced theory of the trace formula.* Annals of Math. Studies, Princeton U. Press, 1989.

[Ba] Barbasch, D.: Fourier inversion for unipotent orbital integrals. Transactions A.M.S 249, no. 1 (1979) 51-83.

[B] Bernstein, J., Deligne, P., Kazhdan, D., and Vigneras, M.-F.: *Représentations des groupes réductifs sur un corps local.* Hermann, Paris, 1984.

[BZ] Bernstein, J. and Zelevinsky, A.: Induced representations of reductive p-adic groups, I. Annales Sc. E.N.S., 4° sér. 10 (1977), 441-472.

[BR_1] Blasius, D. and Rogawski, J.: Fundamental lemmas for $U(3)$ and related groups. In [Mo].

[BR_2] _______: Tate-classes and arithmetic quotients of the two-ball. In [Mo].

[BR_3] _______: Galois representations for Hilbert modular forms. Bulletin A.M.S. (1) **21** (1989), 65-69.

[Bo] Borel, A.: Automorphic L-functions. In [Cor], Part II.

[BW] Borel, A. and Wallach, N.: *Continuous cohomology, discrete subgroups, and representations of reductive groups.* Annals of Math. Studies, Princeton U. Press, 1980.

[Ca] Cartier, P.: Representations of p-adic groups. In [Cor], Part I.

[Ca_1] Casselman, W.: Characters and Jacquet modules. Math. Ann., 230 (1977), 101-105.

[Ca_2] _______: The Hasse-Weil zeta function of some moduli varieties of dimension greater than one. In [Cor], Part II.

[C_1] Clozel, L.: Changement de base pour les représentations tempérees des groupes réductifs réel. Annales Sc. E.N.S., 4° Serie, t. 15 (1982), 45-115.

[C_2] _______: Characters of non-connected, reductive p-adic groups. Can. J. Math. 39 (1987) 149-167.

[C_3] _______: On multiplicites of discrete series representations. Inventiones Math. 83 (1986) 265-284.

[C_4] _______: Orbital integrals on p-adic groups: a proof of the Howe conjecture. Annals of Math. 129, no. 3 (1989), 237-251.

[C_5] _______: Motifs et formes automorphes: applications du principe de fonctorialité. In: *Automorphic forms, Shimura varieties, and L-functions.* Ed. L. Clozel and J. Milne. Academic Press, 1990.

[CD] Clozel, L.: and Delorme, P.: Le Théorème de Paley-Wiener invariant pour les groupes de Lie réductifs. Inventiones Math., 77 (1984), 427-453.

[Cor] *Automorphic forms, representations, and L-functions.* Proc. Symp. Pure Math. XXXIII, Parts I and II, A.M.S. (1979).

[DL] Duflo, M. and Labesse, J.-P.: Sur la formule des traces de Selberg. Annales Sc. E.N.S. 4 (1971), 193-284.

[GJ] Gelbart. S. and Jacquet, H.: A relation between automorphic forms on GL_2 and GL_3. Proc. Nat. Acad. Sci. USA, 73, (1976), 3348-3350.

[GP] Gelbart, S. and Piatetski-Shapiro, I.: Automorphic forms and L-functions for the unitary group. In: *Lie group representations II.* SLN 1041, Springer Verlag, Berlin-Heidelberg-New York (1984), 141-184.

[GK] Gelfand, I. and Kazhdan, D.: Representations of $\mathrm{GL}(n, K)$. In: *Lie groups and their representations*, Summer school on group representations, Budapest, 1971. Halsted Press, NY, 1975, 95-118.

[Hl] Hales, T.: An approach to the fundamental lemma. In preparation.

[Ha] Harder, G.: Bericht über neuere resultate der Galoiskohomologie halbeinfacher grüppen. Jahresbericht d. DMV 70 (1968), 182-216.

[H_1] Harish-Chandra: Admissible invariant distributions on reductive p-adic groups. Queen's papers in pure and applied math. 48 (1978), 281-347.

[H_2] ________: Harmonic analysis on real reductive groups, I. J. Funct. Analysis 19 (1975), 104-204.

[H_3] ________: Fourier transform on a semi-simple Lie algebra, I. Amer. J. Math. 79 (1957), 193-257.

[H_4] ________: *Harmonic analysis on reductive p-adic groups.* SLN 162 Springer Verlag, Berlin-Heidelberg-New York (1970).

[Hn] Henniart, G.: La conjecture de Langlands pour GL(3). Sem. Delange-Pisot-Poitou, 1980-1981.

[H] Howe, R.: The Fourier transform and germs of characters. Math. Ann. 208 (1974), 305-322.

[HP] Howe, R. and Piatetski Shapiro, I.I.: A counterexample to the "generalized Ramanujan conjecture". In [Cor], Part I.

[JL] Jacquet, H. and Langlands, R.P.: *Automorphic forms on* GL(2). SLN 114 Springer Verlag, Berlin-Heidelberg-New York (1970).

[JS] Jacquet, H. and Shalika, J.: On Euler products and the classification of automorphic forms, I and II. Amer. J. Math. 103, no. 3 and no. 4, (1981), 499-558 and 777-815.

[K] Kazhdan, D.: Cuspidal geometry of p-adic groups. J. D'analyse Math. 47 (1986), 1-36.

[Ky] Keys, D.: Principal series of special unitary groups over local fields. Compositio Math. **51** (1984), 115-130.

[KyS] Keys, D. and Shahidi, F.: Artin L-functions and normalization of intertwining operators. Annales Sc. E.N.S., 4e série, 21 (1988), 67-89.

[Kp] Knapp, A.: *Representation theory of semisimple groups.* Princeton U. Press, Princeton, New Jersey (1986).

[Kn] Kneser, M.: Hasse Principle for H^1 of simply connected groups. In: *Algebraic groups and discontinuous subgroups.* Proc. Symp. Pure Math. IX, A.M.S. (1965).

[Kt_1] Kottwitz, R.E.: Base change for unit elements of Hecke algebras. Compositio Math. 60 (1986), 237-250.

[Kt_2] ________, Rational conjugacy classes in reductive groups. Duke Math. J. 49 (1982), 785-806.

[Kt_3] ________, Stable trace formula: cuspidal tempered terms. Duke Math. J. 51, no. 3 (1984), 611-650.

[Kt_4] ________, Stable trace formula: Elliptic singular terms. Math. Ann. 275 (1986), 365-399.

[Kt_5] ________, Sign changes in harmonic analysis on reductive groups. Trans. A.M.S. 278, no. 1 (1983), 289-297.

[Kt_6] ________, Tamagawa numbers. Annals of Math. 127 (1988), 629-646.

[KR] Kottwitz, R.E. and Rogawski, J.D.: The distributions in the invariant trace formula are supported on characters. Preprint.

[KS] Kottwitz, R.E. and Shelstad, D.: in preparation.

[La] Labesse, J.-P.: La formule des traces d'Arthur-Selberg. Sem. Bourbaki 1984/85, exp. 636, Asterisque 133-134 (1986), 73-88.

[LL] Labesse, J.-P. and Langlands, R.P.: L-Indistinguishability for SL(2). Can. J. Math. 31 (4) (1979), 726-785.

[Lai] Lai, K.F.: Tamagawa number of reductive algebraic groups. Compositio Math. 41 (2) (1980), 153-188.

[L] Landherr, W.: Äquivalenze Hermitscher formen über einem beliebigen algebraischen zahlkörper. Abh. Math. Sem. Hamb., 11 (1936), 245-248.

[L_1] Langlands, R.P.: *Base Change for* GL(2). Annals of Math Studies, Princeton U. Press, 1980.

[L_2] _______, *Les débuts d'une formule des traces stable.* Pub. Math. de I'Univ. Paris VII 13, 1982.

[L_3] _______, Stable conjugacy: definitions and lemmas. Can. J. Math. 31 (4) (1979), 700-725.

[L_4] _______, Modular forms and l-adic representations. In: SLN 349 Springer Verlag, Berlin-Heidelberg-New York (1973), 362-499.

[L_5] _______: Remarks on Igusa theory and real orbital integrals. In [Mo].

[L_6] _______: Eisenstein series, the trace formula, and the modern theory of automorphic forms. In *Number theory, trace formulas, and discrete groups.* Academic Press, 1989.

[L_7] _______: On the zeta functions of some simple Shimura varieties. Can. J. Math. **31** (1979), 1121-1216.

[L_8] _______: On the classification of irreducible representations of real algebraic groups. In *Representation theory and harmonic analysis on semisimple Lie groups.* Ed. P. Sally and D. Vogan, A.M.S., 1989.

[L_9] _______: Eisenstein Series. In: *Algebraic groups and discontinuous subgroups.* Proc. Symp. Pure Math. IX, A.M.S. (1965).

[LS_1] Langlands, R.P. and Shelstad, D.: On the definition of transfer factors. Math. Ann. 278 (1987), 219-271.

[LS_2] _______: Orbital integrals on forms of SL(3), II. Can. J. Math. 41, no. 3 (1989), 480-507.

[MW] Moeglin, C., et Waldspurger, J.-W.: Le spectre residuel de GL(n). Annales Sc. E.N.S., 4° série, 22 (1989), 605-674.

[Mo] Proceedings of a workshop on Shimura varieties. Centre de recherches mathématiques, Université de Montréal, 1988. To appear.

[M] Morning seminar on the trace formula (Clozel, Labesse, Langlands). IAS notes, 1984.

[Mu] Mueller, W.: The trace class conjecture in the theory of automorphic forms. Annals of Math. 130, no. 3 (1989), 473-529.

[RZ] Rapoport, M. and Zink, Th.: Über die lokale zetafunktion von Shimuravarietäten. Inventiones Math. 68 (1982), 21-101.

[R_1] Rogawski, J.D.: An application of the building to orbital integrals. Compositio Math. 42 (1981), 417-423.

[R_2] _______, Representations of GL(n) and division algebras over a p-adic field. Duke Math. J. 50, no. 1 (1983), 161-196.

[R_3] _______, Trace Paley-Wiener theorem in the twisted case. Transactions AMS 309, no. 1 (1988), 215-229.

[R_4] _______, Thesis, Princeton (1980).

[R_5] _______, Analytic expression for the number of points mod p. In [Mo].

[SW] Sally, P. and Warner, G.: The Fourier transform of invariant distributions. Proc. Conf. on Harmonic Analysis, College Park, Maryland, 1971. SLN 266 Springer, Berlin-Heidelberg-New York (1972) 297-320.

[Se] Serre, J.-P.: *Local Fields.* Springer, Berlin-Heidelberg-New York (1979).

[S_1] Shelstad, D.: Orbital integrals, endoscopic groups, and L-indistinguishability for real groups, in *Journées Automorphes.* Pub. Math. de l'Univ. Paris VII, 15, 1983.

[S_2] ________, Base change and a matching theorem for real groups. In: SLN 880, Springer, Berlin-Heidelberg-New York (1982), 425-482.

[S_3] __________, L-indistinguishability for real groups, Math. Ann. 259, (1982), 385-430.

[Sh] Shimura, G.: Arithmetic of unitary groups. Annals of Math. 79 (2) (1964), 369-409.

[Si] Silberger, A.: *Introduction to harmonic analysis on reductive p-adic groups.* Princeton U. Press, Princeton, N.J., 1979.

[St] Steinberg, R.: Endomorphisms of linear algebraic groups. Memoirs AMS 80, AMS (1960).

[T] Tate, J.: The cohomology groups of tori in finite Galois extensions of number fields. Nagoya Math. J. 27 (1966), 709-719.

[Ti] Tits, J.: Reductive groups over local fields. In [Cor], Part I.

[V] Varadarajan, V.: *Harmonic analysis on real reductive group.* SLN 576, Springer, Berlin-Heidelberg-New York, 1977.

[W] Wallach, N.: On the Selberg trace formula in the case of compact quotient. Bull. AMS 82, no. 2 (1979), 171-195.

[We] Weil, A.: *Basic Number Theory.* Springer, Berlin-Heidelberg-New York, 1973.

[Z] Zelevinsky, A.: Induced representations of reductive p-adic groups, II. Annales Sc. E.N.S., 4° série, 13 (1980), 165-210.

Subject Index

Notation Index

Milton Keynes UK
Ingram Content Group UK Ltd.
UKHW040740121224
452420UK00001B/86

9 780691 085876